Springer Textbooks in Law

Springer Textbooks in Law compiles high-quality educational content aimed at undergraduate and graduate students in all areas of law. All self-contained volumes are authored by accomplished academics and suitable for use in class as well as individual study. Many of them include chapter abstracts, definitions of technical terms, cases and self-assessment exercises, as well as recommended reading sections. This series is an invaluable resource for students and lecturers alike and spans the full range of topics in international and European law, including fundamentals of law and comparative law. Special attention is paid to current and emerging topics such as IT law, intellectual property, human rights as well as dispute resolution, mediation, arbitration – and many more.

Takis Vidalis

The Emergence of Biolaw

The European Experience and the Evolutionary Approach

Springer

Takis Vidalis
Hellenic National Commission for Bioethics
and Technoethics
Athens, Greece

ISSN 2509-999X ISSN 2510-0009 (electronic)
Springer Textbooks in Law
ISBN 978-3-031-02358-3 ISBN 978-3-031-02359-0 (eBook)
https://doi.org/10.1007/978-3-031-02359-0

This Springer imprint is published by the registered company Springer Nature Switzerland AG
The registered company address is: Gewerbestrasse 11, 6330 Cham, Switzerland

To Manuela, Dimitris, and Costis

Preface

Over the last decades, the term "Biolaw" (Biodroit, Biorecht) constantly appears in academic papers, international conferences, workshops, and even programs of specific studies in law schools. This term aims to cover a rapidly growing amount of legal rules and judicial decisions relevant to the management of the biological structure and functions of living organisms, or in other words, the management of life as a natural phenomenon. Other terms have expressed that aim also ("bioethics and law," "law of genetics," "biotechnology law," etc.) more or less successfully, proving an evident interest of legal scholars for exploring a vast new area of regulation, mostly unknown until recently. Nowadays, this development of specific legislation and case-law clearly indicates the emergence of a new field in contemporary law.

Nothing is surprising in this assumption if we consider the enduring progress of technology in Medicine and Life Sciences, leading to impressive applications that may produce big changes in everyday life and even determine the prospects of contemporary societies. It is sufficient to highlight the interest of the global economy for these applications, in terms of investments, along with informatics, telecommunications, and artificial intelligence. Moreover, the influence of modern biomedicine and biotechnology on almost every scientific discipline (from chemistry and engineering to economics and social anthropology), as well as on expressions of modern art (from literature and cinema to video art and comics), is now significant and characteristic of our culture.

This book introduces Biolaw as a distinct field of law, arguing that the legal regulation of life as a natural phenomenon should be addressed under the light of an integrated approach, capturing all life's expressions, human and non-human. This view also explains why that simple term is more appropriate than the various mentioned terms, which tend to focus on certain expressions of life's management only. Several studies in the international legal bibliography paved the path for this attempt; I would refer to the relatively recent book by Roberto Andorno as a characteristic and very inspiring example.

I decided to focus primarily on the material of rules and decisions that form the common European legal experience, based on the activity of the EU's and the Council of Europe's legislative organs and courts, during the last 30 years. Currently, Europe presents the most detailed legal regulation in relevance, covering the biggest part of Biolaw in its integrated meaning. Another significant part is still governed by national laws and the case-law of national courts, indicating the vital role that local traditions, ethical standards, and specific social circumstances continue to play here, particularly regarding the management of human life. The analysis underlines this role wherever needed. Nevertheless, the fact that an extensive common binding regulation characterizes the European Biolaw illustrates a clear tendency beyond the influence of national particularities, which corresponds actually to the globalized conditions of the scientific and technological development in Biomedicine and Biotechnology.

The study's content is divided into two parts; the first analyzes the legal regulation of human life in all its expressions, while the second explores the regulation of non-human life, including microorganisms, plants, and animals. This structure is justified by the rationale of contemporary law, which clearly distinguishes the value of the human species from all other living organisms. Still, an integrated perspective of Biolaw cannot avoid critical questions precisely regarding this rationale. That is why I decided to open a discussion *de lege ferenda* that goes beyond the description and analysis of current regulation. This is what I call the "Evolutionary approach." On the one hand, it includes a necessary conciliation of the law's rationale with the multiple findings of life sciences confirming the strength and accuracy of the Theory of Evolution and thus suggesting to overcome the strict distinction between humans and the other living organisms in terms of values. On the other hand, this perception causes reflections on the moral status of non-biological entities, products of artificial intelligence, creating new challenges for Biolaw even beyond biology since intelligent machines may develop functions characterising living organisms.

I also intended to make this book suitable for educational purposes. Biolaw is nowadays characterized by certain standards, forming the basis of specific legal education. Therefore, a section of "further studying" is annexed to each chapter, where students will find selected parts from the case law in relevance and also fictional problems that summarize the essential elements of regulation. I must say that this section is not designed for the use of lawyers and legal students exclusively. The knowledge of Biolaw is nowadays necessary for professionals or students with non-legal backgrounds (medical doctors, biologists, pharmacologists, agriculturalists, research professionals, environmental scientists, sociologists, political scientists, etc.) acting in the relevant fields and needing specific guidance regarding regulatory aspects that influence now more than ever their work.

It would be impossible for me to undertake and finish this project without the valuable input of so many colleagues and friends from diverse academic and research backgrounds. The benefit they offered me is not limited to necessary information about legal and technological developments; it is mostly the opportunities they provided me to test arguments, theories, beliefs, assumptions, and interpretative proposals in conferences, workshops, seminars, and informal discussions for more than 10 years of research. Some of them influenced my elaboration on specific parts of the study significantly, even if they may disagree with my positions and conclusions, for which the responsibility is obviously mine.

First, I am grateful to two distinguished legal scholars with a pioneering contribution to the development of Biolaw in Europe: Herman Nys, Emeritus Professor of Medical Law at the University of Leuven, and Tade Matthias Spranger, Professor of Public Law at the University of Bonn, for their generous support and inspiring comments on specific chapters of the study.

I owe important information and methodological remarks, mainly regarding an interdisciplinary view based on references to Life Sciences and Medicine to Frangiskos Kolissis, Emeritus Professor of Molecular Biology at the National Tech University of Athens; Babis Savakis, Emeritus Professor of Biology at the University of Crete; Vassilis Tarlatzis, Emeritus Professor of Medicine at the Aristotle University of Thessaloniki; and Dr. Vicky Mollaki, Geneticist, Senior Researcher at the Hellenic National Bioethics Commission for Bioethics and Technoethics (HNCBT) and an EC expert.

Thanks to Roberto Andorno (University of Zurich), Assya Pascalev (Brown University), Pascal Borry (KU of Leuven), and Tom Douglas (Oxford University), I had the opportunity to revisit some specific topics related to law and neurosciences, advance directives, genetic data, and human enhancement. These discussions influenced my work significantly.

Stavroula Tsinorema, Professor of Philosophy and Bioethics at the University of Crete, and Ismini Kriari-Katrani, Emeritus Professor of Constitutional Law at the Athens Panteion University provided significant input regarding the ethics/law relationship during our long collaboration, that inspired the relevant parts and the methodology of this study.

Thanks to my postgraduate students at the Hellenic International University, and the participants of the annual seminar at the HNCBT , I tested thoughts that I develop in several chapters, particularly those more challenging in legal interpretation. I am grateful to them also.

I very much appreciated the comments and support from my reviewer, which gave me incentives for analyzing a few more important topics in a largely unexplored area of legal interest.

Dr. Brigitte Reschke backs from the first moment the promotion of this work through Springer Publications. Her comments, patience, and continuous assistance in editing were precious to me. I owe her nothing less than my communication with an international interdisciplinary audience in this exciting new field of law.

Athens, Greece
January 2022

Takis Vidalis

Contents

About the Author

Takis Vidalis (1963) received his Ph.D. from the Law School at the University of Athens. In 2001, he was elected a senior researcher and legal advisor to the Hellenic National Commission for Bioethics and Technoethics.

Author of 5 books and numerous papers on biolaw, bioethics, medical ethics and law, human rights, philosophy of law, AI ethics and law, and environmental law, he presented original research in more than 100 international and national conferences and workshops.

Takis is the founder of the BioMedLex service (www.biomedlex.gr), specializing in the law of biomedical research and red biotechnology.

Currently, he teaches law and ethics in advanced technologies at the International Hellenic University, the Law School at the University of Athens, the Medical School at the University of Thrace, and the Open University of Cyprus.

President of the Research Ethics Committee of the leading research Institute in Greece "Democritus," Takis is now a member of the European Group on Ethics in Science and New Technologies (EGE).

A General Introduction

1

During the last decades, the astonishing progress in life sciences has offered enormous information about living Nature. Based on that information, the ongoing development of new technological applications explores ways of manipulating life functions in various sectors.

That process produces a significant influence on modern societies. It is sufficient to mention achievements in medical technology regarding transplantations, artificial support of life, or medically assisted reproduction, applications in human genetics involving methods of extensive processing of human genetic data, or the multiple examples of genetic engineering applications, including gene therapy in humans and the deliberate creation of genetically modified or transgenic microorganisms, plants, and animals.

The use of information related to the biological composition of living species generates hope and fear for the future. This spontaneous, uncomfortable reaction of modern societies inevitably leads to the necessity of ethical considerations. Following the rapid technological progress, ethical reflection is developed widely, paving the way for modern Bioethics, a distinct sector of Applied Ethics.

1.1 Bioethics: Reflecting on the Society/Nature Relationship

The term "Bioethics" became known in the early 70s, for characterizing ethical considerations about the position of humans within the surrounding living environment. Bioethics also covered all ethical questions relevant to humans themselves as biological entities, particularly questions that emerged in the field of medical ethics.[1]

Developments in Genetics and Life Sciences in general, particularly after the first successful attempts of DNA recombination and the rise of modern Biotechnology, guided the interests of Bioethics mostly to the investigation of ethical issues

[1] See, in relevance, Callahan (1995), pp. 278 seq., Durand (1999), pp. 18 seq.

T. Vidalis, *The Emergence of Biolaw*, Springer Textbooks in Law,
https://doi.org/10.1007/978-3-031-02359-0_1

associated with the various forms of species' genome processing, a promising field of current genetic technology for the future.[2] However, modern Bioethics continues to reflect on more traditional topics, such as abortion, transplantations, or animal handling, appraising new data from technology, that affect the relevant ethical approach significantly.

As it happens in all areas of Ethics, Bioethics represents a field of exchange of different opinions, deriving from various philosophical and religious origins. The most important modern philosophical trends and the main religions, including all dogmas and sects, provide their own approach, offering answers and solutions as close as possible to their general doctrine. Even if differences and disagreements continuously occur, this dialogue often reveals some hidden dimensions of critical social values, under the influence of scientific applications.

The open expression of any ethical approach related to the understanding of relationships between humans and other species or of human biological nature itself has, as such, a crucial value; this ensures, at least, the free dialogue and review of arguments, even if entirely different ethical attitudes are involved. A public expression of thoughts makes possible the practical control and rationalization of social needs supposed to be served by modern biological and medical applications. This is a critical matter in order to avoid an exclusive dominance of economic interests in relevance, reflecting nowadays an undoubted reality. At the institutional level, a typical expression of the Bioethics reasoning is the establishment of various ethics committees, international or national, state, or non-governmental. Ethics committees are mostly interdisciplinary advisory bodies of independent experts, frequently with the participation of laypersons representing "average" social attitudes and opinions.

Nevertheless, an inducement of relativistic approaches always occurs as a real danger in that institutional context. A simple presentation of different attitudes without comparison and appraisal of their arguments leads inevitably to deadlocks regarding the best possible solution to a certain ethical question. Such an approach would disapprove of any prioritization of pertinent ethical and social values, assuming that any such attempt has only "relative" significance, accepted only from its supporters' subjective viewpoint. Apparently, this assumption is incapable of offering practical answers to ethical questions with the aspiration to gain consent even from those who do not share common ethical views. The deadlock is evident if, for instance, we encounter controversies between, on the one hand, religious attitudes rejecting the supposed therapeutic use of embryonic stem cells and, on the other, attitudes that defend the prioritization of scientific research in that concrete field, regardless of the expected benefit from its outcome. Following a relativistic approach, no comparison between these two opposed but equivalent, in terms of ethics, attitudes would be possible. Thus, no practical solution is expected in cases of such extreme disagreements, for the interest and benefit of society. From that point, exactly, Bioethics leaves an open field to the competence of Biolaw.

[2] Regarding the question on whether Bioethics constitutes a distinct discipline, see Jonsen (1998), pp. 325 seq.

1.2 Biolaw

1.2.1 Basic Material

The necessity to provide practical solutions for crucial problems raised by the development of biological applications leads gradually to the emergence of a new branch of law, defined nowadays as "Biolaw".

So far, legislators play the leading role in shaping the content of the new branch. During the last decades, intense legislative activity is developed at the level of international law, E.U. law, and national legal systems. On the one hand, this mobility illustrates evidence of the vast social interest regarding new technologies associated with an anxious effort to "prevent" their fantastic potential; on the other, it is evidence of a particular caution when dealing with unprecedented scientific realities affecting substantially social relationships and stances.

Nevertheless, the courts also have started to contribute to the development of Biolaw.[3] It is worth noting that the judicial system's relevant cases attracted significant public interest, particularly concerning end-of-life and assisted reproduction issues. In the near future, two elements seem to be essential for the courts' role enlargement. Firstly, insofar as new applications are multiplied, it is realistic to expect a proportional increase of litigation and case-law, as demonstrated by the examples of assisted reproduction and patents in biological applications. Secondly, current legislation on Biomedicine already affects the field of medical liability, the content of which is grounded mostly on case law. The development and further elaboration of that field seem inevitable, as new cases of medical malpractice involving the use of advanced technologies appear before the courts, particularly in the U.S.[4]

Also, decisions, directives, recommendations, reports, and guidelines produced by ethics committees and administrative authorities (like, for instance, the data protection or the human fertilization authorities) cover, to a certain extent, the general need for interpretative guidance when specific provisions of Biolaw are to be implemented in particular circumstances. This practice leads gradually to normative standards that contribute to a solid structuring of the new branch.

1.2.2 The Epistemological Status of Biolaw: Life as a Distinct Object

We can define the object of Biolaw as *the legal management of life considered as a natural phenomenon* that includes *all legal rules aiming at the regulation of deliberate interventions in the biological structure and functions of species.*

---

[3] Cf. Deharo (2006), pp. 33 seq. On the necessity for a separate jurisdiction pertinent to Bioethics, see Michaud (2006), pp. 61 seq.

[4] Kachalia et al. (2005), pp. 416 seq.

These interventions may change the biological structure of an organism or collect data from it. Change of the structure includes an extensive range of interventions, such as those related to the conservation or end of life (abortion, euthanasia in humans or animals), genome modification in various species, and all conventional or experimental therapeutic medical acts (transplantations, assisted reproduction, enhancement). Data collection includes a variety of genetic testing for identifying characteristics of the DNA in humans, animals, plants, and microorganisms, as well as all methods of medical data collection and processing for preventive, therapeutic, or experimental purposes. There are also some "satellite" topics related to the conditions of biological interventions, such as, in particular, the issue of biological material and data commercialization, which also includes the question of patenting in biological applications.

All the above topics are characterized by an epistemological coherence with two fundamental elements, the first of descriptive and the second of normative nature. The descriptive element is the common biological origin of all forms of life, including the human species (a). The normative element recognizes life as a phenomenon of the natural world with superior value demanding special protection (b).

(a) As we know, the common origin of life has been demonstrated by the Darwinian theory of Evolution and confirmed at a molecular level with the discovery of the DNA structure; based on that discovery, any comparison of different species' genomes reveals extensive similarities between them.
(b) At the normative level, our understanding of that community of species leads us to consider life a phenomenon with special and superior significance in all its forms (human or non-human). Certainly, we adopt also a hierarchical order of value between various life forms, presupposing though the normative significance of all of them.[5] Thus, life is distinguished from the rest natural world, meaning that the management of its various forms must be ethically and legally different from the management of the non-living objects that surround us. International and constitutional instruments explicitly express this normative priority of life as a natural phenomenon.

The emergence of these two fundamental elements of Biolaw is undoubtedly due to modern technology. Indeed, since we are able to design and perform substantial interventions, with direct effects on the structure and functions of living species, including human life, we take real risks for public goods, such as the natural environment and the public or individual health. The distinct normative significance

[5] On the primacy of human dignity in Biolaw, see Duprat (2004), pp. 45 seq. Yet, the author adopts a limited content of Biolaw, related only to applications pertaining to human life. Also, see the approach of Biolaw as an extension of human rights in the field of Biomedicine, in Andorno (2013), esp. sec. II, and Carlos Loureiro (2000), pp. 73–74. However, that issue became crucial during the preparation of the UNESCO's Bioethics Declaration (2005), when particular criticism had been expressed against the primacy of human dignity. See Hayry and Takala (2005), pp. 225 seq., particularly p. 232.

of life plays a crucial role in preventing those risks; until recently, this was not the case, since the lives of natural beings were developing independently, not affected by human interventions in their molecular composition and deriving vital functions.

1.2.3 Sources of Biolaw

Starting from the above general remarks, we can now examine the current image of Biolaw in Europe, as revealed from a variety of legal instruments and case-law, that covers a vast range of applications concerning human life and the life of other species.

Regarding their binding force, these legal instruments present differentiations. Briefly, there are the main distinct categories of binding and soft-law instruments. In the first approach, instruments of soft-law have much richer content, providing more normative tools that facilitate the interpretation of binding instruments and eventually make regulation more efficient. In the following, we expose the most important sources of Biolaw, applicable commonly in Europe, that is, in the member states of the Council of Europe and the E.U. member states.

I. *The international law:* Basic instruments of current Biolaw, in terms of their content, belong to international law.[6]
   i. Regarding applications of Biomedicine, having human life as their object, the existing instruments with binding force are the Council of *Europe's Convention on Human Rights and Biomedicine* of 1997 (known as the *Oviedo Convention*), and its additional Protocols on the *Prohibition on Human Cloning* (1998), on *Tissue and Organ Transplantations* (2002), on *Biomedical Research* (2005) and on *Genetic Data for Health Purposes* (2008).[7] Important soft-law texts are the UNESCO *Declarations on Bioethics and Human Rights* (2006), on *Human Genome and Human Rights* (1997), and on *Genetic Data Protection* (2003).

The Oviedo Convention includes provisions for the protection of fundamental rights in the broader field of Medicine by adopting modern medical ethics principles, particularly the "informed consent" principle. Provisions regarding applications of Genetics and Biotechnology in human life, clinical research, transplantations, research on embryos, and commercialization of the human body are also included. The Convention's Protocols contain detailed regulations in most of these topics.

It should be noted that the Oviedo Convention has been adopted by most, but not all, member states of the Council of Europe. Member states developing advanced biomedical research, such as the United Kingdom and Germany, have not even

---

[6]See in relevance, Dahl Rendtorff and Kemp (2000), pp. 281 seq., 341 seq., using the term "bioright," and Andorno (2004), pp. 131 seq., which indicates four main characteristics in Biolaw: Human dignity as a central notion, its development within the context of human rights, the flexibility of its rules, and the vague content of them. See also, Byk (2004), pp. 151 seq.

[7]Belanger (2004), pp. 75 seq. (also referring to the role of the CDBI—the ethics committee of the Council of Europe).

signed the Convention due to substantial objections regarding concrete provisions. This demonstrates the difficulty of reaching common European attitudes in such a sensitive area for the protection of human rights. However, currently, the majority of member states have signed and ratified the Convention and its Protocols, accepting thus a European minimum of binding legal rules, regardless of national ethical or religious differences.[8]

The UNESCO Declarations have more comprehensive content. The Universal Declaration on Bioethics and Human Rights (2006), a legal framework for Bioethics in general, accepts the priority of human dignity and fundamental rights, regardless of cultural differences. Based on that, the Declaration recognizes several principles, like those of personal autonomy, medical beneficence, informed consent, respect for personal integrity, respect for privacy, equal and fair treatment, non-discrimination and non-stigmatization, solidarity and cooperation, promotion of health, equal access to scientific research benefits, protection of future generations, protection of the environment, biosphere and biodiversity.[9]

The Declaration on the Human Genome (1997) includes general principles regarding the relationship of human dignity with the human genome, safeguards for the protection of rights of persons undergoing genetic testing or genetic interventions (highlighting, in particular, the principle of "prior, free and informed consent"—art. 5 b), pertaining to the accepted limits of genetic research, as well as the essential requirements of scientific activity in that field.

Specific provisions for protecting genetic and other biological human data, as sensitive personal data, are included in the UNESCO Declaration on Genetic Data (2003). The most important of them refer to the rules of the data subject's informed consent, protection of privacy, non-genetic discrimination and stigmatization, and the safe use, storage, and transfer of data.

There is also a U.N. Declaration on Human Cloning (2005), calling the U.N. member states to ban *"all forms of human cloning, including cloning for medical treatment [known as "therapeutic cloning"], as incompatible with human dignity and the protection of human life*."

Other known international instruments for the protection of human rights are also relevant to topics of Biolaw, such as, in particular, the *European Convention on Human Rights* (*ECHR*—particularly its articles 2 on the right to life, 3 on the prohibition of torture, 8 on the protection of private and family life, 14 on the prohibition of discriminations) and the *EU Charter of Fundamental Rights* (particularly, articles 1 on human dignity, 2 on the right to life, 3 on personal integrity, 4 on the prohibition of torture, 8 on data protection and 13 on the freedom of research).

ii. Two binding international instruments govern applications of Biotechnology in non-human species: The U.N. *Convention on Biodiversity* (*CBD - Rio de Janeiro Convention,* 1991) and its Protocols on *Biosafety* (*Cartagena Protocol*, 2000),

[8]Regarding the Convention's influence on national legislation, cf. Nys et al. (2007), pp. 223–235.

[9]Cf. Ten Have (2018), pp. 31–39.

and on *Access to Genetic Resources and the Fair and Equitable Sharing of Benefits Arising from their Utilization* (*Nagoya Protocol*, 2010).

The Rio Convention does not refer in detail to genetic modification of microorganisms or organisms; nevertheless, it recognizes certain principles for the management of biodiversity pertinent to GMMs and GMOs, in particular the sustainable development principle, the precautionary principle, and the priority of sovereign national rights regarding the management of genetic resources.

The Cartagena Protocol is more specific, referring to the safe transfer and use of GMOs for the protection of the natural environment and public health.

Animal handling (meaning, any possible intervention in their organism) is governed by Conventions of the Council of Europe, such as the *Convention for the Protection* of *Vertebrate Animals* used for *Experimental* and other Scientific Purposes (1986), the *Convention* for the Protection *of Animals during International Transport.* (revised text, 2003), the *Convention for the protection of Animals kept for Farming Purposes* (1976), *the Convention for the Protection of Animals for Slaughter* (1979), and the *Convention for the Protection of Pet Animals* (1987).

iii. The international Biolaw also includes instruments related to patent rights. The recognition of patents for biotechnological applications in humans or other species is a highly controversial issue affecting advancements in innovative technology directly. Patents represent the most important motivation for attracting funds to cutting-edge technologies that demand significant investments while they involve a high risk of failure.

The international instruments in relevance are (a) The *Agreement on Trade-Related Aspects of Intellectual Property Rights* (*TRIPS Agreement*, WTO 1995), particularly its article 27, regarding exceptions from patenting, (b) the *European Patent Convention* (Munich, EPO 1973), particularly its articles 52 para 4 and 53, on the same issue, and (c) *the Budapest Treaty on the International Recognition of the Deposit of Microorganisms for the Purposes* of *Patent Procedure* (1977).

II. *The EU law:* From the late 80s already, the European Communities proceeded to the first legislative steps on Biolaw.[10]

In 1990, we had the enactment of the basic Directives 1990/219/EC *"on the contained use of genetically modified micro-organisms"* and 1990/220/EC *"on the deliberate release into the environment of genetically modified organisms."* Both Directives were relevant to genetic engineering in non-human species, setting up preventive control systems to manage these biotechnological products in order to eliminate potential risks to the environment and health. This regulation was transferred into the member states' internal legal order, and its actual content remains still

[10]Dubos (2004), pp. 101 seq. (referring mostly to the European Group on Ethics, as the basic advisory organ at the level of the European Commission).

in force. Due to technological developments, modifications to that content are related mostly to GMMs and GMOs' technical characteristics, as they are described in relevant annexes. Besides, the Directives included an explicit provision for their review's necessity in their original form after technology advances. Currently, *Directives 2001/18* on GMOs and *98/81* on GMMs incorporate new developments, although there are still doubts on whether they apply to the recently developed genome editing methods (CRISPR Cas9, and similar).[11]

Two Regulations also govern the area of GMOs on *GM food and feed* (1829/2003) and on *traceability and labeling* of these products (1830/2003).[12]

Furthermore, the EU law encompasses specific regulations on patenting in biotechnology, regarding applications in humans and non-human species. *Directive 98/44/EC* had triggered controversial opinions, regarding the extent of patentable applications, especially in humans. Nevertheless, in 2001, the EC Court rejected a relevant application of the Netherlands (supported by other member states as well) against the entrance in force of that Directive.

Applications of Genetics in humans are only partially regulated at the level of the EU law. Most relevant is the *General Regulation on Data Protection* (*GDPR* - 679/2016), which also covers the collection and processing of genetic data as a "special category" of data (art. 9).

In the field of clinical research, the *Clinical Trial Regulation* (536/2014) has replaced the previous legislation of Directives 2001/20/EC and 2005/28 (Good Clinical Practice/GCP Directive), regarding interventional clinical studies with pharmaceutical products in humans. That legislation governs the most sensitive area of biomedical experimentation involving the administration of experimental drugs into the human organism.

Directives 2004/23/EC and *2006/17/EC* regulate all issues related to the safe management (conservation, storage, and transfer) of human biological material for transplantation use, assisted reproduction, or other clinical use (but not research use), governing, mostly, the function of tissue and gamete banks.

In the area of animal welfare, *Regulation 1/2005* concerning the protection of animals during transport and related operations, establishes control measures ensuring conditions of welfare similar to those required by the Council of Europe's Convention of 2003. Also, *Regulation 1099/2009*, and *Directive 98/58* on farm animals are based on the Council of Europe's relevant Conventions that we mentioned previously.

Regarding preclinical experimentation, *Directive 2010/63* governs the use of vertebrate animals in biomedical research, providing a more detailed regulation compared to that of the Council of Europe's Convention mentioned above.

---

[11] See, in relevance, judgement C-528/16 (Confédération paysanne and Others v. Premier ministre and Ministre de l'Agriculture, de l'Agroalimentaire et de la Forêt) of the European Court of Justice/ECJ (2018), that accepts the applicability of the current legislation.

[12] Reg. 1830/2003 also contains amendments to Dir. 2001/18. Reg. 1946/2003 regulates the cross-border transfer of genetically modified organisms (GMOs) following the Cartagena Protocol.

In the EU law, we do not yet have specific instruments pertinent to essential areas of Biomedicine, such as assisted reproduction and end-of-life decisions. These areas are covered, at present, with national legislation in most EU countries.

1.2.4 The Position of Biolaw in the Legal System

We can now attempt to locate Biolaw within the modern legal system's general classification, based on the above presentation.

First, we need to avoid confusion with two relevant branches of law:

- a) the Medical Law, the object of which is more restricted, as it focuses on the patient/physician relationships, but also, it is more extensive since it covers all organizational aspects of the health system (including public funding, insurance, hospital and clinic organization, and administration, etc.)
- b) the environmental law, which is interested in legal regulation concerning the human societies/natural environment relationships, encompassing also the impact of human activity on non-living environmental elements but does not cover issues related to the modification of the biological nature of living species, particularly in research contexts.

Furthermore, as happens with other new law branches also, Biolaw represents a comprehensive corpus of legal rules, affecting almost every traditional branch either of Public or of Private Law.

Rules of Biolaw are closely associated with Public Law since:

i. They regulate situations of conflict between interests representing aspects of fundamental rights (i.e., privacy v. research, in the field of genetic data processing). Also, they establish limits to the exercise of rights, under the view of fundamental legal principles (i.e., limits to research for protecting human dignity in the field of clinical trials, or limits to the economic freedom for protecting the environment in the area of genetically modified organisms' development). From that aspect, the rules of Biolaw are associated with Constitutional Law, in particular with the Law of Human Rights.

ii. They often provide general regulation, at a regional or global level, regardless of national legal orders, for controlling advanced technological developments not restricted within national borders. Therefore, rules of Biolaw are included in international instruments, with binding or no legal force, and thus fall within the interest of Public International Law.

iii. In many cases, they involve the state to guarantee their implementation and efficiency by adopting preventive or coercive controls at various levels (licensing, sanctioning systems). These rules affect the establishment and function of research and business agents or individual experts, both in the public and the private sector (universities, research institutes, fertility clinics, hospitals, biobanks, etc.). In that sense, Biolaw is associated with Administrative Law.

iv. Sometimes, rules of Biolaw regulate sensitive issues with vital socio-ethical significance, involving also criminal sanctions (i.e., in abortion, assisted reproduction, euthanasia, assisted suicide, transplantations, etc.); they fall therefore within the interest of Criminal Law.

On the other hand, rules of Biolaw are associated with branches of Private Law also:

i. They regulate relationships between individuals acting in terms of free will, with no presence of state authority. Even if in private contracts frequently one of the contracting parts holds a dominant position (like, for example, the physician over the patient, or a food or drug company over consumers, etc.), there is no doubt that the legal basis for these contracts is the autonomy of will, as considered in the framework of classic Civil Law.
ii. They encounter, to a significant extent, issues related to intellectual property, particularly concerning the recognition of patent rights related to biotechnological applications. The patenting system functions as a major motivation for developing research policy in modern Biotechnology; Biolaw thus is associated with Commercial Law and Patent Law.
iii. In some cases, they affect the arrangement of relationships in the workplace and private insurance (regarding, for instance, the collection and processing of medical and genetic data), having thus a close contact with Labor and Insurance Law.

That intricate image makes doubtful an approach of "isolated" branches and reveals the integrated nature of modern law and Biolaw in particular.

1.3 Introducing an Evolutionary Approach

Bearing in mind the above presentation, in the following, we will try to analyze the significant issues of Biolaw, with reference mainly to the positive European law, as articulated by both the international legislation and the EU legislation in relevance, as well as the case-law of the European Court of Human Rights and the EU Courts.

These issues are systematically classified into two parts. The first includes those related to human life management, while the second refers to all forms of management of the non-human species' life (microorganisms, plants, and animals). The distinction is legally justified since current law recognizes humans as the only "subjects" enjoying a supreme value *per se*, whereas all other living beings are considered legal "objects" having only relative value meant mostly for serving the interests and needs of humans.

Still, this study defends an evolutionary view of Biolaw, according to which we need to consider a closer connection of positive law with both the findings of life sciences and the developments in artificial intelligence and robotics. This approach criticizes the absolute correspondence of legal "subjects" with the human species, suggesting an open meaning of that fundamental legal concept that may encompass other beings as well.

This argumentation is grounded on two considerations. On the one hand, human life cannot exist without the support of non-human organisms, which, in terms of values, means that the value of the latter is, in fact, not separate but an integral part of the human value itself. On the other, it is not the material of living organisms (namely their specific structure with the presence of DNA) but rather the functions they can develop that really matter for the law since these functions influence our social life and are subject to regulation. Following these considerations, a different legal status should be observed not only for some non-human organisms but also for some "intelligent machines," although this conclusion presupposes future significant changes in the existing provisions of Biolaw.

References

Andorno R (2004) Towards an international bioethics law. Journal International de bioéthique 15: 131–149

Andorno R (2013) Principles of international biolaw. Seeking common ground at the intersection of bioethics and human rights. Bruylant, Brussels

Belanger M (2004) Le système d' Oviedo et l' établissement d' un nouvel ordre Européen de la bioéthique. Journal International de Bioéthique 15:73–87

Byk C (2004) International law and life sciences: a new battlefield of power or a new temple for justice and peace? Journal International de Bioéthique 15:151–162

Callahan D (1995) Bioethics. In: Post SG (ed) Encyclopedia of bioethics, vol I, 3rd edn. Macmillan Reference, New York 2004, pp 278–287

Carlos Loureiro J (2000) The Kemp principles: a bio-legal perspective. In: Dahl Rendtorff J, Kemp P (eds) Basic ethical principles in European bioethics and biolaw, vol II, Partners' Research, Centre for Ethics and Law, Institut Borja de Bioética, Copenhagen – Barcelona

Dahl Rendtorff J, Kemp P (2000) Basic ethical principles in European bioethics and biolaw, vol I, Autonomy, dignity, integrity and vulnerability. Centre for Ethics and Law, Institut Borja de Bioética, Copenhagen – Barcelona

Deharo G (2006) La science est-elle un instrument de pouvoir concurrent dans l' exercice de l' activité juridictionnelle? Journal International de Bioéthique 17:33–54

Dubos O (2004) Droit communautaire et bioéthique: Étude des internormativités à travers les avis du Group Européen d' Éthique. Journal International de Bioéthique 15:101–127

Duprat JP (2004) Le biodroit, un phénomène global sans principe unificateur? Journal International de Bioéthique 15:37–50

Durand G (1999) Introduction générale à la Bioéthique, Fides, Cerf, Montreal

Hayry M, Takala T (2005) Human dignity, bioethics, and human rights. Dev World Bioeth 5:225–233

Jonsen AR (1998) The birth of bioethics. Oxford U.P., Oxford, New York

Kachalia A, Choudhry NK, Studdert DM (2005) Physician responses to the malpractice crisis: from defense to offense. J Law Med Ethics 33:416–428

Michaud J (2006) De la procédure bioéthique: Experience judiciaire, pratique des comités. Journal International de Bioéthique 17:61–68

Nys H, Stultiens L, Borry P, Goffin T, Dierickx K (2007) Patient rights in E.U. Member States after the ratification of the Convention on Human Rights and Biomedicine. Health Policy 83:223–235

Ten Have H (2018) The Universal Declaration on Bioethics and Human Rights as a landmark in the development of global bioethics. In: Caporale C, Pavone IR (eds) International biolaw and shared ethical principles. The Universal Declaration on Bioethics and Human Rights. Routledge, London, and New York, pp 31–39

Part I

Persons

2 Introduction: The Concept of Biological Autonomy

At the very moment when newborns are released from the maternal organism, they acquire a unique quality characterizing their entire future life. At that moment, precisely, the newborn becomes a "person," the exclusive "subject" in legal terms. As such, the person enjoys rights, representing a privileged individual compared to any other living or nonliving being of the natural world.

The crucial element of that superior position of persons in the legal world is autonomy, a space of free existence and activity in which no interventions of others are allowed without the person's permission. In Biolaw especially, the most exciting dimension of personal autonomy is what we can call *"biological autonomy"*: Any intervention of others in the person's biological constitution eventually affects that unique dimension of freedom.

Therefore, with "biological autonomy," we define *any person's ability to control all functions and data of their organism as a mere biological entity.* This is not an absolute ability; it is determined by Nature since many functions and data of the human organism remain unknown, and also many known functions are not dependent on our will and control (like the function of the heart and other vital organs). Nonetheless, besides these natural limitations, our body's control may be considered part of our self-determination's free space. It is an integral expression of our autonomy, the most important in fact, since all other expressions of autonomy, through the exercise of fundamental rights, seem to presuppose it.[1]

Being a dimension of personal autonomy, biological autonomy is extended in terms of time between the beginning and the end of a certain person. We need to examine these two crucial limits as they indicate the importance of human biology for the law.

---

[1] On the significance of human body in the conceptualization of human dignity, see Moutouh (1999), pp. 167, 172.

T. Vidalis, *The Emergence of Biolaw*, Springer Textbooks in Law,
https://doi.org/10.1007/978-3-031-02359-0_2

2.1 The Beginning of Persons

The first question in relevance is whether the birth of a human individual represents the accepted "beginning" of a person, in legal terms. Why this moment and not a previous or a later one is considered legally suitable for attributing the qualities of "person" and "subject" to a living human being[2]? The basic instruments of the European Biolaw (ECHR, Oviedo Convention) are not quite helpful to provide an answer,[3] since they mention terms like "human value," "human life," and "human being," without guidance about their proper use, before or after birth.

Nevertheless, acknowledging the moment of birth, legally, as the only suitable for the beginning of a person, does not seem absurd, precisely because all meaningful expressions of autonomy, from the elementary to the most complex ones, become certain in the real world only if presupposing birth. Indeed, the moment of birth illustrates the coexistence of a biological and a cognitive element in the newborn's nature. On the one hand, it is the fact of release from the maternal organism, and subsequently the suspension of any further biological dependence; on the other, it is the existence of a cognitive "infrastructure" (a developed brain and neural system) that enables new individuals to gradually exercise actions of autonomy, starting from actions that express their basic instincts.

Before birth, the human fetus is absolutely dependent organically on the maternal organism, not enjoying biological individuality. At that stage, self-determination is impossible, also given that no cognitive element occurs yet since the brain and the neural system are still under development.[4] After birth, the newborn has acquired biological independence, a signal of its biological autonomy. Simultaneously, the cognitive condition also exists, as brain and neural development have been completed. Spontaneous expressions of basic instincts, demanding feeding, etc., indicate sufficiently the ability of the individual's self-determination, destined to last an entire life.

Two necessary remarks should be mentioned here: a) The fact that the human fetus before birth is not recognized as a "person" does not mean that it remains with no legal value. On the contrary, several provisions existing in all modern legal systems, attribute to the unborn human life special status of protection, supporting its potential to be developed into a subject of law. However, that status does not emanate from the "rights" or "interests" of the fetus itself.,[5] since a crucial condition for acknowledging them lacks, that is the quality of a "person." b) Certainly, quite

---

[2] On the concepts in relevance and emerging ambiguities, see Palazzani (2007), pp. 89–98.

[3] See ECtHR, Vo v. France (2004), concluding that the "beginning" of the right to life (that is, the beginning of a person) is a matter of national legislation.

[4] Burgess and Tawia (1996), pp. 1 seq., Trevarthen (1987), pp. 101 seq. Also, Wiesemann (2006), pp. 117 seq.

[5] Schulz (2001), pp. 527–528. For a similar interpretation of the UN Convention on the Rights of the Child, see Alston (1990). Contra, Mathieu (1999), pp. 452–454. See also, Andorno (1994), pp. 151–152, arguing that the necessity for the embryo protection derives from the ambiguity regarding its status.

long after birth, humans remain incapable of surviving without assistance from other persons. In particular, newborns are unable to exercise acts of self-determination, at least beyond a very elementary degree. Still, this does not mean that autonomy does not exist, to conclude that the quality of a "person" is disputable.[6] On the contrary, precisely because newborns have acquired that quality, they become able to experience socialization, to learn how to exercise self-determination acts, an ability that goes in parallel with the knowledge of the external world; if the ability of self-determination were absent, no socialization would be possible, as evident.

2.2 The End of Persons

Death seals the legal end of a person. The termination of the human organism's functioning corresponds to removing both the biological and cognitive coexisting elements that determine autonomy. In that sense, a deceased human being cannot be considered a legal "subject" anymore.[7] In contrast to situations where a person, although the holder of rights, remains temporarily unable to exercise acts of self-determination (due to specific social conditions, like imprisonment or hospitalization, or conditions of health, like a temporary loss of conscience), in the case of death, the autonomy loss is irreversible; this makes meaningless an assumption considering deceased humans as subjects of rights or legal subjects in general.

The definitive loss of self-determination transforms these previous persons into "objects" of law, placing them at an inferior level in terms of value. The fact that legal systems require certain respect for the dead human body,[8] is not inconsistent with that quality of a legal "object." To attribute specific value to objects does not change their legal nature. That value is always relative and serves the respective rights of existing persons, by "reflecting" their own personality. In the specific case of dead humans, all legal provisions requiring respect for the dead body, the exclusive treatment involving burial, funeral acts, etc., are generated by the fundamental rights to personality, to private and family life of third existing persons, namely, the relatives, friends or the members of the community in which the now dead lived in the past. These real rights include feelings for the dead, and this is where respect for the body is embedded, contrary to an approach arguing for an "extending" conception of "rights" and "personhood" even after death.

[6]Contra, Tooley (2001), pp. 124–125.

[7]Luchaire (1994), p. 1650, referring to the French Constitution.

[8]See art. 18 of the Add. Protocol on Transplantations of the Oviedo Convention.

2.3 Advanced Control Over the Human Organism

Not only do birth and death determine the extent of our biological autonomy. Progress in technology indicates that many functions of the human organism are now subjected to biological autonomy; that is, the person can intentionally control them, and, in that sense, they are not guided by the teleology of Nature anymore.

Although being realized only over the last decades, this development had been prepared throughout the whole course of Medicine. Indeed, since its very first steps, Medicine has expressed nothing but the relief of the human organism from natural deterioration caused by disease or other pathology; Medicine has introduced, in fact, alternatives for the condition of our health, other than what Nature dictates.

Moreover, due to modern developments in Biomedicine, namely a variety of biological applications to the human organism, our control has, nowadays, a meaning beyond health restoration. Modern technology makes possible the artificial prolongation of life, supports human reproduction, and provides methods for enhancing our natural characteristics and abilities, or even for "designing" our offspring. In that constantly enlarged area of our biological self-control, the exercise of biological autonomy is not undisputable. Just the opposite: Our decisions on biological self-determination may be contested by the intervention of other persons, experts, family members, or agents of the public interest. Are there any limits to such interventions, directly affecting the condition of our biology?

Given that framework, the first Part of this study intends to cover the legal aspects of the essential expressions of a person's biological autonomy. Pathology, biomedical research, death, data management, reproduction, and enhancement are topics where biological autonomy remains always a matter of controversy. In an additional chapter, we will examine the question of the human body's ownership as a necessary dimension of the biomedical part in Biolaw.

2.4 Further Studying...

ECtHR: Vo v. France, 8 July 2004 (Grand Chamber)

"Owing to a mix-up with another patient with the same surname, the applicant's amniotic sack was punctured, making a therapeutic abortion necessary. She maintained that the unintentional killing of her child should have been classified as manslaughter.

The Court held that there had been no violation of Article 2 (right to life) of the Convention. It found that it was not currently desirable or possible to rule on whether an unborn child was a person under Article 2 of the Convention.

(continued)

And, there was no need for a criminal law remedy; remedies already existed allowing the applicant to prove medical negligence and to seek compensation”.[9]

Question

Comment on the following paragraphs, under the light of the ECHR and the Oviedo Convention:

84. At European level, the Court observes that there is no consensus on the nature and status of the embryo and/or foetus (...), although they are beginning to receive some protection in the light of scientific progress and the potential consequences of research into genetic engineering, medically assisted procreation or embryo experimentation. At best, it may be regarded as common ground between States that the embryo/foetus belongs to the human race. The potentiality of that being and its capacity to become a person – enjoying protection under the civil law, moreover, in many States, such as France, in the context of inheritance and gifts, and also in the United Kingdom (...)– require protection in the name of human dignity, without making it a “person” with the “right to life” for the purposes of Article 2. The Oviedo Convention on Human Rights and Biomedicine, indeed, is careful not to give a definition of the term “everyone”, and its explanatory report indicates that, in the absence of a unanimous agreement on the definition, the member States decided to allow domestic law to provide clarification for the purposes of the application of that Convention (...). The same is true of the Additional Protocol on the Prohibition of Cloning Human Beings and the Additional Protocol on Biomedical Research, which do not define the concept of “human being” (...). It is worth noting that the Court may be requested under Article 29 of the Oviedo Convention to give advisory opinions on the interpretation of that instrument.

85. Having regard to the foregoing, the Court is convinced that it is neither desirable, nor even possible as matters stand, to answer in the abstract the question whether the unborn child is a person for the purposes of Article 2 of the Convention (“personne” in the French text). As to the instant case, it considers it unnecessary to examine whether the abrupt end to the applicant’s pregnancy falls within the scope of Article 2, seeing that, even assuming that that provision was applicable, there was no failure on the part of the respondent State to comply with the requirements relating to the preservation of life in the public-health sphere. With regard to that issue, the Court has considered whether the legal protection afforded the applicant by France in respect of the loss of the unborn child she was carrying satisfied the procedural requirements inherent in Article 2 of the Convention.

[9] ECtHR, Press Unit.

References

Alston P (1990) The unborn child and abortion under the draft convention on the rights of the child. Hum Rights Q 12:156–178

Andorno R (1994) Les droits nationaux Européens face a la procréation médicalement assistée: primauté de la technique ou primauté de la personne? Revue Internationale de Droit Comparé 46:141–152

Burgess JA, Tawia SA (1996) When did you first begin to feel it? Locating the beginning of human consciousness. Bioethics 10:1–26

Luchaire F (1994) Le Conseil constitutionnel et l'assistance médicale à la procréation. Revue de Droit Public et de la Science Politique en France et a l'Étranger 6:1647–1658

Mathieu B (1999) La recherche sur l'embryon au regard des droits fondamentaux constitutionnels. Recueil Dalloz Sirey de doctrine de jurisprudence et de législation 41:451–456

Moutouh H (1999) La dignité de l' homme en droit. Revue de Droit Public et de la Science Politique en France et a l'Étranger 1:159–196

Palazzani L (2007) Person and human being in bioethics and biolaw. In: Weisstub DN, Diaz-Pintos G (eds) Autonomy and human rights in health care. An international perspective. UCLM (Universidad de Castilla- La Mancha) Springer, Dordrecht, pp 89–98

Schulz S (2001) Schleichende Harmonisierung der Stamm-zellforschung in Europa? Europarat und Europäische Ethik-Räte könnten Ansätze zu gemeinsamen Standards bieten. Zeitschrift für Rechtspolitik 34:526–529

Tooley M (2001) Personhood, in H. Kuhse – P. Singer (eds) A companion to bioethics, Blackwell, Oxford, pp. 117-126

Trevarthen C (1987) Brain development. In: Gregory RL, Zangwill OL (eds) Oxford companion to the mind. Oxford UP, Oxford, New York, pp 101–110

Wiesemann C (2006) Relational ethics and the moral status of the embryo. In: Deltas C, Kalokairinou EM, Rogge S (eds) Progress in science and the danger of hubris. Genetics, transplantation, stem cell research. Waxmann, Munster, pp 117–126

Pathology

3

3.1 General

The most common experience for understanding biological autonomy is human pathology, when diseases or injuries distract the organism's regular function. Regardless of a pathological condition's seriousness, two standards should be considered in relevance; first, biological disorders obstruct to a certain degree the person's self-determination, and second, the restoration of the latter presupposes restrictions to some of its expressions, that is, to individual freedoms. In that sense, biological autonomy is affected twice; on the one hand, by pathology itself, on the other, temporarily, by treatment's requirements.

Biological autonomy reflects our freedom to address pathological conditions. This does not necessarily mean a positive outcome, namely the full recovery of our organism (which is not possible in chronic diseases, for example). It means also attempts purposing to relieve pain and suffering or moderation of symptoms. Addressing pathology belongs, thus, to our self-determination.[1] It cannot be imposed on us, as an obligation, by the state or others. Every person is the only one entitled to decide "if" and "how" he/she will cope with his/her organism's unhealthy condition, which means that we are not obliged by anybody to be (or, rather, to try to be) "healthy." Of course, we can imagine some limitations to that freedom, particularly for the sake of other persons' rights, but on the condition that its core remains inalienable. An opposite argument would undermine self-determination as a principle, leading to a sort of person's instrumentalization, even if health is necessary for enjoying most of our rights; in a democratic society, any paternalistic

[1] See ECtHR, Scialacqua v. Italy (1998), in which the right to Life is associated with the social right to Health. According to the Strasbourg Court, the patient autonomy is grounded on the right to Private Life (art. 8 ECHR). The Oviedo Convention has no specific provision in relevance besides that protecting human dignity (art. 1).

T. Vidalis, *The Emergence of Biolaw*, Springer Textbooks in Law,
https://doi.org/10.1007/978-3-031-02359-0_3

imposition of healthcare decisions, is inconsistent with the very concept of personal freedom.[2]

Given that, the social right to Health, as declared by the European Social Charter.,[3] has the sole meaning of a legal guarantee requiring the state to provide a minimum of adequate infrastructure and health services, for enabling persons to encounter pathological conditions freely; the social right to Health does not imply, thus, some kind of public "duty" for persons to ensure their individual good health.

3.2 The Concept of Health

With "pathology," we understand some kind of restriction imposed by Nature on our autonomy due to biological reasons. These may be only temporary and disturb only specific expressions of self-determination when, for instance, we suffer from non-serious diseases or injuries (like a cold or a hand fracture), or they may result in extensive damage to autonomy (like in cases of psychiatric diseases, or paralysis). Moreover, some diseases lead to death, that is, the definite loss of autonomy.

The opposite state is "health." Its concept, in some situations, is not easily determined.[4] It may depend on social perceptions or even it may be confused with personal differentiations in lifestyle. This means that, sometimes, a typical situation of the human organism may be considered "pathological" just because it is difficultly adapted to certain strict social conditions imposing concrete habits, preferences, behavior, etc. The most characteristic example comes, of course, from the history of Psychiatry, revealing many cases of human behavior perceived as "unhealthy" just because they were deviating from specific "normal" social standards, including political attitudes, particularly in pre-modern societies or totalitarian regimes.

Nevertheless, we cannot deny that the concept of health has a core not determined by social perceptions or conditions. We can define that content as *the biological state of a human organism in which its structure and functions remain unaffected by random external factors that may produce biological harm or decline*. If this condition is not met, we face a situation of pathology.

Some may argue that this definition of health is insufficient. According to that approach, health is not merely the absence of pathology, but something much more: It is the condition in which the human organism is presented in a place to develop its abilities fully. However, this latter definition of health, adopted nowadays by the

[2]The issue of intentional harm to the human organism (that may result even in the person's death) needs to be addressed under the view of biological autonomy also. Thus, the practice of force-feeding in hunger strikers violates their biological autonomy, no matter if one may consider this as a measure that protects their health. This is because that practice entails the person's instrumentalization since his/her autonomous decision to deny food is not in conflict with another person's special right or the public interest. In some occasions, force-feeding may entail physical or psychological pain and, as such, falls within the concept of torture (art. 3 ECHR).

[3]Article 11.

[4]On the history of social perceptions of disease, see Engelhardt (1974), pp. 234 seq.

WHO,[5] cannot avoid two problems. First, a logical problem occurs since, even if we define health as an ideal situation that we are trying permanently to reach, we still need to contrast it to the condition of pathology; an intermediate situation between health and pathology would be logically meaningless. Under this view, by setting an optimal condition of health, we enlarge the concept of pathology, which leads us to a second essential question: If health is an ideal practically never achieved, are we all not considered "patients," or at least "unhealthy," needing some kind of treatment? Once again, we meet here an approach imposing a paternalistic model, denying biological autonomy and health as an object of individual freedom. In extremis, that view may inspire hostile social discriminations against those not accepting "treatments" for "improving" their supposed "imperfect" health.

These problems lead us to adopt the first "narrow" definition of health, stressing as a crucial element the absence of physical or mental harm, as more compatible with the modern law's principles and rationale, for not compromising personal autonomy.

3.3 The Patient/Doctor Relationship

Addressing pathological conditions involves clinical acts performed by medical doctors having specific knowledge for intervening in the human organism usually with the support of other health professionals (nurses, laboratory personnel, etc.).

The contact of patients (or persons asking for medical care in general) with medical doctors and supporting experts inevitably affects aspects of personal autonomy. Thus, a patient discloses sensitive elements of private or family life, necessary for the attending physician to act on the basis of that information. Moreover, a hospitalized patient accepts strict restrictions on the freedom of movement, and patients undergoing a surgical operation risk their corporal integrity. However, all the above are nothing but the intentional exercise of biological autonomy by persons willing to accept temporary limitations to general autonomy in order to recover from pathology. Thus, patients proceed to a specific balancing between different dimensions of self-determination; they take risks by restricting temporarily the exercise of certain fundamental rights, as a necessary condition for achieving a complete restoration of their overall autonomy, damaged by pathology.

From the European Biolaw's point of view, that restriction of patients' self-determination is justified if three conditions are met: respect for human dignity, informed consent of the person interested, and protection of medical secrecy.

[5] *"Health is a state of complete physical, mental and social well-being and not merely the absence of disease or infirmity"* (Preamble to the Constitution of WHO as adopted by the International Health Conference, New York, 1946, Official Records of WHO, no. 2, p. 100).

3.3.1 Human Dignity

Respect for human dignity is an essential condition for a legitimate medical act. It is acknowledged already from the Hippocratic tradition of western medical ethics, in which doctors have full control and responsibility for the patients' care, while patients played a relatively "passive" role.

Indeed, the patient/doctor relationship was governed in this paternalistic model by doctors' dominant position, permitting an almost absolute discretion for deciding and performing medical acts. Patients had only the freedom to choose the doctor they trusted, based on personal preferences and supposed skills. Still, trust was not ungrounded. Doctors were morally obliged to act only for the benefit of patients: *"Beneficence or non-maleficence"* represents the core moral principle of the classic model in Medicine, and, as such, it reveals, in terms of modern ethics and law, respect of the patient's human dignity.[6] For that reason, this principle continues to characterize medical ethics in the modern era of Medicine also.

From the aspect of modern law, this means that no medical act is legitimate if it ignores the patients' fundamental rights since it violates their human dignity by transforming them into a simple instrument of uncontrolled handling by the physician;[7] no matter how serious the need for healthcare, this is an absolute limit regarding any medical act.

The human dignity principle finds its explicit expression in the European Biolaw, particularly in article 1 of the EU Charter of Fundamental Rights and the same article of the Oviedo Convention.[8]

3.3.2 Informed Consent

By exercising biological autonomy, patients and healthy persons asking for medical services have the last word before undergoing medical acts of diagnosis, prevention, or treatment. This is expressed in modern medical ethics and law by an additional condition that characterizes medical care; a medical act presupposes the person's consent after information about its potential consequences.

The "informed consent" principle represents a novel element in medical ethics, established just in the second half of the twentieth century.[9] Declared explicitly for the first time in the field of biomedical research (after the experience of the Nazi

[6] Affecting even the patient's right to Life. See ECtHR, Center of Legal Resources on behalf of Valentin Campeanu v. Romania (2014). Also, ECtHR, D. v. UK (1997), mentioning "degrading treatment" (art. 3 ECHR).

[7] Meaning that, in principle, self-determination of patients never stops, contrary to the argumentation of. Ackerman (1982), pp. 80 seq.

[8] In soft-law, see art. 3 para 1 of the UNESCO's Bioethics Declaration.

[9] See, in detail, Beauchamp and Childress (2001), pp. 77 seq.

atrocities regarding experimentation on prisoners), soon after it covered any field of medical practice.

Informed consent illustrates the role of a biologically autonomous recipient of medical care; that person is no more a "passive" factor in the fundamental medical relationship, as it happened under the traditional paternalistic model.[10] The current European Biolaw[11] recognizes this role explicitly. Thus, according to article 5 of the Oviedo Convention:

> An intervention in the health field may only be carried out after the person concerned has given free and informed consent to it.
>
> This person shall beforehand be given appropriate information as to the purpose and nature of the intervention as well as on its consequences and risks.
>
> The person concerned may freely withdraw consent at any time.

Exceptions to that rule exist in persons unable to provide consent and in emergency situations, as we will see in detail later. However, even in these cases, the Oviedo Convention (art. 6–9) provides guarantees of protection for avoiding inappropriate treatment of the person undergoing medical care.

Indeed, what ensures the "informed consent" rule is that the interested person always holds direct control of any medical act and its consequences. More specifically, direct control means that the person always[12] has the freedom to consider and balance the need for temporary autonomy restrictions aiming to support health recovery, not leaving this area of decisions to the attending physician.[13] Three remarks are necessary here:

a) Consent needs to be harmonized with the subject's human dignity. It is valid only on that condition since, by definition, human dignity represents the ground of any free will. This is also expressed by article 1 of the Oviedo Convention, recognizing dignity as an objective of superior protection. Under this view, for example, the informed consent that accepts the person's instrumentalization by transforming the latter into a means of experimentation would not be valid.
b) The patient either provides "consent" to a unique proposed medical act or makes a "choice" between alternative acts that the physician proposes ("informed

[10]The forcible medical intervention or treatment falls within the content of art. 3 ECHR (prohibition of inhuman and degrading treatment). See ECtHR, Jalloh v. Germany (2006), R.S. v. Hungary (2019).

[11]See art. 3 para 2 a of the EU Charter of Fundamental Rights, art. 5 of the Oviedo Convention. Also, see art. 6 of the UNESCO's Bioethics Declaration.

[12]Meaning that consent should be considered a continuous act, characterizing at every particular moment the patient/physician relationship. See on that, Di Paolo et al. (2019).

[13]Regarding patient self-determination, see ECtHR Y. F. v. Turkey (2003), Glass v. U.K. (2004). Yet, the Court has also accepted restrictions to patient consent, for public interest reasons (art 8 para 2 ECHR); see Sandren v. Sweden (1996)—control of disability for determining pension rights—and J.R., G.R., R.R., Y.R. v. Switzerland (1995), for the protection of public health.

choice"[14]). No other form of decision is legally accepted, which means that patients cannot ask for "preferred" medical acts not included in the doctor's information and proposal. For, if patients could "dictate" their medical care to doctors, no scientific independence would be acknowledged to the latter, which is absurd given their expertise. That problem of "preferred" medical acts emerges in the field of enhancement, particularly regarding aesthetic interventions.

c) Consent is considered free only if the previous information is "appropriate." Indeed, for understanding medical interventions and their consequences, common experience is not sufficient. Appropriate information provided by an expert (the attending physician) includes necessarily specific knowledge and should guarantee that the interested person's consent will be severe, that is, wholly conscious and independent, analyzed in autonomous decisions of vital significance. This is a kind of "enhanced" process of individual decision-making, compared to simple decisions of everyday life. Nevertheless, this task is not easily accomplished. Since the freedom of will presupposes a process of information by an expert, a risk of misleading (not necessarily intentional) is always a possibility.[15] We need then to presuppose certain conditions for ensuring genuine information suitable for real consent.
   The critical element of that information is the use of language understandable to the patient, not an extensive presentation of pathology's characteristics or technical details of a particular medical intervention, exposed in scientific terminology.[16] Bearing that in mind, the physician must decide the manner of information that fits a particular case, whether, for example, the information should be formal or not, needing not only oral communication but also a written form. Thus, it is evident that, while for administering a conventional drug the written information provided in its packaging could be sufficient, for a surgical operation, an extensive discussion between the physician and the patient, along with a formal informed consent process, are necessary.[17]

Nevertheless, exceptions are also recognized by the Oviedo Convention to the rule of informed consent. These exceptions refer to persons unable to provide consent, namely minors and adults suffering from a cognitive disability or mental disease.[18]

In these cases, the exercise of biological autonomy by its subject is impeded due to natural reasons, which cause risks for human dignity if medical intervention is to

[14] On differences between these notions, see Nys (2012), pp. 328–329.

[15] Beauchamp and Childress (2001), p. 78.

[16] See in the U.S., Canterbury v. Spence, 464 F2D 762 (D.C. Cir. 1972), regarding "patient-oriented" information.

[17] Following the ECtHR case, X v. Danemark (1983), the patient's consent is sufficient even with no previous information. That position undermines, though, the need for genuine consent, that always presupposes appropriate information. A different issue is the exercise of the "right-not-to-know" by the patient (see *infra*), which is a legitimate option, on the necessary condition that information is available.

[18] See also, art. 7 of the UNESCO's Bioethics Declaration.

be performed. However, the need for such intervention makes necessary an alternative protection of the person concerned.

The law provides, in this regard, two relevant guarantees: a) the involvement of the person's legal representative or a competent organ entitled to provide informed consent for the medical act, b) a requirement according to which the opinion of the patient, if expressed, must be taken into account, as a minimal manifestation of biological autonomy.[19]

Still, protection of health as a legal interest, in some instances, justifies medical interventions even without the guarantee of a legal representative's consent. According to article 8 of the Oviedo Convention, these are situations of emergency in which a timely performance of the needed medical act is vital for avoiding a serious deterioration of health or even death.[20]

Again, two conditions should be observed in these situations: a) The emerged pressure of time must impede absolutely an informed consent procedure, either of the able to consent patient or of any of the legal representatives in charge, in the sense that any delay will lead to irreversible health damage. b) The requirement for respect the patient's dignity by the physician performing the needed intervention remains a legally binding prerequisite. As the attending physician is expected to act alone, the risk of inappropriate medical treatment in such situations (leading to the violation of the patient's human dignity) makes necessary strict evidence for applying this special provision of the Oviedo Convention. This is also the ground for determining the relevant liability of the physician in charge.

Article 9 of the Convention also provides a conditional guarantee for medical interventions in emergency situations, if advance directives of the patient exist. Advance directives, including living wills, formulate a person's previously expressed wishes regarding preferred medical treatment if informed consent cannot be provided. We will examine extensively their consequences in the chapter relevant to end-of-life decisions, where they present particular significance. At this point, it is sufficient to underline that, according to article 9 of the Convention, advance directives are not binding instruments, but they need only to be taken into account by the attending physician. Hence, they are not considered a strict condition for guiding medical intervention in emergency situations, like the other two conditions. In any case, advance directives cannot be equated to a timely obtained informed consent; they cannot "restrain" medical liability, given that eventually, it is the attending physician who decides about their implementation.

---

[19] For children, in particular, see art. 24 para 1 of the EU Charter of Fundamental Rights.

[20] See ECtHR, Mehmet Senturk and Bekir Senturk v. Turkey (2013). However, pregnancy is not considered an "emergency" as a natural process, when no implications in health occur. See Stohl (2018).

3.3.3 Medical Secrecy

The confidential nature of personal information that the patient discloses to the attending physician, and of medical data collected after a series of relevant medical acts, also represents an essential guarantee for patients' autonomy,[21] explicitly acknowledged in article 10 of the Oviedo Convention.

Personal information and medical data are of sensitive nature, meaning that any disclosure of them to third persons may significantly affect the exercise of certain fundamental rights of the patient.[22] It may also affect others' rights, particularly of family members, even after the patient's death.

The guarantee of medical secrecy does not only refer to the patient's health data; it also includes other data of the patient (lifestyle, professional, etc.), and even personal data of third persons that the patient has disclosed to the doctor. That guarantee produces mutual binding force strictly between a concrete patient and a concrete physician, no matter if medical acts will be performed; secrecy is binding for the physician, even if the patient does not decide to undergo medical care, or the physician refuses to undertake or stops medical treatment for scientific or conscious reasons. Moreover, since confidential information may always affect third persons even after the patient's death, for the physician secrecy remains binding for life.[23]

Nevertheless, medical secrecy is also subjected to limitations, following relevant judgments of the Strasbourg Court, either for reasons of public interest or protection of other persons' compelling rights. On such occasions, a relevant appraisal of the conflicting interests is necessary under the concrete circumstances that need to be considered.[24]

3.4 Informed Consent and Medical Liability

The patients' informed consent, as a legal condition required for accepting any medical act, changes the approach of medical liability significantly.

[21] Included already in the Hippocratic Oath. See in relevance, ECtHR, Konovalova v. Russia (2014), regarding students' unauthorized presence during the birth of a child.

[22] Medical secrecy is also considered as a condition of the doctor's reliability. See ECtHR, Z v. Finland (1995), also P. and S. v. Poland (2012). On patient access to medical records, see ECtHR, K.H., and Others v. Slovakia (2009).

[23] Medical secrecy is deeply influenced by new health data processing technologies, particularly with applications of electronic health records (EHR). On the advantages and risks of EHR systems, under a human rights view, and the necessary cooperation of EU countries in that field, cf. Rynning (2007), pp. 105–112. On the issue of medical liability that emerges in relevance, cf. Mangalmurti et al. (2010), pp. 2060–2067.

[24] See ECtHR, Z v. Finland (1995), that accepted data disclosure to criminal court, on the ground of art. 8 para 2 ECHR, and M.S. v. Sweden (1997). In the U.S., see Tarasoff v. Regents of the University of California, 51 P. 2D 334 (Cal. 1976) regarding disclosure of non-medical data belonging to the context of medical secrecy.

An obvious question in relevance is whether patients undertake part of responsibility concerning medical acts decided after their informed consent, even for acts of high risk for their health. In other words, we need to answer whether the legal recognition of patients as active factors of medical treatment means that doctors do not carry the burden of decisions in total; therefore, they do not have to pay the same attention when performing medical acts, as it happened under the paternalistic model of the past. If so, a further problem is whether it is legitimate to transfer part of the responsibility to presumably non-expert patients, even informed, for decisions presupposing specific expertise.

In a first approach, the law seems indeed to accept a specific patient's responsibility (even of the non-expert patient) since it recognizes an active role through the informed consent rule. Nevertheless, we need to bear in mind that the formation of the patient's will depends directly on the nature and content of the information provided by the expert physician; this is where medical liability is embedded mostly in modern law. A patient's consent for undergoing medical acts with no guarantees of previous appropriate information by the medical doctor is in legal terms invalid and cannot restrain the responsibility of the latter for these acts.

In a closer view, we can distinguish two different dimensions of medical liability.

In the first, we assume that the doctor provides appropriate information and obtains genuine consent from the patient, but the performance of the relevant act does not comply with *leges artis* and provokes damage to the patient's health. This is an example of "traditional" liability (criminal, civil, and professional), already existing before the era of informed consent, since the patient's decision does not affect the medical malpractice.

In the second dimension, we assume that the information provided is misleading or false, or inappropriate and ultimately manipulates the patient's consent. There, we address two possibilities; the doctor performs either a false act following the patient's wrong decision or the right act deviating from that decision. In both cases, the doctor remains fully responsible if damage to the patient's health occurs after the intervention; in the first case, because of the nature of information on which the patient's consent was based, in the second, for deviating from the patient's consent.

Following the above analysis, in reality, there is no question for "restraining" the extent of medical liability, even under the model of informed consent; since any decision of the patient remains strictly dependent on the information provided by the doctor, there is no free space for transferring part of responsibility to the patient. The only case in which no medical liability could be demonstrated if the medical intervention damages the patient's health is when this act is performed a) in compliance with genuine consent after adequate information and b) in compliance with *leges artis*. These two conditions must coexist in the model of informed consent, and that is the only difference from the traditional model of medical paternalism.

In addition to what we have already mentioned about the content of information, we can elaborate further on some remarks under the stricter view of medical liability. Any "appropriate" information includes the aim and the nature of the medical intervention in relevance, as well as its implications or potential risks for the patient's health. Even if these elements exist, an information that manipulates the patient's will, not leaving a free margin for reflection, should be defined as inappropriate.

Obviously, this is a difficult balance and cannot be determined without considering the concrete conditions that characterize any particular case. However, a general rule is that information needs to be necessary and not excessive for enabling the patient to decide consciously. Excessively detailed information that states, for example, risks rarely occurred according to relevant scientific evidence, may be misleading for the non-expert patient by causing unreasonable fear, resulting in refusal of consent even for a necessary medical intervention. The same is possible when detailed information is provided in standard form regardless of the patients' educational level or their real ability for understanding. Considering the specific nature of each case, that is, the necessity for adapting respectively the information provided, is a crucial feature of the doctor's responsibility; in most medical acts, information cannot be restricted within a certain formality, but it is ensured only with a direct personal communication between the doctor and the patient.

This intricate balance that the doctor must guarantee during the information process explains why medical liability is not, in reality, minimized under the modern law of informed consent. There are indeed attempts to minimize the extent of medical liability. Under the pressure of a considerable number of medical malpractice cases before the courts occurred some decades ago, particularly in the U.S., the legal requirement of informed consent has been the basis for the emergence of the so-called "defensive medicine".[25] This trend supports the practice of excessive information on the part of the doctors in order to transfer responsibility to patients for coping with the possibility of malpractice. Eventually, defensive medicine promotes the legal protection of physicians instead of the needed care for patients. This represents an apparent deviation from the concept of "appropriate" information, leading to a bureaucratic exercise of medical practice that includes multiple formalities of informed consent, even for routine medical acts. However, since it is the quality and not the quantity of patient information that matters legally, that practice cannot remove the attending physician's responsibilities regarding medical care.

3.5 The "Right-Not-to-Know"

A consequence deriving from the acknowledgment of biological autonomy is the "right-not-to-know," namely the will of the person to deny information about their state of health.[26] The informed consent rule is a necessary condition for any medical

[25] See Kachalia et al. (2005), p. 418.

[26] See, in general, Chadwick (1997), pp. 17 seq. In defense of that right as a matter of principle, see Davies (2020) and, contra, Harris (2020), arguing that other rights prevail (freedom of Expression, etc.).

intervention, but this does not mean that information should be "imposed" on the patient as an obligation.[27]

This is an integrated approach of the individual right to Health based upon our biological autonomy; consent is needed not only for a proposed medical intervention but also for the information to be provided in relevance. Indeed, it is not evident that this information would always be beneficial for the person concerned. Regarding diagnostic findings of severe diseases, that information may disturb the person's life, if leading to uncertainty for the future, influencing family relationships, even causing psychological or psychiatric disorders.[28]

Considering that dimension of biological autonomy, the Oviedo Convention recognizes the right not-to-know (art. 10 para 2) explicitly, as part of the intimate core of the patient's privacy.

An interesting aspect of the right not-to-know emerges when a patient has not expressed a relevant will, and family members disagree with the doctor on whether the patient should be informed anyway, particularly on the occurrence of severe diseases (cancer, etc.).[29] Here, there is a need to balance, on the one hand, the right not-to-know of the patient and, on the other, the physician's medical duty to protect that patient's health. Still, the physician has a legal responsibility to ask the patient whether he/she wants to exercise the right not-to-know or, alternatively, to be informed about the condition of his/her health. In any case, a necessary medical intervention for protecting the patient's health is lawful if the doctor has fulfilled that obligation.[30]

3.6 Treatment Refusal

Since a necessary condition for any medical act is informed consent, the patient's explicit refusal for consenting either to start or to continue undergoing a specific treatment stops the latter, even if the doctor or family members disagree. This is again a logical consequence of informed consent, by which a person exercises the fundamental right to Health negatively.

A patient has a plain reason to refuse or cease medical treatment when believing that the expected benefit would be trivial compared to other benefits that he/she enjoys as an autonomous person (for example, the freedom of movement, aspects of the right to privacy, the right to work, etc.). In other words, by refusing treatment, the patient makes choices between different aspects of self-determination.

[27] On "negatively informed consent," based on patients' trust and requiring only limited information about the aims and not the methodology or risks of medical acts, cf. Kihlbom (2008), pp. 146–149.

[28] Regarding genetic information, in particular, see Malauzat (2000), pp. 339 seq.

[29] On the equation of truth-hiding and lying, see Higgs (1985), p. 512. See, also, Lipkin (1979), pp. 85 seq., Gordon and Daugherty (2003), pp. 142 seq.

[30] See, Fraisseix (2000), pp. 396–397.

However, according to the current European law, a limit to treatment refusal is recognized for protecting the patient's life.[31] As we will see in the chapter on end-of-life decisions, the European Court of Human Rights does not acknowledge a "right to die" as a dimension of the right to life under the European Convention on Human Rights (article 2). Yet, although at the European level human life is protected in this absolute sense, the Court also admits a margin of appreciation for national legislators to recognize end-of-life decisions (like assisted suicide or euthanasia), which means to accept refusal of treatment even if death is a possible outcome. Therefore, the limit imposed on the patient's consent denial (represented by the current legal protection of life), is relative only, insofar as national legislators in Europe may opt for a "right to die".

In any other case, when treatment refusal does not present life-threatening risks, ceasing treatment is a legal obligation of the doctor, in terms of medical liability, regardless of the potential health damage that may occur.[32] The doctor has a legal duty to intervene, in such cases, only in a situation of emergency, and on the condition that informed consent (or refusal) cannot be expressed by the patient or (if unable) by legal representatives,[33] according to the provision of article 8 of the Oviedo Convention.[34]

3.7 Autonomy and Public Health Issues

Public health issues represent a crucial field of interest for Bioethics and Biolaw, raising questions about potential restrictions on individual biological autonomy.[35]

[31] Refusal of blood transfusion for religious reasons is a classic example in relevance. See Kleinman (1994), and for Europe, Loix et al. (2020). In legal terms, that refusal is valid only if end-of-life decisions are explicitly recognized by law, and therefore no medical liability can be deduced. If this is not the case, physicians should proceed to transfusion if considered appropriate (*lege artis*) for saving the patient's life, no matter if refusal has been expressed by the patient or their legal representatives. Regarding the question of compensation by the physician for declining to comply with an expressed transfusion refusal, see McQuoid-Mason (2021).

[32] Following article 5 of the Oviedo Convention. Contra, in ethical terms, Murphy (2005), pp. 189 seq.

[33] However, in the case of mature minors the role of legal representatives is questionable. See in relevance, Moritz and Ebbs (2021). For psychiatric patients, see Pinals et al. (2017).

[34] Several important cases recognize the option of treatment refusal in the U.S. (see in Ch. 5, ref. 6). From the French case-law, see Cour de Cassation (Crim., 3.1.1973) and Conseil d' Etat (6.3.1981, 27.1.1982).

[35] According to article 11 of the European Social Charter, the contracting parties have responsibilities inter alia *"to remove as far as possible the causes of ill health; to provide advisory and educational facilities for the promotion of health and the encouragement of individual responsibility in matters of health; 3 to prevent as far as possible epidemic, endemic and other diseases"*. Also, article 35 of the EU Charter of Rights states that *"Everyone has the right of access to preventive health care and the right to benefit from medical treatment under the conditions established by national laws and practices. A high level of human health protection shall be ensured in the definition and implementation of all the Union's policies and activities"*.

On certain occasions, dangers to public health threatening more persons and possibly an entire population, may require immediate action with preventive measures affecting personal decisions directly. Since decisions on personal health taken by an individual may influence the health condition of others, we address a situation that justifies restrictions on the individual autonomy.

I. *Vaccinations:* The protection of public health is dependent, first, on the prevention of infectious diseases. Vaccinations for healthy adults or children are nowadays the most crucial part of this policy, as they promise high levels of efficiency for addressing risks of many serious diseases. Nevertheless, vaccinations are interventional medical acts, and, as such, they may challenge the status of biological autonomy. The question, here, is whether restrictions on our freedom to accept vaccination are legitimate, especially when public health faces the danger of epidemics or pandemics.

As a medical act, vaccination presupposes the informed consent of the person concerned. The general information provided by public authorities to the population regarding the occurrence of infectious diseases and necessary measures to be observed, although useful in the framework of organized national health systems, does not substitute the prerequisite of individual information and consent; it only encourages access to vaccination on the basis of the free will of citizens. The question of compulsory vaccinations of a particular population emerges in that context. The recent COVID – 19 pandemic on the global scale illustrates the issue characteristically.

In European law, public health is considered a legitimate reason for imposing restrictions on the fundamental right of movement, as stated explicitly by the ECtHR (art. 5). Indeed, in cases of infectious diseases, measures affecting the exercise of that right may be appropriate for preventing the disease's spread to the population. Therefore, prohibitions of access to public areas or even quarantine measures fall within the scope of this article's reservation. However, restrictions on the freedom of movement need to comply with the general criteria of the proportionality principle and respect for the essence of that right; as we know, besides the Human Rights theory, these criteria are now explicitly stated by the EU Charter of Rights (art. 52 para 1), which means that permanent monitoring of the condition of danger needs to be in place for updating the restrictive measures.

On the other hand, in article 26, the Oviedo Convention justifies broader personal autonomy restrictions on the ground of public health protection as a general rule. This stipulation accepts exceptions to that rule only for a limited number of personal rights, that is, for non-genetic discriminations, the rights related to clinical research, those related to transplantations, and the non-commercialization of the human body.

It seems then that the essential safeguards of personal autonomy in health decisions, namely the informed consent principle, the right not-to-know, and the rights related to the person's privacy (medical secrecy, data protection), are considered as potential subjects of restrictions on that ground. In that context, when public health is at stake due to an epidemic situation, and vaccines for the population are available, we need to consider the possibility to accept even measures of compulsory vaccinations for protecting public health. Can we force persons to be vaccinated despite the informed consent prerequisite, on the basis of article 26 of the Oviedo Convention? "Compulsory" means, here, that a person has a public obligation to be vaccinated, under a threat of legal sanctions in case of non-compliance.

The answer to that question should be negative. Again, we must underline that any restrictions on fundamental rights are subjected to strict scrutiny based on compliance with the proportionality principle and non-violation of the right's essential content. Even if the informed consent prerequisite (corresponding to the fundamental right to Health) is not excluded by article 26 of the Oviedo Convention explicitly, it is difficult to imagine which compulsory medical act could be performed without violating the essential content of the right to Health. It is true that in emergency situations, the Oviedo Convention accepts the performance of medical acts with no previous consent (art. 8), but on the condition that the attending physician cannot obtain that consent timely, that is, without jeopardizing the health of the person concerned. Nevertheless, this exceptional case refers to a therapeutic and not a preventing medical intervention; moreover, it does not exclude the informed consent prerequisite in advance but only sets a condition of time for its fulfillment. These are sufficient reasons for concluding that the specific regulation of emergencies respects the essential content of the informed consent principle and thus of the individual right to Health. This is not the case with compulsory vaccinations, where a general obligation backed with sanctions is imposed on all healthy persons with no present harm to their individual health as a public policy preventive measure.[36]

However, we must note that vaccination for certain diseases may be imposed on individuals as a prerequisite for exercising one or more fundamental rights, particularly the right to Movement (for example, to visit a particular country) or the right to Work (for healthcare professionals in hospitals, etc.). In such cases, the person concerned needs to choose between these rights and the right to Health as expressed by the informed consent principle. Even if that choice may be difficult, the person is not threatened by any legal sanctions in the strict sense of this term, as it happens in compulsory vaccinations; therefore, no infringement of the right to Health exists.

Special attention should be paid to regular vaccinations of children. These are preventive medical acts with no necessary relation to the emergence of danger for public health; they aim at protecting the vulnerable health in childhood, against the manifestation of pediatric and other infectious diseases, particularly in environments of necessary stay (kindergarten, schools, etc.).[37] Acting as legal representatives of the child, parents are usually those entitled to consent to vaccination, according to article 6 of the Oviedo Convention, on the basis of the child's best interest. This

---

[36] *Contra,* see ECtHR, Solomakhin v. Ukraine (2012), that accepted compulsory vaccinations, for preventing the spread of infectious diseases. Nevertheless, the Ukrainian law in relevance did not include any sanctions, in case of disobedience.

[37] See on that issue, Ciolli (2008), pp. 129–137, presenting the experience in the U.S., where differentiation in states' legislation exists. Also, ECtHR, Vavřička and Others v. Chech Republic (2021), that accepted vaccinations as a requirement for admission to nursery schools (non-compulsory education).

means that, in principle, parents evaluate the need for certain vaccines, after receiving appropriate information from the pediatrician, like in any medical act. For that reason, compulsory vaccination cannot be a legitimate requirement for the inscription of children in schools of compulsory education, even if the benefit of vaccination is evidence-based. On the other hand, schools need to be informed about non-vaccinated children in order to observe restrictive measures in case of an infectious disease danger for protecting them. Thus, in the event of measles epidemics, or meningitis incidents, a school is justified to exclude non-vaccinated children from class attendance as a prevention measure.

II. *Stigmatization:* The proportionality of measures is also associated with the danger of social stigmatization for certain vulnerable groups of the population. Relevant experience with persons positive for HIV indicates that the characteristics of concrete infectious diseases are crucial for deciding restrictive measures since they determine the standards of proportionality.

It is necessary to clarify that discriminative treatment in social relationships for persons positive for HIV or suffering from AIDS is legally allowed, but only in exceptional conditions. Since infection from HIV is comparatively difficult, in principle, no restriction on personal autonomy is legitimate for protecting public health or even the individual health of the person concerned, including restrictions on the freedom of movement, the right to privacy, the rights of assembly, and association, etc.

Nevertheless, a social prejudice regarding AIDS still accepts the patients' isolation and negative discriminations.[38] This element needs to be taken into account along with the medical secrecy guarantee (see *supra*) for avoiding stigmatization of persons without compromising public health protection.

III. *Allocation of health resources:* Furthermore, the exercise of biological autonomy raises the more general issue of health resources' allocation, which is a crucial aspect of the social right to Health. The allocation of resources refers to public health in the sense of the state's obligation to ensure equal access to available means and services for health care.

Dangers to public health, like the manifestation of epidemics or pandemics, often entail a shortage of preventive or therapeutic means (vaccines, tests, drugs, etc.) and facilities (ventilators, ICUs, etc.), even a shortage of available health professionals, particularly when new diseases emerge. Shortage means that resources are insufficient for covering a particular population; therefore, access criteria in terms of justice need to be observed.

The existing law does not refer to any criteria of prioritization in access under such conditions. The Oviedo Convention recognizes only a general principle of equitable access to health care (art. 3) as an obligation of states "taking into account health needs and available resources." Similarly, the European Social Charter (art. 11) and the EU Charter of Fundamental Rights (art. 35) refer only to the traditional social right to health care. Therefore, a legal approach of "triage" (which is a classic

[38] See ECtHR, Kijutin v. Russia (2011), Novruk and Others v. Russia (2016).

topic in Bioethics) should be associated directly with the meaning of other legal values, particularly the human dignity and the right to life.

A general standard on prioritization is that access to means and services must be ensured, first, for persons experiencing direct threats to their life, due to severe diseases or injuries, in the context of article 2 ECHR, since the life's protection supports their overall autonomy. Dangers for health-threatening human dignity, that is, the ability of self-determination (in cases of serious accidents resulting in multiple injuries, etc.), also require set priority on access, particularly if urgent treatment is needed. Apart from these cases, criteria in access may differ according to the existing circumstances, and they include the priority of vulnerable persons, children, young persons, or healthcare personnel for ensuring the availability of services for the rest population. Priority in ICUs should also take into consideration the time of a particular patient's recovery, in order again to make feasible the treatment of more incidents within the shortest possible time, as a measure guaranteeing equitable access for more patients.

Prioritizing in access results necessarily in limited health care for certain persons (e.g., healthy adults waiting for vaccination or less severely injured, etc.). This is a legally acceptable situation, in terms of equitable treatment, on conditions a) that the state makes reasonable efforts to ensure full healthcare coverage for these persons too, and b) that these persons will have immediate access to the necessary means and services if meeting determined criteria of priority, due to a gradual deterioration of their health.

3.8 Specific Issues

3.8.1 Compulsory Hospitalization of Psychiatric Patients

The patient/doctor relationship must be specifically examined in the case of psychiatric diseases when the patient's hospitalization for undergoing treatment cannot be depended on conditions of free will, which means that the interested person cannot provide informed consent.[39]

Indeed, regardless of the vast differentiation in mental disorders and psychiatric diseases, their common characteristic is a confirmed inability of the person involved to form and express rational will. Here, the absence of will makes problematic the condition of informed consent as a guarantee for the patient's biological autonomy; thus, a psychiatric treatment presents direct risks of jeopardizing the latter.

It is worth noting that, in principle, mental disorders are not excluded from the general application of informed consent, which involves necessarily a legal representative to provide consent on behalf of the person concerned. Even so, this latter must be heard to the extent possible during the informed consent procedure.

[39] See ECtHR, Luberti v. Italy (1984) and, in the U.S. law, O' Connor v. Donaldson, 422 U. S. 563 (1975).

This is the legal approach, according to article 6 of the Oviedo Convention, which also includes a specific article regarding serious mental disorders (article 7), stating that:

> Subject to protective conditions prescribed by law, including supervisory, control and appeal procedures, a person who has a mental disorder of a serious nature may be subjected, without his or her consent, to an intervention aimed at treating his or her mental disorder only where, without such treatment, serious harm is likely to result to his or her health.

A question here is whether, according to article 7, no informed consent is required, either from the patient or from a legal representative. A consistent interpretation is that informed consent of the latter is always necessary, except for an emergency, when "*serious harm is likely to result in his or her [the patient's] health.*" Article 7, thus, excludes informed consent only under that exceptional situation regarding severe mental disorders, following the approach of article 8 on emergency situations, potentially occurring in any pathology.

However, even when the guarantee of informed consent stands, and the legal representative makes the relevant decision, we address just an indication for the interested person's protection, not comparable to a genuine consent of the latter specifically referred to psychiatric treatment conditions of hospitalization. Consequently, the protection of biological autonomy depends on the emotions and ethical behavior of third persons (acting as legal representatives) and, legally, needs to be ensured by a system of adequate requirements and controls for avoiding absurd treatments.

In this respect, firstly, the consent ability of the person concerned must be appraised by an independent expert opinion, that is, by medical specialists.[40] This is a critical requirement because, if inability is confirmed, the legal representatives are entitled to make decisions on medical treatment, including the performance of compulsory medical testing and, furthermore, compulsory hospitalization. The independent expert opinion must refer to these steps explicitly, in the sense that they need to be mentioned to the necessary information of legal representatives before consenting. Secondly, the essential procedural guarantee for a person's compulsory hospitalization requires that the relevant decision should be taken by a competent judicial or administrative organ and not by family members (possibly involved in a conflict of interest situation).[41]

From the moment of hospitalization, the psychiatric patient is treated by the attending physician and supporting health professionals, and undergoes medical interventions with no previous informed consent. This intensifies medical liability, especially regarding standard medical care. On the other hand, concerning serious medical interventions, the attending physician cannot act in an informed consent

---

[40] See, ECtHR, Winterwerp v. the Netherlands (1979), Varbanov v. Bulgaria (2000).

[41] On the conflict of interests regarding decisions of legal representatives, see Hardwig (1993), pp. 187 seq. As an appropriate standard for decision-making, the presumed will of the patient is suggested by Brody (1993).

"vacuum" and needs to obtain specific relevant permission from the legal representatives, according to the general rule of article 6 of the Oviedo Convention.

Finally, compulsory hospitalization stops, regardless of the patient's will, in contrast to what happens with non-psychiatric hospitalized patients or consensually hospitalized psychiatric patients. In compulsory hospitalization, the patient is considered unable to control the treatment process; thus, a return to normal life is always dependent on the judgments of third persons. For that reason, the law must ensure the reliability of these judgments, particularly with provisions requiring them to be based on independent expert opinions.

The unstable situation of a psychiatric patient regarding the exercise of fundamental rights, including the right to Health, is due to the lack of will ability, which substantially restrains biological autonomy. The compulsory hospitalized person, in particular, is exposed to extensive medical interventions for which consent needs to be obtained only in exceptional cases and never by the patient concerned. It is clear that the legal guarantees mentioned above represent a substitute to the lacking personal control, involving inevitably third persons and competent organs, the mobilization of which resulting, often, in complicated legal disputes.

3.8.2 The Case of HIV/AIDS

A particular case regarding the patient/doctor relationship is that of HIV/AIDS. Here, we address issues of medical data protection of persons either positive to HIV or patients. Disclosure of a person's contamination may raise problems of social stigmatization and negative discrimination due to persistent social attitudes of fear regarding the disease (see *supra*). Therefore, the question of medical data confidentiality is of high importance, being the only guarantee for a normal personal and social life for people with HIV or AIDS.[42] Can this affect medical liability?

Protection of medical data is a physician's legal duty, in the framework of medical secrecy recognized by the Oviedo Convention (article 10), as we already mentioned. Doctors are legally obliged not to disclose to other person elements from the medical file of patients, including the result of medical tests (usually standard tests before blood donation) confirming contamination by the HIV.[43]

However, the fact that AIDS is an infectious disease that may be transmitted under certain circumstances of bodily contact with others, partially changes the question of medical secrecy. Indeed, to prevent the virus transmission for the sake of public health, certain exceptions regarding the protection of medical secrecy and the corresponding responsibility of doctors are ethically justified. Thus, the attending physician may disclose the fact of contamination to family members of the person concerned, for considering their own health situation in timely, by undergoing tests and even therapeutic care if needed. Still, this is only an ethical duty of doctors and

[42] See ECtHR, Z v. Finland (*supra*).

[43] See ECtHR, Armonas v. Lithuania, and Biriuk v. Lithuania (2008).

not a legal one; since no medical contract with these persons has been concluded, their information cannot be considered a matter of medical liability. It is worth mentioning that, on the occasion of that disclosure on the part of the doctor, the protection of the patient's medical data is not entirely removed. Indeed, only part of them become known to thirds, and it also happens that these persons, as family members, share with the patient a common space of privacy, protected by specific provisions of law (article 8 ECHR). This may be seen as a minimum guarantee for keeping confidentiality, even in these very sensitive circumstances.

3.8.3 Transplantations from Living Donors

Persons with seriously damaged tissues or vital organs can replace them with other persons' biological material. So far, Medicine has not achieved the use of artificial replicas of kidneys, liver, heart, etc. that might ensure a similar level of performance when functioning in the human organism as natural human organs, despite the significant progress occurring in that sector. Indeed, nowadays, we have relevant efforts in the field of medical devices' engineering, with the development of artificial implants (heart pacers, implants for hearing or vision activating specific areas of the neural system, etc.), including the use of nanotechnology that permits the development of relevant devices at a minimal size.[44] Furthermore, similar efforts occur in the field of biotechnology, aiming at the development of tissue and organs with the use of human stem cells, the development of 3d bioprinted material,[45] or the alteration of animal organs for achieving their compatibility and functionality in the human organism (xenotransplantations).[46]

However, at present, the only available source of transplants is the body of other humans, either living or deceased. This raises complex issues related to the nature of a human person that needs to be resolved at the level of law.

At this point, we will examine specifically the case of living donors of tissue and organs for transplantation purposes. According to the Oviedo Convention (Chapt. VI) and its Additional Protocol on Transplantations, the use of such biological material from living persons is allowed only if no therapeutic alternative of comparable efficiency exists and no transplants from deceased donors are available.

In terms of biological autonomy, the option of obtaining tissue or vital organs from living donors involves risks for their own health inevitably, due to the medical intervention that needs to be performed on their healthy organism. In other words, a healthy person exercises his/her biological autonomy by undergoing a serious

[44] The use of implants, either for enhancement or data collection purposes, are also included here. See European Group on Ethics (2005) and Wood et al. (2003), pp. 16–17, 21.

[45] On the ethical and legal issues in relevance, see Vidalis (2020), pp. 51–68.

[46] See, in general, on the ethical issues occurred in that field, Caplan (1992), pp. 404 seq. From a legal aspect: Byk (1992), pp. 543 seq.

medical act not intending for a certain preventive, diagnostic, or therapeutic goal for individual benefit, but for serving the healthcare of another person.[47]

The Oviedo Convention states certain presuppositions for the protection of donors (art. 19). Therefore, the removal of organs or tissue from a living person is allowed only

i) For the therapeutic benefit of the recipient,
ii) Where no suitable organ or tissue from a deceased person and no other alternative therapeutic method of comparable effectiveness exists, and
iii) Upon the donor's explicit and specific informed consent either in written form or before an official body

The donor needs to be able to consent. Unable to consent persons (minors, psychiatric patients, etc.) are excluded explicitly (art. 20) unless if

i) There is no compatible donor able to consent,
ii) The recipient is a sibling of the donor,
iii) The donation has the potential to be life-saving for the recipient,
iv) The authorization provided by the legal representatives is specific and in writing, "with the approval of the competent body",
v) The potential donor does not object

The Additional Protocol on Transplantations mentions the possibility for a donor to have no family relationship with the receiver, on conditions that the national legislation specify that relationship and after obtaining approval by an independent national organ (art. 10). This is an interesting issue because potential donors may have either motivation of solidarity or a utilitarian motivation for deciding to exercise in this way their biological autonomy, namely to accept the risk of transplant removal.

The option provided by the Protocol must be interpreted in conjunction with the explicit prohibition of financial gain from the commodification of the human body (art. 21 of both the Oviedo Convention and the Protocol). This excludes clearly financial motives, but not necessarily a possibility of personal interdependence between donor and recipient, which may also be harmful to the latter's human dignity. That possibility makes crucial the conditions that national legislators require in order to exclude motives other than solidarity.[48] In the same sense, appropriate

[47] On the influence that relatives may exercise on decisions of putative donors, see Franklin (2003), pp. 56–57.

[48] See, moreover, the case Strunk v. Strunk 445 S. W. 2D 145 (KY. 1967) in the U.S., where organ donation to a sibling was considered a benefit for the donor as well.

guarantees need to be in place for ensuring the independent role of the organ providing the relevant approval.[49]

Nowadays, we know that the illicit trafficking of organs has created an international money-driven market where actual commodification of the human body clearly undermines these provisions.[50] The Council of Europe's Convention on Action against Trafficking in Human Beings and the Directive 2011/36/EU provide a binding legal framework destined to tackle this pressure. The relevant provisions of these instruments that need to be implemented by national authorities should also be taken into consideration when national organs are expected to grant approval for organ donation in the above framework of the Transplantations Protocol.[51]

3.8.4 Gene Therapy: Interventions of Genetic Engineering in Humans

Gene therapy in patients aims to radicalize the treatment of genetic pathologies by targeting not their phenotypic manifestations but the genetic cause directly, through adequate alteration of "responsible" genes' functions.

The first issue that we need to address, here, is whether such radical interventions in the human organism, being irreversible, eventually reduce personal autonomy. Our genome defines a biological framework within which our organism develops its characteristics; in that sense, it determines, at least roughly, the whole biological process of our life. Considering this, one can wonder if "designing" a human genome would be against the freedom of the person concerned by imposing irreversible artificial changes on a given natural condition.

The second issue refers to genetic interventions that may affect the offspring of the person concerned and therefore future generations, or even to interventions in the genome of embryos or fetuses before implantation or during pregnancy; that perspective is largely debated nowadays, due to the CRISPR Cas9 and the other genome editing methods' development.

The current European Biolaw has adopted an attitude that accepts interventions in the human genome if performed for health reasons exclusively. Certainly, insofar as

---

[49] Use of human biological material for health purposes is, also, that of primordial hemopoietic cells obtained by the umbilical cord or the placenta, after birth, for replacing similar cells of the donor, family member, or other people to address serious blood-related diseases (leukemia, etc.), if needed during the lifespan of the recipient. These cells are stored in special banks for that reason, but their use must be distinguished from conventional tissue and organ transplantations; this is because the newborn/donor does not contribute with biological material extracted from the body, therefore there is no issue concerning his/her biological autonomy. Still, the newborn is considered the legal owner of that material, which means that any use presupposes the consent of parents as legal representatives (see, also, in Chap. 9 on "self-ownership"). Cf. Gunning (2003); Vidalis (2011), pp. 119–126.

[50] See United Nations (2013).

[51] At the level of non-legal regulation, see the Declaration of Istanbul (2008) of the Transplantation Society and International Society of Nephrology. Cf. Muller et al. (2019).

our knowledge about the functions of genes in the human genome remains limited,[52] gene therapies involve a great deal of uncertainty about their potential outcomes in terms of safety and efficiency. Nevertheless, we address this problem under their general therapeutic scope, representing an ultimate hope for persons suffering from serious genetic pathologies.

The Oviedo Convention includes a particular provision for these genetic interventions (art. 13), according to which:

> An intervention seeking to modify the human genome may only be undertaken for preventive, diagnostic or therapeutic purposes and only if its aim is not to introduce any modification in the genome of any descendants

Genome modifications may entail limited influence only on the organism of persons undergoing genetic interventions, or they may affect these persons' descendants as well. When the intervention points to the DNA of common diploid "somatic" cells that are not involved in the organism's reproduction, only limited influence on genetic characteristics occurs. In contrast, when a particular gene therapy targets the DNA of gametes or the cells that produce them, we expect genetic modifications to descendants since they are derived from the modified reproductive material. Similar consequences will be manifested in the descendants' offspring and future generations also.

Article 13 of the Oviedo Convention clearly accepts only interventions in "somatic" cells, that is, limited intentional genetic modifications only to the organism of the person concerned. As mentioned, this reservation is so far justified, given our relative ignorance about the exact role of genes in the human organism and the subsequent necessity to protect future persons from unforeseen and irreversible harms that a genetic intervention might incite.

However, it is worth mentioning that uncertainty about the exact interactions between genes does not exclude risks regarding the transmission of genetic modifications to the offspring, even for persons undergoing gene therapy in somatic cells. Indeed, even if the target is somatic cells, the function of their genome may affect the germ cells of the organism as well.[53] A similar risk also exists when conventional therapies are performed (i.e., chemotherapies) that may yield genetic alterations to germ cells, which will be transmitted to descendants.[54] Still, the Oviedo Convention accepts that possibility, considering more vital the expected health benefit for the person undergoing gene therapy.[55]

---

[52] See, on that, in Chap. 6 on "Data".

[53] Cf. Salvi (2003), pp. 34 seq.

[54] See the explanatory report on art. 13 of the Oviedo Convention.

[55] A *de lege ferenda* analysis, regarding in particular genetic interventions in the context of human enhancement, see in Chap. 8.

3.8.5 Medical Acts in the Brain: The Objective of Neurolaw

The rapid progress in neurosciences over the last decades, thanks to developments in diagnostic technologies and related applications, raises ethical and legal questions in two areas; first, regarding the handling of medical data obtained from the neural system of a patient or healthy person, and then, regarding invasive medical acts for brain stimulation with the use either of drugs or other medicinal products and methods (implants, electric stimuli, etc.). A new topic of reflection also emerges with the development of artificial brain organoids, that is, biological entities capable to express advanced neural functions.

Questions in relevance are essential not only when prevention or treatment are observed but also when the aim is to enhance brain functions, as we will see in the relevant chapter. Nowadays, the terms "neuroethics" and "neurolaw" illustrate this new field of medical and research performance characteristically.

Brain Data The extensive use of neuroimaging techniques (electroencephalography/EEG, functional magnetic resonance imaging/fMRI, in particular), along with the identification of genetic causes for various brain functions, offer a valuable amount of data that may guide preventive and therapeutic medical acts. These methods provide evidence over correlations between specific biomarkers and reactions of the brain, that is, of human behavior. In that sense, they reveal key elements of the sentient and cognitive abilities of a person, particularly regarding the origin and formation of these.[56]

In ethical and legal terms, this intrusive performance into the very intimate area of the human mental and cognitive world affects the core of autonomy; data from the brain indicates the formation of a person's will and even biological limitations on free thought and expression. As necessary acts for addressing severe pathologies of the neural system and psychiatric diseases, including dementia, psychosis, Parkinson's disease, depression, etc., the collection and processing of such data are covered by the current legislation under the safeguards of medical secrecy (Oviedo Convention and EU Charter of Rights) and sensitive data protection (GDPR, as analyzed in Chap. 6, *infra*).

In this regard, we must pay special attention to the management of incidental findings, that is, important data for the health of a person undergoing brain diagnosis, the presence of which is not expected initially. The attending physician must evaluate the nature of incidental findings and decide upon the necessity of the person's information. Nevertheless, this information should follow the informed consent rule in conjunction with the person's right-not-to-know. This means that the person needs to receive information on the possibility of incidental findings prior to undergoing the diagnostic act in order to provide or deny consent on their disclosure by the attending physician. In terms of the current law, this procedure is necessary given the informed consent and the right-not-to-know explicit stipulations,

[56] Cf. Ienca and Andorno (2017), pp. 2–4.

meaning that medical liability could also be proved when a physician acts differently.

Interventions in Brain Functions The issue of medical interventions in brain functions raises safety problems, given the largely unknown nature of these in biological terms. The most important question here is that of possible risk of altering essential elements of the individual's personality.

Besides the general framework that ensures good medical practice, the European law does not contain specific regulations for addressing safety questions in that field. The basic rule of the Oviedo Convention regarding the necessity of compliance with "professional obligations and standards" (art. 4) needs then to be interpreted under the light of generally accepted protocols pertinent to the performance of brain stimulation, with the use of drugs, implants or other methods, including these of transcranial direct current stimulation (tDCS), transcranial magnetic stimulation (TMS) and deep brain stimulation (DBS).[57]

At the legal level, we do not expect to find any guidance for avoiding the risk of altering elements of the patient's personality other than those considered pathological, needing medical intervention. We can only refer to the fundamental rights related to an individual's inner world, particularly those of the freedom of thought, the freedom of expression, and the right to privacy, which guarantee our overall personal autonomy, as recognized by the European legal instruments (ECHR, EU Charter of Rights, Oviedo Convention). These rights form a normative framework for the legal evaluation of medical interventions in brain functions involving high risk for the personality, even if compliant with relevant protocols; it is a physician's legal duty to assess the implementation of such a protocol in a way that ensures respect of these fundamental rights, and, on that ground, evidence for medical liability cannot be excluded.

However, in terms of legal certainty, the adoption of specific individual rights either explicitly or through interpretation of the above fundamental rights seems suitable to ensure that the legal system remains aware of the rapid developments in neuroscience and the relevant technology. In this regard, there is already a theoretical proposal of four distinct rights to "cognitive liberty", "mental privacy", "mental integrity", and "psychological continuity"[58] that shapes the field for further legal elaboration.

Brain Organoids The development of brain organoids raises issues of a different nature but still falling within the general interests of Neuroethics and Neurolaw.

Brain (or cerebral) organoids are artificial products of embryonic pluripotent stem cells cultivation that imitate functions of the natural human brain.[59] As we will see in the next chapter on "research," embryonic stem cells have the potential to

[57] Ibid., p. 5.

[58] Ibid., pp. 10–23.

[59] See, Di Lullo and Kriegstein (2017), pp. 573–584.

differentiate into various types of human tissues, including those of the neural system. With the use of techniques for the growth of stem cells in vitro (in gel droplets) and then their transformation into neural tissue (with a bioreactor's assistance), the final product is an organoid of relatively small size but with brain-alike biological structure and functions. These organoids enable researchers to perform in vitro research on brain characteristics without the limitations that occurred in research with animal or natural human brains, for studying injury or disorders, including stroke, degenerative diseases, the effects of substances, etc., as well as for testing drugs. Moreover, cerebral organoids may be used as a brain tissue source for transplants in clinical practice.

From an ethical point of view, besides problems related to embryonic stem cell use (see in the next chapter), a question emerges regarding the nature of brain organoids.[60] As brain-alike products, organoids are sensitive to reacting to external stimuli, and possibly they develop characteristics of human consciousness, although they are not part of a complete human organism. Are these characteristics sufficient to attribute a certain ethical status to these artificially created biological products? Moreover, in ethical terms, can we equate an "advanced" brain organoid, capable of developing a conscience, to a natural person? This problem is not absurd; even assuming that a natural person lives in a state of complete paralysis, with only brain functions remaining undamaged, his/her ethical status would not be questionable. This could be an argument for defending an equal ethical status for an isolated artificial brain as well.

In the context of Biolaw, that problem is critical. It is worth noting that the existing legal instruments avoid to identify the natural person/subject of human value and fundamental rights with a corporally "complete" human organism. Although corporal integrity is the object of a fundamental right according to the EU Charter of Rights (art. 3 para 1), this pertains to all persons by definition, regardless of their bodily structure or functions. In that sense, in principle, there is no legal basis to exclude a legal status for brain organoids similar to that of natural persons, on the condition that these artificial products have the potential to develop conscious abilities and, if supported with artificial equipment, even to act autonomously. In that context, an unconditional creation and availability of brain organoids for experimental purposes becomes problematic and needs to be regulated with specific legislation.

[60] See in general, Lavazza and Massimini (2017), pp. 606–610.

3.9 Further Studying...

ECtHR: K.H. and Others v. Slovakia, 28 April 2009

"The applicants, eight women of Roma origin, could not conceive any longer after being treated at gynaecological departments in two different hospitals, and suspected that it was because they had been sterilised during their stay in those hospitals. They complained that they could not obtain photocopies of their medical records.

The Court held that there had been a violation of Article 8 (right to private and family life) of the Convention in that the applicants had not been allowed to photocopy their medical records. It found that, although subsequent legislative changes compatible with the Convention had been introduced, that had happened too late for the applicants".[61]

Question

Comment on the following paragraphs, under the light of the ECHR, the Oviedo Convention and the GDPR:

48. It can be accepted that it is for the file holder to determine the arrangements for copying personal data files and whether the cost thereof should be borne by the data subject. However, the Court does not consider that data subjects should be obliged to specifically justify a request to be provided with a copy of their personal data files. It is rather for the authorities to show that there are compelling reasons for refusing this facility.

49. The applicants in the present case obtained judicial orders permitting them to consult their medical records in their entirety, but they were not allowed to make copies of them under the Health Care Act 1994. The point to be determined by the Court is whether in that respect the authorities of the respondent State complied with their positive obligation and, in particular, whether the reasons invoked for such a refusal were sufficiently compelling to outweigh the Article 8 right of the applicants to obtain copies of their medical records.

50. Although it was not for the applicants to justify the requests for copies of their own medical files (see paragraph 48 above), the Court would nevertheless underline that the applicants considered that the possibility of obtaining exclusively handwritten excerpts of the medical files did not provide them with effective access to the relevant documents concerning their health. The original records, which could not be reproduced manually, contained information which the applicants considered important from the point of view of their moral and physical integrity as they suspected that they had been subjected to an intervention affecting their reproductive status.

[61] ECtHR, Press Unit.

51. The Court also observes that the applicants considered it necessary to have all the documentation in the form of photocopies so that an independent expert, possibly abroad, could examine them, and also in order to safeguard against the possible inadvertent destruction of the originals are of relevance. As to the latter point, it cannot be overlooked that the medical file of one of the applicants had actually been lost (see paragraph 27 above).

52. The national courts mainly justified the prohibition on making copies of medical records by the need to protect the relevant information from abuse. The Government relied on the Contracting States' margin of appreciation in similar matters and considered that the Slovak authorities had complied with their obligations under Article 8 by allowing the applicants or their representatives to study all the records and to make handwritten excerpts thereof.

ECtHR: Jalloh v. Germany, 11 July 2006 (Grand Chamber)

"This case concerned the forcible administration of emetics to a drug-trafficker in order to recover a plastic bag he had swallowed containing drugs. The drugs were subsequently used as evidence in the criminal proceedings against him. The applicant claimed in particular that he had been subjected to inhuman and degrading treatment as a result of having been forcibly administered the emetics in question.

The Court reiterated that the Convention did not, in principle, prohibit recourse to a forcible medical intervention that would assist in the investigation of an offence. However, any interference with a person's physical integrity carried out with the aim of obtaining evidence had to be the subject of rigorous scrutiny. In the present case, the Court held that the applicant had been subjected to inhuman and degrading treatment contrary to Article 3 (prohibition of inhuman or degrading treatment) of the Convention."[62]

Question

Comment on the following paragraphs, under the light of the ECHR and the Oviedo Convention:

75. At the outset the Court notes that in the Government's view the removal of the drugs from the applicant's stomach by the administration of emetics could be considered to be required on medical grounds, as he risked death through poisoning. However, it is to be observed that the domestic courts all accepted that, when ordering the administration of emetics, the authorities had acted on the basis of Article 81a of the Code of Criminal Procedure. This provision entitles the prosecuting authorities to order a bodily intrusion to be effected by a doctor without the suspect's consent in order to obtain evidence, provided that there is no risk of damage to the suspect's health. However, Article 81a does not cover measures taken to avert an imminent danger to a person's health. Furthermore, it is

[62] ECT HR, Press Unit.

undisputed that the emetics were administered in the absence of any prior assessment of the dangers involved in leaving the drug bubble in the applicant's body. The Government also stated that emetics are never administered to juvenile dealers unless they are suspected of selling drugs on a commercial basis. Juvenile dealers are, however, in no less need of medical treatment than adults. Adult dealers, for their part, run the same risks to their health as juvenile dealers when administered emetics. Consequently, the Court is not satisfied that the prosecuting authorities' decision to order the impugned measure was based on and required by medical reasons, that is, the need to protect the applicant's health. Instead, it was aimed at securing evidence of a drugs offence.

76. This finding does not by itself warrant the conclusion that the impugned intervention contravenes Article 3. As noted above (see paragraph 70 above), the Court has found on several occasions that the Convention does not, in principle, prohibit recourse to a forcible medical intervention that will assist in the investigation of an offence. However, any interference with a person's physical integrity carried out with the aim of obtaining evidence must be the subject of rigorous scrutiny, with the following factors being of particular importance: the extent to which forcible medical intervention was necessary to obtain the evidence, the health risks for the suspect, the manner in which the procedure was carried out and the physical pain and mental suffering it caused, the degree of medical supervision available and the effects on the suspect's health (compare and contrast also the criteria established by the United States courts in similar cases – ...). In the light of all the circumstances of the individual case, the intervention must not attain the minimum level of severity that would bring it within the scope of Article 3. The Court will now examine each of these elements in turn.

77. As regards the extent to which the forcible medical intervention was necessary to obtain the evidence, the Court notes that drug trafficking is a serious offence. It is acutely aware of the problem confronting Contracting States in their efforts to combat the harm caused to their societies through the supply of drugs (see, in particular, D. v. the United Kingdom, 2 May 1997, § 46, Reports 1997-III). However, in the present case it was clear before the impugned measure was ordered and implemented that the street dealer on whom it was imposed had been storing the drugs in his mouth and could not, therefore, have been offering drugs for sale on a large scale. This is reflected in the sentence (a six-month suspended prison sentence and probation), which is at the lower end of the range of possible sentences. The Court accepts that it was vital for the investigators to be able to determine the exact amount and quality of the drugs that were being offered for sale. However, it is not satisfied that the forcible administration of emetics was indispensable in the instant case to obtain the evidence. The prosecuting authorities could simply have waited for the drugs to pass through his system naturally. It is significant in this connection that many other member States of the Council of Europe use this method to investigate drugs offences.

...

79. As to the manner in which the emetics were administered, the Court notes that, after refusing to take the emetics voluntarily, the applicant was pinned down by four police officers, which shows that force verging on brutality was used against him. A tube was then fed through his nose into his stomach to overcome his physical and mental resistance. This must have caused him pain and anxiety. He was subjected to a further bodily intrusion against his will through the injection of another emetic. Account must also be taken of the applicant's mental suffering while he waited for the emetics to take effect. During this time he was restrained and kept under observation by police officers and a doctor. Being forced to regurgitate under these conditions must have been humiliating for him. The Court does not share the Government's view that waiting for the drugs to pass through his body naturally

would have been just as humiliating. Although it would have entailed some invasion of privacy because of the need for supervision, such a measure nevertheless involves a natural bodily function and so causes considerably less interference with a person's physical and mental integrity than forcible medical intervention (see, mutatis mutandis, Peters, cited above, and Schmidt, cited above).

80. As regards the medical supervision of the administration of the emetics, the Court notes that the impugned measure was carried out by a doctor in a hospital. In addition, after the measure was executed the applicant was examined by a doctor and declared fit for detention. However, it is a matter of dispute between the parties whether an anamnesis of the applicant was obtained prior to the execution of the measure in order to ascertain whether his health might be at risk if emetics were administered to him against his will. Since the applicant violently resisted the administration of the emetics and spoke no German and only broken English, the assumption must be that he was either unable or unwilling to answer any questions that were put by the doctor or to submit to a prior medical examination. The Government have not submitted any documentary or other evidence to show otherwise

ECtHR: Novruk and Others v. Russia, 16 March 2016

"All five applicants wished to obtain residence permits in Russia. To complete their application, they were required to have a medical examination which included a mandatory test for HIV infection. After they tested positive for HIV, the migration authorities refused their applications by reference to the Foreign Nationals Act, which prevents HIV-positive foreign nationals from obtaining residence permits. The applicants alleged in particular that they had been discriminated against because they were HIV-positive.

The Court held that there had been a violation of Article 14 (prohibition of discrimination) read together with Article 8 (right to private life and family) of the Convention. It notably noted that the legislation aimed at preventing HIV transmission, which was used in the present case to exclude the applicants from entry or residence, had been based on an unwarranted assumption that they would engage in unsafe behaviour, without carrying out a balancing exercise involving an individualised assessment in each case."[63]

Question

Comment on the following paragraphs, under the light of the ECHR and the Oviedo Convention:

98. Once an applicant has shown that there has been a difference in treatment, it is then for the respondent Government to show that the difference in treatment can be justified (see Chassagnou and Others v. France [GC], nos. 25088/94, 28331/95 and 28443/95, §§ 91-92, ECtHR 1999-III). Such justification must be both objective and reasonable or, in other

[63] ECtHR, Press Unit.

words, it must pursue a legitimate aim and there must be a reasonable relationship of proportionality between the means employed and the aim pursued. The Contracting States enjoy a margin of appreciation in assessing whether and to what extent differences in otherwise similar situations justify differing treatment. The scope of this margin will vary according to the circumstances, subject matter and background (see Burden, § 60, and Carson and Others, § 61, both cited above).

99. The existence of a European consensus is an additional consideration relevant in determining whether the respondent State should be afforded a narrow or a wide margin of appreciation (see Dickson v. the United Kingdom [GC], no. 44362/04, § 81, ECtHR 2007-V, and S.L. v. Austria, no. 45330/99, § 31, 9 January 2003). Where there is a common standard which the respondent State has failed to meet, this may constitute a relevant consideration for the Court when it interprets the provisions of the Convention in specific cases (see Kiyutin, cited above, § 65; Tănase v. Moldova [GC], no. 7/08, § 176, ECtHR 2010; and Demir and Baykara v. Turkey [GC], no. 34503/97, § 85, ECtHR 2008).

100. If a restriction on fundamental rights applies to a particularly vulnerable group in society that has suffered significant discrimination in the past, then the State's margin of appreciation is substantially narrower and it must have very weighty reasons for imposing the restrictions in question. The reason for this approach, which questions certain classifications per se, is that such groups were historically subject to prejudice, with lasting consequences resulting in their social exclusion. Such prejudice could entail legislative stereotyping which prohibits the individualised evaluation of their capacities and needs (see Kiyutin, cited above, § 63, and Alajos Kiss v. Hungary, no. 38832/06, § 42, 20 May 2010). The Court has found that people living with HIV have to face a whole host of problems, not only medical but also professional, social, personal and psychological, and to confront deeply rooted prejudice even from among highly educated people (see I.B. v. Greece, cited above, § 80). The prejudice was born out of ignorance about the routes of transmission of HIV/Aids, and has stigmatised and marginalised those who live with the virus. Consequently, the Court has held that people living with HIV are a vulnerable group and that the State should be afforded only a narrow margin of appreciation in choosing measures that single out this group for differential treatment on account of their health status (see I.B. v. Greece, § 81, and Kiyutin, § 64, . . .).

Provide legally appropriate solutions

1. Jim A. suffers from prostate cancer. His attending physician is aware of two treating chemotherapy options. The first promises a better outcome but involves serious side-effects, while the second has limited side effects but with no certain outcome for all patients. The physician starts the first treatment for Jim A.

- Is that decision legally valid?
- Has the physician a legal duty to inform the wife and the adult children of Jim A. instead of the patient himself?
- Has the physician a legal duty to inform the wife and children of Jim A. along with the patient himself?
- What should the physician do if these relatives disagree with Jim A.?
- What if the relatives express different opinions and the patient remains reluctant to decide?

- Could the physician miss out on the treatment's side effects from the information provided for "not terrifying" the patient?
- Would your answer be different if these effects were reported only in very few cases of patients?
- If Jim A. refuses both treatment options, what should the physician do?

2. Under the threat of an Ebola pandemic, the competent authorities suggest patient access to experimental drugs just after reports encouraging pre-clinical trials in animals with no testing in clinical trials so far.

1. Comment on this guidance under the view of the current European Biolaw
2. How do you legally judge a measure of compulsory isolation for an unlimited time period of persons identified as having contact with patients?

References

Ackerman TF (1982) Why doctors should intervene (Hastings Center Report 12: 14-17). In: Mappes ThA, Degrazia D (eds) Biomedical ethics, 5th edn. McGraw-Hill, New York 2001, pp 80–85

Beauchamp TL, Childress JF (2001) Principles of biomedical ethics, 5th edn. Oxford U.P.

Brody B (1993) Hardwig on proxy decision making (J. of Clin. Ethics 4: 66-67). In: Mappes TA, Degrazia D (eds) Biomedical ethics, 5th edn. McGraw-Hill, New York, pp 196–198

Byk C (1992) Les aspects juridiques des xenotransplantations, Semaine Juridique-édition générale 50: I 3633

Caplan AL (1992) Is xenografting morally wrong? (Transplantation Proceedings 24: 722-727). In: Kuhse H, Singer P (eds) Bioethics. An anthology. Blackwell, Oxford 1999, pp 404–411

Chadwick R (1997) The philosophy of the right to know and the right not to know. In: Chadwick R, Levitt M, Shickle D (eds) The right to know and the right not to know. Ashgate Publ. Co, Aldershot, pp 13–22

Ciolli A (2008) Mandatory school vaccinations: the role of tort law. Yale J Biol Med 81:129–137

Davies B (2020) The right not to know and the obligation to know. J Med Ethics 46:300–303

Di Lullo E, Kriegstein AR (2017) The use of brain organoids to investigate neural development and disease. Nat Rev Neurosci 18:573–584. https://doi.org/10.1038/nrn.2017.107

Di Paolo M, Gori F, Papi L, Turillazzi E (2019) A review and analysis of new Italian law 219/2017: 'provisions for informed consent and advance directives treatment'. BMC Med Ethics 20:1–7

Engelhardt TH (1974) The disease of masturbation: values and the concept of disease. Bull Hist Med 48:234–248

European Group on Ethics (EGE) (2005) Ethical Aspects of ICT Implants in the human body, Opinion No 20, file:///C:/Users/My/Downloads/Ethical_aspects_of_ICT_implants_in_the_human_body___opinion_presented_to_the_Commission_by_the_European_Group_on_Ethics.pdf

Fraisseix P (2000) La protection de la dignité de la personne et de l' espèce humaines dans le domaine de la biomédecine : l' exemple de la Convention d' Oviedo. Revue Internationale de Droit Comparé 52:371–413

Franklin P (2003) The recipient's perspective. In: Morris P (ed) Transplants. Council of Europe Publ., Strasbourg, pp 51–62

Gordon EJ, Daugherty CK (2003) "Hitting you over the head": oncologists' disclosure of prognosis to advanced cancer patients. Bioethics 17:142–168

Gunning J (2003) A worldwide study of umbilical cord cell banking, in European Group on Ethics in Science and New Technologies to the European Commission (EGE), Opinion No 19. Ethical Aspects of Umbilical Cord Blood Banking, European Communities 2004
Hardwig J (1993) The problem of proxies with interests of their own: toward a better theory of proxy decisions (J. of Clinical Ethics 4: 20-27). In: Mappes ThA, Degrazia D (eds) Biomedical ethics, 5th edn. McGraw-Hill, New York 2001, pp 187–196
Harris J (2020) Is there a right not to know? J Med Ethics 46:414–415
Higgs R (1985) On telling patients the truth (M. Lockwood - ed., Moral dilemmas in modern medicine, Oxford U.P.: 507-512). In: Kuhse H, Singer P (eds) Bioethics. An anthology. Blackwell, Oxford 1999, pp 87–93
Ienca M, Andorno R (2017) Towards new human rights in the age of neuroscience and neurotechnology. Life Sci Soc Policy 13:2–4. https://doi.org/10.1186/s40504-017-0050-1
Kachalia A, Choudhry NK, Studdert DM (2005) Physician responses to the malpractice crisis: from defense to offense. J Law Med Ethics 33:416–428
Kihlbom U (2008) Autonomy and negatively informed consent. J Med Ethics 34:146–149
Kleinman I (1994) Written advance directives refusing blood transfusion: ethical and legal considerations. Am J Med 96:563–567
Lavazza A, Massimini M (2017) Cerebral organoids: ethical issues and consciousness assessment. J Med Ethics 44:606–610. https://doi.org/10.1136/medethics-2017-104555
Lipkin M (1979) On lying to patients (Newsweek, June 4: 13). In: Mappes ThA, Degrazia D (eds) Biomedical ethics, 5th edn. McGraw-Hill, New York 2001, pp 85–87
Loix S, Henin P, Descamps O, Reusens I (2020) Jehovah's witnesses and transfusion: where do we stand in Europe? ISBT Sci Ser 15:212–220
Malauzat M-I (2000) Le droit façe aux pouvoirs des données génétiques. Presses Univ. D' Aix-Marseille, Aix-en-Provence
Mangalmurti SS, Murtagh L, Mello MM (2010) Medical malpractice. Liability in the age of electronic health records. N Eng J Med 363:2060–2067
McQuoid-Mason DJ (2021) Refusal for religious reasons of a blood transfusion for her husband by wife as legal proxy ignored by the doctors–can the wife claim damages? S Afr Med J 111:1172–1173
Moritz D, Ebbs P (2021) Consent and refusal of treatment by older children in emergency settings. Emerg Med Australas 33:168–171
Muller E, Dominguez-Gil B, Martin D (2019) The Declaration of Istanbul on organ trafficking and transplant tourism (2018 Edition) introduction. Transplantation 103:217
Murphy P (2005) Are patients' decisions to refuse treatment binding on health care professionals? Bioethics 19:189–201
Nys H (2012) The right to informed choice and the patients' rights directive. Eur J Health Law 19: 328–329
Pinals DA, Nesbit A, Hoge SK (2017) Treatment refusal in psychiatric practice. In: Rosner R, Scott C (eds) Principles and practice of forensic psychiatry, 3rd edn. CRC Press, Boca Raton, pp 185–194
Rynning E (2007) Public trust and privacy in shared electronic health records. Eur J Health Law 14: 105–112
Salvi M (2003) What is wrong in modifying the human germ line. Journal International de Bioethique 14:33–42
Stohl H (2018) Childbirth is not a medical emergency: maternal right to informed consent throughout labor and delivery. J Leg Med 38:329–353
United Nations (special report) (2013) Trafficking in persons, especially women and children, New York, https://www.ohchr.org/Documents/Issues/Trafficking/A-68-256-English.pdf
Vidalis T (2011) A matter of health? Legal aspects of private umbilical cord blood banking. Eur J Health Law 18:119–126
Vidalis T (2020) 3D printing, and bioprinting: ethical and legal issues. In: Tsouflas G, Bangeas PI, Suri JS (eds) 3D printing. Applications in medicine and surgery, vol I. Elsevier, pp 51–68
Wood S, Jones R, Geldart A (2003) The social and economic challenges of nanotechnology. ESRC, London

4 Research

4.1 General

Experimentation on human beings is an extreme challenge for biological autonomy and a matter of specific interest for Biolaw.[1] Every new therapeutic means (drug, medical device, therapeutic method) needs to be tested in humans before becoming available to the general public.

Clinical trials represent the most common and characteristic example of that testing. If clinical trials did not exist, we could not expect any progress in medical research leading to improvements in treatment methods and means.[2] The organization of clinical trials under certain institutional conditions is necessary, not only for practical reasons (i.e., the best possible scientific control of experiments) but also for protecting all the persons involved, that is, those undergoing experimentation and the future consumers of the final product.

The law acknowledges three sorts of experimentation involving deliberate interventions in human gametes, human embryos, and human persons, establishing different frameworks of protection. Their common characteristics are a) that they need to aim at scientific purposes (basic or applied), therefore, no experiments are justified for pleasure, curiosity, etc.; b) that they need to follow specific strict rules of conduct.

Experimentation on embryos is mostly related to collecting and processing embryonic stem cells from early embryos in vitro (up to two weeks after the

[1] Particularly, after the experience of such experimentation in Nazi's Germany and in Japan. See McNeill (2001), pp. 370–371. See also, in the U.S., the case-law regarding experimentations involving large numbers of unaware humans: Jaffee v. the United States, 456 U.S. 972, 102 S. Ct. 2234, 72 L. Ed. 2d 845 (1982), Begay v. the United States, 591 F. Supp 991 (1984), the United States v. Stanley, 483 U. S. 669, 107 S. Ct. 3054, 97 L. Ed. 2d 550 (1987), CIA v. Sims, 471 U. S. 159, 105 S. Ct. 1881, 85 L. Ed. 2d 173 (1985). The most known case in the U.S. is the notorious experimentation with syphilis in Alabama. See Curran and Hyg (1973), pp. 730–731.

[2] See Hellman (1995), p. 436.

T. Vidalis, *The Emergence of Biolaw*, Springer Textbooks in Law,
https://doi.org/10.1007/978-3-031-02359-0_4

fertilization of oocytes, usually) for research or therapeutic purposes. Questions on the appropriate handling of this biological material are certainly raised here, and we will discuss them in the relevant paragraph. However, these questions do not refer directly to the human dignity principle since embryos are not considered subjects of law and biological autonomy, as argued before. Moreover, the same is true for research on human gametes, since no living organism exists before fertilization.

In contrast, in clinical trials where persons are involved, the human dignity principle requires special safeguards. Indeed, experimentation on persons must comply with two additional presuppositions, not existing in the other sorts of experimentation mentioned above: a) the necessary informed consent of the persons involved, and b) an expectation of health benefit for these persons. Both presuppositions are associated with the human person's legal status, and they are directly linked to his/her biological autonomy; the lack of any of them would compromise that special legal status by transforming a person into a mere object of experimentation. Nevertheless, in some instances, exceptions occur for both, raising compliance issues with the principle of human dignity.

Apart from these basic considerations, the complex organization of clinical trials is subjected to specific legal rules, that distinguish trials from conventional medical interventions.[3]

4.2 Phases of Clinical Trials

Clinical trials are interventional medical studies, that is, experimental methods involving the administration of the substance to be tested on the human organism in order to monitor its effects in terms of safety and efficiency. In contrast, non-interventional clinical studies monitor an approved drug's long-term effects when administered to patients according to its standard prescription. These studies intend to ensure safety levels (pharmacovigilance) that the respective clinical trial cannot appraise due to its limited duration.[4]

Clinical trials of drugs usually follow three distinct phases that ensure integrated control. Each phase presupposes the successful results of the previous one. The selection of putative participants (recruitment) is based on standard methodology

[3] An issue emerges on whether the patient/investigator relationship is similar to the classic relationship between the patient and the attending physician, or differs, given the research purpose's importance. On the ethical dilemma that faces the doctor/investigator in such cases, and the need for maintaining a certain equipoise, see Hellman and Hellman (1991), pp. 253 seq. See also Freedman (1987), justifying a reasonable uncertainty regarding the best treatment in clinical trials, from the part of doctors, Litton and Miller (2005), pp. 566 seq. and (regarding legal implications) Cranley Glass and Waring (2005), pp. 575 seq. On the primacy of patient treatment in that context, see Helsinki Declaration, art. 4, 8, 14.

[4] See in relevance, Helsinki Declaration, art. 6, according to which: *"Even the best proven interventions must be evaluated continually through research for their safety, effectiveness, efficiency, accessibility, and quality"*.

with inclusion/exclusion criteria, ensuring mitigation of statistical errors. In phase I, which follows the successful testing of the experimental substance in pre-clinical trials (performed on laboratory animals), the aim is to control safety for the human organism (toxicity, etc.) when the substance is administered to a limited number of healthy volunteers.[5] Phase II intends to study the substance's efficiency when administered to a limited number of patient volunteers. During the trial, these patients may stop receiving their conventional drugs following the plan of experimentation. In phase III, the drug's effects are tested in large numbers of patients to reach statistically valid results. Upon successful completion of this last phase, the tested substance may be subject to patent rights and obtain a license for marketing. A post-approval phase IV may follow to detect the new drug's effects for other medical use that the original prescription may not cover.

In both phases II and III, trials are based on comparisons of the experimental substance's efficiency either with the conventional drug that the patients receive or with placebo, that is, a substance with no therapeutic value. This latter case is medically justified when no conventional drug exists or when the placebo may have some therapeutic effects (in psychiatric patients, for example). Patient volunteers are allocated to separate groups, based on randomization, namely the random selection of persons following specific algorithms, according to their gender, race, age, etc. (stratification). Volunteers, and in some cases investigators, are not previously informed about the substance assigned for each group (the experimental or the conventional one) to avoid bias (single/double-blind trials). All these measures intend to ensure the scientific reliability of a trial.

This complex organization of clinical trials raises questions regarding the protection of volunteers, that is, related to their biological autonomy, which needs to be encountered not only in terms of medical ethics but also in terms of Biolaw.

4.3 The Law of Clinical Trials

The Oviedo Convention, its Protocol on Biomedical Research, and Regulation 536/2014 of the EU[6] constitute the common European legal framework for clinical trials. However, essential guidance for the interpretation of their stipulations can be found in generally accepted rules of medical ethics, as expressed in internationally known texts, like the Nuremberg Code, the Belmont Report, and mostly the Helsinki Declaration.[7]

[5] Exceptionally only, Phase I may involve seriously ill patients.

[6] Cf. Nys (2012), pp. 405–414.

[7] Byk (2004), pp. 17 seq. A comparison between the European and the U.S. law is presented by Muttin (2018, Chapt. Five), showing almost identical details in regulation, which facilitates international collaborations in that area.

4.3.1 A Special Consent

The fundamental legal requirement for conducting clinical trials, regardless of the phase they belong to, is to obtain the informed consent of the person involved as in any medical intervention.[8] This term makes the difference between experimentation with persons and other sorts of research involving living beings. Consent in clinical trials is correlated to a high degree of risk for the person's biological autonomy. For that reason, consent needs specific guarantees confirming its genuine nature (i) and, mostly, it is considered a necessary but not sufficient requirement for involving a volunteer in a trial; the law refers to additional requirements that need to be met in this respect (ii).

(i) Apart from its general characteristics pertaining to all medical interventions,[9] the consent for clinical trials must be "explicit" (that is undoubtful), "specific" (that is, related to a concrete intervention), and "documented," a feature usually ensured by written form.[10]

(ii) Additional requirements for participation in trials are considered necessary guarantees due to conditions of relative uncertainty for health that occur when a person undergoes experimental treatment. Legal provisions in relevance require that a) no alternative method of research exists, b) the expected benefits for the person's health prevail over potential risks,[11] and c) a previous approval of the research protocol needs to be obtained by competent organs. The latter term refers to both the scientific and the ethical review of the research protocol by organs with a status ensuring independent opinion for avoiding conflict of interests.[12] According to the EU Regulation, the scientific review of clinical trial protocols in the EU is assigned to central organs, while the ethical review remains to national ethics committees.[13]

These additional requirements mean that the mere expression of the individual's consent, as a dimension of their biological autonomy, is not sufficient for the participation in experimental procedures; consent is activated only within the above restrictive normative framework, functioning as a guarantee of human dignity.

---

[8] See art. 16 iv, v of the Oviedo Convention. Also, art. 7 of the UN International Covenant on Civil and Political Rights.

[9] According to art. 5 of the Oviedo Convention. The withdrawal of consent is crucial in experimental procedures to guarantee the volunteer's biological autonomy, given their vulnerable situation. Nevertheless, exercising that option does not always ensure the patient's best interest, notably when appropriate information is missing. See Edwards (2005), pp. 112 seq.

[10] See, in conjunction, art. 16 v of the Oviedo Convention, art. 29 para 1 of the Clinical Trial Regulation.

[11] See Helsinki Declaration, art. 16–18.

[12] Art. 16 iii of the Oviedo Convention, art. 4 of the Regulation.

[13] Art. 4 of the Regulation.

What happens when the ability of consent does not exist? Can we accept, ethically, persons' involvement in experimentation when the fundamental guarantee for maintaining their biological autonomy lacks? As in conventional medical interventions, the answer here is affirmative, even if we address conditions of uncertainty regarding the therapeutic result or the potential effects on the person's health. This is because there is still a possibility to benefit an unable to consent person with his/her participation in experimental procedures when conventional therapies have been proved inefficient. It is worth mentioning that, in certain diseases (pediatric or psychiatric), this participation is necessary for improving therapeutic means and consequently for expecting direct benefit to persons involved. In legal terms, then, these persons' participation in clinical trials still expresses an aspect of their fundamental right to Health, as it happens with patients able to provide consent.

Article 17 of the Oviedo Convention requires additional conditions for the participation of persons unable to consent. These persons may undergo clinical experimentation if the requirements mentioned above pertaining to all volunteers are met, and, also, a) research may conclude to "real" and "direct" benefit for the health of the person involved,[14] b) research of similar efficiency cannot be performed with the participation of able to consent volunteers, c) an explicit and written authorization by the legal representative of the person involved exists, and, if possible, with the assent of the latter,[15] and d) the person involved does not object to participating.

We can, therefore, conclude that according to the Oviedo Convention: a) research with unable to consent persons is legitimate as an exception, only when alternative trials with able volunteers cannot be performed, b) the lack of consent has as a counterpart a reasonably expected direct benefit for the person involved, c) in any case, the objection of that person, regardless of his/her maturity or mental condition, is considered a valid expression of will, and functions as an absolute veto terminating participation.

For certain vulnerable groups of persons, participation in clinical trials is ethically questionable.[16] This refers, mostly, to the participation of persons facing difficulties to express free will due to their submission to special conditions of hierarchy, like in cases of military personnel and detainees. In the past, clinical research involved extensively such persons since no informed consent conditions were required. The new model of biological autonomy is incompatible with that practice.[17]

Hospitalized patients are also subjected to special regulations. In their case, it is crucial to ensure the continuation of their standard treatment if they object to

---

[14] On the determination of the best treatment in concrete circumstances, see Freedman (1987), pp. 429 seq.

[15] See in the U.S. the case Grimes v. Kennedy Krieger Institute Inc., 366 Md. 29, 782 A. 2d 807 (2001), in which the court considered that, in such experimental procedures, parental consent is not sufficient, and a judicial approval would be necessary.

[16] See in relevance, Helsinki Declaration, art. 20.

[17] See in relevance, Penneau (1998), p. 388.

participating (or withdraw participation) in clinical trials in which the investigator is their attending physician.

Finally, the participation of pregnant or breastfeeding women is accepted on strict conditions of safety since the experimental substance's administration may entail implications for the embryo or fetus. The EU Regulation requires, in particular, an expected direct benefit for the woman or the fetus. However, it also allows participation with no such benefit, on similar conditions as those required for persons unable to consent (see *infra*). Within this legal context, we can accept that only in exceptional cases pregnant or breastfeeding volunteers can be recruited, e.g., when no alternative therapeutic method exists, or the trial is relevant to specific drugs for pregnancy, and putative participants are facing risks of miscarriage.[18]

From a medical view, a distinction between conventional and urgent clinical trials is usually defended. Conventional trials are those performed under normal conditions when their completion does not involve time pressure. Urgent trials are performed in specific circumstances demanding rapid completion in order to ensure new means of prevention or therapy (vaccines or drugs) and, thus, to avoid a present danger to the individual health of participants or even the public health (in case of epidemics, etc.).

The EU Regulation covers the case of urgent trials performed in intensive care units when patients are not able to provide consent, and there is no access to legal representatives for obtaining their informed consent timely. In such cases, the Regulation accepts starting of the trial without consent, on conditions a) that the patient had not objected in a previous time, b) the experimentation involves minimal risk for the patient's health, and significant improvement is expected, and c) the patient will approve his/her participation afterwards, when retrieving the consent ability and being informed adequately.[19]

This is a legal acceptance of a kind of "presumed consent" facilitating the research procedure as obvious. In that view, the legislator equates urgent trials with conventional medical acts performed in emergency situations, where also previous informed consent is not a necessary condition (article 8 of the Oviedo Convention). Nevertheless, a significant difference between the two cases is that conventional acts are associated with concrete scientific evidence regarding their expected effects; therefore, the degree of uncertainty is not comparable to any clinical research uncertainty. This is a crucial guarantee for the protection of patients, which means that, in terms of human dignity, the option provided by the Regulation with urgent trials should be followed only when no alternative exists for the treatment of the person concerned.[20]

[18] Art. 33 of the Regulation.

[19] Art. 35 of the Regulation. Also, Helsinki Declaration, art. 30.

[20] Regarding the importance of consent against alternative methods of participation (involving "waiver of consent requirements" or "incomplete information"), see Murdoch and Caulfield (2018).

4.3.2 The Person's Benefit and the Research Purposes

Since participation in trials ensures a better expectation for treatment, in excellent conditions of hospitalization and with no economic burden, it usually attracts more candidate volunteers. Their selection, then, must be based upon scientific criteria of inclusion/exclusion on the one hand and upon a standard of justice requiring equal review of candidatures.[21] The EU Regulation does not recognize a "right to recruitment," although it provides access to information for the general public about ongoing clinical trials.[22]

Thus, the exclusion of patient volunteers is legitimate, only if their concrete profile differs to the extent that it may yield statistic errors or because of random elements of the protocol (if, for instance, the number of volunteers is completed). Under the general rule of equitable access to health care (art. 3 of the Oviedo Convention), no other reason for exclusion is legally accepted, even if not jeopardizing the trial's scientific planning.

Both the Oviedo Convention and the EU Regulation mention exceptional clinical research circumstances, where research purposes seem to prevail over other values. In its article 17 para 2, the Convention recognizes an option for national legislators to allow participation in clinical trials of persons unable to consent, even if no expectation of direct benefit for their health occurs, on condition that all other requirements in relevance are met, and, also:

a) The research intends to an improvement of scientific understanding of pathology, disease or disorder of the patient, in order to yield a benefit either for the person involved or for other persons addressing similar problems of health.
b) That research involves minimal risk and burden for the person concerned.

Even if all other requirements for participation in clinical trials are fulfilled, here the risk for biological autonomy is evident due to the lack of both consent ability and potential benefit. Indeed, the margin of discretion for national legislators to accept clinical trials serving the exclusive benefit of third persons seems incompatible with the principle of human dignity; it looks like a measure of "mandatory" social solidarity that inevitably transforms persons unable to consent into mere objects of experimentation.

However, there are two guarantees for interpreting that provision in compliance with biological autonomy and human dignity. First, as mentioned above, no research can be performed if the person concerned objects, even being unable to consent. Moreover, research presupposes the person's legal representative's explicit authorization, in accordance with the best interest of that person. This means that the legal

[21] See on that, Chastain et al. (2020), regarding problems occurred with certain population groups' representation in COVID-19 trials.

[22] See art. 81 para 4, on public access to the EU database. Nevertheless, open access to clinical trials data raises some concerns regarding the protection of intellectual property and privacy. See Minssen et al. (2019).

representative has no authority to consent to participation in clinical trials not expected to benefit the represented person. However, at least for minors, the concept of "benefit" should not be considered ethically as related to health exclusively; it may be associated with the wider purpose of the minor's socialization if contributing to the development of personality by cultivating an input of social solidarity. Under this view, the minors' participation in clinical trials, even if not expected to yield concrete benefit for their own health, may serve that wider purpose of socialization. This seems the only interpretation fitting the principle of human dignity, as expressed in article 1 of the Oviedo Convention and, thus, recognizing clinical trials in minors for the necessary development of pediatric treatments.[23]

Furthermore, it should be noted that the benefit guarantee is meaningless in healthy volunteers of Phase I clinical trials since, for them, no question of health benefit exists. Neither the Oviedo Convention nor the EU Regulation cover this particular issue, which may involve a financial or other motivation for these persons to undergo biomedical experimentation.

More specifically, the Oviedo Convention adopts the informed consent prerequisite for the participation of humans in research procedures (art. 15) but also prohibits financial gain from the use of the human body or part of it (art. 21).[24] Even if the case of healthy volunteers can be captured under the view of article 15, article 21 clearly excludes any financial motivation for their participation. Nevertheless, this should be considered only as an exclusion of rewards, not referred to as financial compensation of any sort, covering direct expenses of participation, or even indirect damages, like losses of worktime wages, etc.

A final issue concerns the exceptional access to the experimental substance of seriously ill patients not participating in a particular trial, before its completion, as an ultimate attempt to protect their health or life.[25] Indeed, these patients seek the last chance, having no time to wait for the new drug's official approval. In terms of biological autonomy, a right to access cannot be excluded, even legally, as an aspect of the fundamental right to Health,[26] on condition that the information provided by the attending physician on potential implications follows the instructions of the ongoing trial's investigators.[27]

---

[23] Regarding non-therapeutic experimentation, see De Castro (2001), pp. 380–381.

[24] See also art. 3 para 2 c of the EU Charter of Rights.

[25] A similar issue is that of patient early access to "breakthrough therapies," suggesting "fewer, smaller, or shorter clinical trials.". Cf. Darrow et al. (2014), pp. 1252–1258. On "unproven interventions" and the responsibility of attending doctors, see Helsinki Declaration, art. 37.

[26] See art. 35 of the EU Charter of Rights, art. 11 of the European Social Charter, art. 14 of the UNESCO's Bioethics Declaration.

[27] *Contra* in the U.S. case-law. Cf. Darrow et al. (2015), pp. 279–286. See also ECtHR, Hristozov, and Others v. Bulgaria (2012).

4.3.3 The Use of Placebo

The use of placebo in clinical trials raises crucial ethical questions.[28] A criticism in relevance insists that the administration of "no-therapy" to patient volunteers is unethical if they stop receiving their conventional therapy during the trial.

In legal terms, if the patient received appropriate information before consenting, which refers explicitly to the use of placebo, this is a sufficient guarantee for the protection of biological autonomy. Indeed, on that condition, the interested person knows that there is a possibility to be included in a "control group" of volunteers. This group will receive placebo instead of the experimental substance if, of course, standard methods of randomized allocation of patients are to be followed, and no negative discrimination between them will occur. Here, the risk for the patient is comparable to the risk undertaken when consenting after appropriate information to a controversial conventional treatment (with proven efficiency only in limited cases), since the possibilities of failure are the same.

A different matter is whether the use of placebo in clinical trials is justified if conventional therapy exists. In Medicine and Medical Ethics, various opinions have been expressed, yet strong criticism is exercised to the use of placebo for economic reasons, that is, for shortening the clinical trial duration, in order to avoid high costs.[29] This matter is related, also, to the liability of doctors providing placebo instead of the conventional therapy to patients included in control groups. This is because that practice may contravene the fundamental medical duty for health care if conventional therapy stops for research purposes and not for clinical reasons.[30]

Nevertheless, given the legal rule of informed consent, the patient volunteer, if informed of the above possibility and subsequent implications for health, still he/she may refuse consciously the conventional therapy that receives for participating in a trial involving the use of placebo. In that case, medical liability refers exclusively to the content of information and not to the patient's enrollment in the placebo group.

4.4 Research on Human Gametes and Embryos

Gametes Research on human gametes involves issues of informed consent and data protection of their donors. No specific legislation at the European level exists in relevance; therefore, the data protection legislation is the only applicable here, likewise in any sort of human biological material extracted from the human organism, as we will see in detail in the relevant Chap. 9.

[28] See Helsinki Declaration, art. 33, and, in general, Rothman and Michels (1994), pp. 394 seq. On uncertainty regarding the approach of the Helsinki Declaration, see Skierka and Michels (2018).

[29] See Rothman and Michels (1994).

[30] See ref. 3.

Embryonic Stem Cells After fertilization of gametes, the embryo creation signifies the appearance of a living organism having the potential to become a complete person under certain circumstances (mostly, the implantation into a woman's womb). As already mentioned, research on embryos is particularly of interest with regard to the use of embryonic stem cells for therapeutic purposes.

During the last decades, biomedical research focuses on the processing of embryonic stem cells in order to produce tissues and even organs suitable for transplantations. These primordial human cells, still non differentiated into specific types of tissue, constitute parts of the first divisions that occurred in fertilized oocytes. Initially, as blastomeres, they are "totipotent", namely, they include the genetic potential to be specified in any type of tissue. Gradually, as they are located in specific areas of the human organism, they lose their original potential, although still maintaining a significant degree of it, and they become "pluripotent." The processing of pluripotent embryonic stem cells is expected to be a reliable alternative in the future for addressing the persistent problem of transplant shortage for therapeutic purposes.

Relevant research may be performed either on already existing "spare" embryos, created for reproductive purposes but never used, or on new embryos originally destined for therapeutic use, created either through conventional in vitro fertilization or through "therapeutic" cloning (if human cloning becomes possible, in the future).

In contrast to the so-called "reproductive" cloning (which we will discuss in Chap. 7), the purpose of "therapeutic" cloning is not related to human reproduction, that is, the creation of "complete" human organisms. The fertility process is developed exclusively in vitro, which means that, after the nucleus transfer from an adult cell to the oocyte and the first divisions of the new "cloned" human cell, it is expected to make possible the collection and use of embryonic stem cells for producing tissue or even vital solid organs. Due to their identical genetic profile with the original human organism's used nucleus, these tissues and organs would be presumably suitable transplants, perfectly histocompatible, for that organism, whenever needed. Practically, if this method proved successful, this would mean that any person could create and store "substitute" tissue and vital organs by fertilizing oocytes with the DNA of their adult cells and collecting stem cells for possible therapeutic use throughout their life.

The current European legislation accepts in principle the collection and use of embryonic stem cells for therapeutic purposes.[31] In this regard, we need to take into account that the timeframe of interest is limited to the first 14 days after the moment of conception that determines the creation of an embryo by in vitro fertilization.[32] During that period, the embryo is not considered a law subject, enjoying rights, although the law recognizes its special value and protects its life. Under this view,

[31] See, on that issue, in German law, Dederer (2002), pp. 4 seq, and, in France, Martin (2002), p. 386.

[32] Although there are also opinions arguing for research beyond that limit. See Hyun et al. (2021).

the Oviedo Convention provides a general framework allowing research on embryos in vitro that may result in therapeutic applications.[33]

According to article 18, the Convention, on the one hand, prohibits the creation of embryos for research purposes, allowing the use of "spare" embryos only, created initially but not used for fertility purposes; on the other, it requires "sufficient protection" of these embryos when used for research. This last clause leaves unclear the meaning of "sufficient protection," given that, so far, research on embryos in vitro, including research intending for the therapeutic use of embryonic stem cells, results necessarily in these embryos destruction.[34] Therefore, the meaning of this condition logically is not associated with the embryos' potential conservation after research for future fertility purposes. The rationale of "sufficient protection" is to ensure a) that "spare" embryos can be used for research or therapeutic purposes, only after the informed consent of the gametes' donors, and b) that research on them is legitimate only when no alternative method for meeting the same research objective is available. Therefore, a research use intending for therapeutic applications that involve embryonic stem cells is legitimate if these two conditions are fulfilled.

In contrast, the absolute prohibition of creating embryos for research purposes impedes any method of "creation" either with the fusion of gametes or with cloning. Article 18 para 2 excludes the option of the so-called "therapeutic cloning" although the First Additional Protocol of the Convention explicitly prohibits only the "reproductive" cloning in humans. Nevertheless, in Europe, countries not belonging to the Oviedo Convention system accept that option, following specific national legislation (UK, Sweden).

De lege ferenda, the absolute prohibition of article 18 para 2 remains questionable. Given that the embryo before implantation has no potential to become a human person, there is no reason to be treated differently compared to any non-reproductive sort of human biological material. Indeed, if reproduction is not the purpose, the embryonic primordial cells share the same status with gametes and even the somatic human cells, no matter the original intention for their creation. Only if we accepted a special value of the embryo in vitro, directly related to the legal status of persons, regardless of reproductive purposes, we can justify the prohibition of article 18 para 2. Yet, under this assumption, for consistency reasons, the absolute prohibition should cover also research on "spare" embryos, which is now allowed, according to the first paragraph of article 18.

Hybrids and Chimeras Stem cell research is also associated with the creation of hybrids and chimeras, that is, organisms derived from a mixture of animal and

---

[33] See a survey of current national legislations demonstrating substantial differences on embryo research, in Matthews and Moralí (2020). The lack of international binding legislation leads to the development of "stem cell tourism", for people seeking therapeutic solutions in countries with less strict regulation. See Gopalan et al. (2018).

[34] See Fraisseix (2000), pp. 412–413.

human biological material. A hybrid is the potential product of the transfer of a human cell nucleus into an animal oocyte. If developed, the embryos would be characterized by one cell line throughout their entire organism, which is formed by a unique human-animal genetic combination.[35] On the other hand, chimeras are products of the coexistence of at least two cell lines derived from different species (animal and human) in the same organism, with the fusion of cells at the early developmental stage of the embryo.[36]

The creation of such organisms through a mixture of animal species exclusively has been proved successful; there are many examples from the distant past already. With the progress of stem cell research, experimentation with human biological material attracts particular scientific interest. This is because the creation of human-animal early embryos is considered a potential source of embryonic stem cells suitable for transplantations (being compatible with the human organism) with no need to address ethical and legal questions related to the use of human embryos.

However, the creation of human-animal hybrids or chimeras raises other doubts concerning the human biological material's use, particularly on whether that use is compliant with the respect of human value. The European Biolaw does not contain specific regulations on that issue, although several national laws in Europe (Italy, Greece, Cyprus etc.) explicitly prohibit the creation of such organisms, no matter if these are expected to serve therapeutic purposes ultimately. The existing provisions of article 18 of the Oviedo Convention, as mentioned previously, are not relevant here since they refer strictly to human embryos *in vitro*.

Nevertheless, in the framework of the European legal instruments, the protection of human dignity could be considered an appropriate basis for addressing the problem, in the sense that it sets the rule of non-instrumentalization of persons. The creation of human-animal hybrids or chimeras does not exclude the appearance of beings having human characteristics and even the ability to develop autonomy deserving the status of persons. Still, this is highly uncertain, since acknowledging this status would not be evident from the moment of birth, as it happens in humans. Due to this uncertainty, for avoiding the risk of treating these beings as mere instruments contrary to the human dignity principle, we need to accept that the creation of human-animal hybrids or chimeras for reproductive purposes is legally excluded. In contrast, there is no reason to refuse this experimentation if intending for research or therapeutic purposes, that is, if there are guarantees that the whole process would be performed *in vitro*, with no further prospect of the early embryo development.[37] Since no question of the potential appearance of an organism deserving a person's status occurs, the fundamental rights to Research and Health justify the creation and use of human-animal hybrids or chimeras as a legally accepted source of embryonic stem cells.

---

[35] See, for example, Chang et al. (2003), pp. 1380 et seq.

[36] See, for example, Brüstle et al. (1998), pp. 1040 et seq.

[37] See in relevance the position of the Human Fertilization and Embryology Authority (HFEA) in the UK, in: HFEA (2007).

4.5 Further Studying...

ECtHR: Hristozov and Others v. Bulgaria, 13 November 2012

"The ten applicants were cancer sufferers who complained that they had been denied access to an unauthorised experimental anti-cancer drug. Bulgarian law stated that such permission could only be given where the drug in question had been authorised in another country. While the drug was permitted for "compassionate use" in a number of countries, nowhere had it been officially authorised. Accordingly, permission was refused by the Bulgarian authorities.

The European Court of Human Rights held that there had been no violation of Article 8 (right to respect for private and family life) of the European Convention on Human Rights. Considering that the restriction in question concerned the patients' right to respect for private life, protected by Article 8 of the Convention, it observed a trend among European countries towards allowing, under exceptional conditions, the use of unauthorised medicine. However, the Court found that this emerging consensus was not based on settled principles in the law of those countries, nor did it extend to the precise manner in which the use of such products should be regulated. The Court further held that there had been no violation of Article 2 (right to life) and no violation of Article 3 (prohibition of torture and of inhuman or degrading treatment) of the Convention in this case."[38]

Question

Comment on the following paragraphs, under the light of the ECHR and the Oviedo Convention:

110. Article 3 of the Convention enshrines one of the most fundamental values of a democratic society. It prohibits in absolute terms torture and inhuman or degrading treatment or punishment. However, to fall under that provision a given form of treatment must attain a minimum level of severity. The assessment of this minimum level is relative. It depends on all the circumstances of the case, such as the duration of the treatment, its physical and mental effects and, in some cases, the sex, age and state of health of the victim (see, as a recent authority, A, B and C v. Ireland [GC], no. 25579/05, § 164, ECtHR 2010-...). In considering whether a treatment is "degrading", the Court will have regard to whether its object was to humiliate and debase the person concerned and whether, as far as the consequences are concerned, it adversely affected his or her personality in a manner incompatible with Article 3 (see, among other auhtorities, Wainwright v. the United Kingdom, no. 12350/04, § 41, ECtHR 2006-X).

[38] ECtHR, Press Unit.

. . .

113. The applicants rather claim that the refusal by the authorities to allow them access to an experimental product which, according to them, was potentially life-saving, amounted to inhuman and degrading treatment for which the State was responsible, as it thereby failed to protect them from the suffering resulting from the final stages of their illness. However, as in Pretty (. . .), the Court considers that this claim puts an extended construction on the concept of inhuman or degrading treatment that it cannot accept. It cannot be said that by refusing the applicants access to a product – even if potentially life-saving – whose safety and efficacy are still in doubt, the authorities directly added to the applicants' physical suffering. It is true that the refusals, inasmuch as they prevented the applicants from resorting to a product which they believed might improve their chances of healing and survival, caused them mental suffering, especially in view of the fact that the product appears to be available on an exceptional basis in other countries. However, the Court does not consider that the authorities' refusal reached a sufficient level of severity to be characterised as inhuman treatment (see, mutatis mutandis, A, B and C v. Ireland, cited above, §§ 163-64). It notes in this connection that Article 3 does not place an obligation on the Contracting States to alleviate the disparities between the levels of health care available in various countries (see, mutatis mutandis, N. v. the United Kingdom, cited above, § 44). Lastly, the Court does not consider that the refusals can be regarded as humiliating or debasing the applicants.

Provide legally appropriate solutions

1. In a popular newspaper, a private hospital publishes an announcement, according to which adolescents of 14–16 years are recruited for a clinical trial of a very promising new drug designed for curing a rare disease. The announcement clarifies that if the new drug is not expected to produce any direct benefit to certain adolescents' health conditions, these volunteers will receive a compensation of 200 euros for their participation.

Comment on the announcement, in the light of the current European legislation.

2. The parents of a minor at the age of 14, which suffers from a serious metabolic disease, receive information about a clinical trial in a US hospital for a new promising drug. The child's life expectancy is low, so the parents ask the attending physician to support the potential recruitment of the child for this particular clinical trial

- Can the physician deny that?
- What if the child rejects the parents' initiative?
- What if the child wishes to be recruited and the parents disagree?
- Can the investigators in the US hospital deny the child's recruitment?
- If recruited, would it be legally acceptable for the child to be included not in the group of patients that will take the drug but in the control group that will take placebo?
- If the parents agree, would it be legally acceptable for the child to be recruited in the trial even without the expectation of any therapeutic benefit?

3. A research team intends to invite gamete donors to create human embryos not for reproductive purposes but for studying the development of embryonic stem cells with the aim to investigate their potential therapeutic use as an alternative to transplantations. During or after that experimentation, the embryos will be destroyed.
Under the light of the ECHR, the EU Charter of Fundamental Rights, and the Oviedo Convention, do you believe that the above practice is legally permitted?

References

Brüstle O, Choudhary K, Karram K, Hüttner A, Murray K, Dubois-Dalcq M, McKay RD (1998) Chimeric brains generated by intraventricular transplantation of fetal human brain cells into embryonic rats. Nat Biotechnol 16:1040–1044

Byk C (2004) La Déclaration d' Helsinki révisée: un nouveau contexte, de nouveaux défis pour la recherche biomédicale. Journal International de Bioethique 15:17–30

Chang KH, Lim JM, Kang SK, Lee BC, Moon SY, Hwang WS (2003) Blastocyst formation, karyotype, and mitochondrial DNA of interspecies embryos derived from nuclear transfer of human cord fibroblasts into enucleated bovine oocytes. In Vitro Fertilization (Fertil Steril) 80: 1380–1387

Chastain DB, Osae SP, Henao-Martínez AF, Franco-Paredes C, Chastain JS, Young HN (2020) Racial disproportionality in Covid clinical trials. N Engl J Med 383:2486–2487

Cranley Glass K, Waring D (2005) The physician/investigator's obligation to patients participating in research: the case of placebo-controlled trials. J Law Med Ethics 33:575–585

Curran WJ, Hyg SM (1973) The Tuskegee Syphilis Study. N Engl J Med 289:730–731

Darrow JJ, Avorn J, Kesselheim AS (2014) New FDA breakthrough-drug category — implications for patients. N Engl J Med 370:1252–1258

Darrow JJ, Sarpatwari A, Avorn J, Kesselheim AS (2015) Practical, legal, and ethical issues in expanded access to investigational drugs. N Engl J Med 372:279–286

De Castro LD (2001) Ethical issues in human experimentation. In: Kuhse H, Singer P (eds) A companion to bioethics. Blackwell, Oxford, pp 379–389

Dederer HG (2002) Menschenwurde des Embryo in vitro? Der Kristallisationspunkt der Bioethik-Debatte am Beispiel des therapeutischen Klonens. Archiv des öffentlichen Rechts 127:1–26

Edwards SJL (2005) Research participation and the right to withdraw. Bioethics 19:112–130

Fraisseix P (2000) La protection de la dignité de la personne et de l'espèce humaines dans le domaine de la biomédecine : l' exemple de la Convention d' Oviedo. Revue Internationale de Droit Comparé 52:371–413

Freedman B (1987) Equipoise and the ethics of clinical research (N. England J. Of Medicine 317: 141-145). In: Kuhse H, Singer P (eds) Bioethics. An anthology. Blackwell, Oxford 1999, pp 429–435

Gopalan N, Nor SNM, Mohamed MS (2018) Global human embryonic stem cell laws and policies and their influence on stem cell tourism. Biotechnol Law Rep 37:255–269

Hellman S (1995) The patient and the public good (Nature Medicine: 400-402). In: Kuhse H, Singer P (eds) Bioethics. An anthology. Blackwell, Oxford 1999, pp 436–440

Hellman S, Hellman D (1991) Of mice but not men: problems of the randomized clinical trial (N. England Journal of Medicine 324: 1585-1589). In: Mappes ThA, Degrazia D (eds) Biomedical ethics, 5th edn. McGraw-Hill, New York 2001, pp 253–258

Human Fertilization and Embryology Authority/HFEA (2007) Hybrids and chimeras: a consultation on the ethical and social implications of creating human/animal embryos in research, London. http://www.hfea.gov.uk/docs/Hybrids_Chimera_review.pdf

Hyun I, Bredenoord AL, Briscoe J, Klipstein S, Tan T (2021) Human embryo research beyond the primitive streak. Science 371:998–1000
Litton P, Miller FG (2005) A normative justification for distinguishing the ethics of clinical research from the ethics of medical care. J Law Med Ethics 33:566–574
Martin R (2002) Les premiers jours de l' embryon (A propos du projet de loi rélatif à la bioéthique). La Semaine Juridique-édition générale 8:383–386
Matthews KR, Moralí D (2020) National human embryo and embryoid research policies: a survey of 22 top research-intensive countries. Regen Med 15:1905–1917
McNeill PM (2001) Experimentation on human beings. In: Kuhse H, Singer P (eds) A companion to bioethics. Blackwell, Oxford, pp 369–378
Minssen T, Neethu R, Bogers M (2019) Clinical trials data transparency & GDPR compliance: what are the effects on data sharing and open innovation? In: Sideri K, Dutfield G (eds) Openness, intellectual property and science policy in the age of data driven medicine, Special Issue of Science and Public Policy. Available at SSRN: https://ssrn.com/abstract=3360302
Murdoch B, Caulfield T (2018) Pragmatic clinical trials and the consent process. Res Ethics 14:1–14
Muttin C (2018) Clinical trials law and policy: human subjects protection and global dynamics (Doctoral dissertation)
Nys H (2012) New European rules regarding the approval of clinical trials, the role of ethics committees, and the protection of subjects. Archivum Immunologiae et Therapiae Experimentalis 60:405–414
Penneau J (1998) Protection de la personne dans le domaine de la recherche médicale. Revue Internationale de Droit Comparé 50:383–399
Rothman KJ, Michels KB (1994) The continuing unethical use of placebo controls. N Engl J Med 331:394–398
Skierka AS, Michels KB (2018) Ethical principles and placebo-controlled trials–interpretation and implementation of the Declaration of Helsinki's placebo paragraph in medical research. BMC Med Ethics 19:1–12

Death

5

5.1 General

Death in humans means, legally, the end of a person. This explains why the most critical challenge of biological autonomy is related to decisions about the end of life. Such decisions may be taken by the person concerned, like suicide and euthanasia, or by others acting as representatives of a person unable to form and/or express free will, like in withdrawal of artificial support of life.

Even after its occurrence, death is also a topic of interest insofar as functions of the deceased person's vital organs are supported by artificial means. In terms of biological autonomy, the question here is whether decisions taken by that person, while living, continue to yield legal consequences after death regarding the destination of these organs; this is the case of removal and use of transplants from a deceased donor.

5.2 A Right to Death?

Suicide is a classic topic in ethics, yet for the law its significance is rather limited. This is because the legal justification of a successful suicide attempt has no meaning for the individual concerned after death. Legal questions are raised when the attempt is failed or in the case of assisted suicide, where third persons are involved in the decision's execution.

From the aspect of biological autonomy, the basic problem is whether self-determination is extended to decisions resulting in the end of its subject, if, in other words, we can acknowledge a legal "right to death".[1]

[1] For a comprehensive presentation of ethical and legal arguments in relevance, see Besirevic (2016), pp. 149–177.

T. Vidalis, *The Emergence of Biolaw*, Springer Textbooks in Law,
https://doi.org/10.1007/978-3-031-02359-0_5

To answer affirmatively, we need to assume that human life itself may be considered the object of a right, freely disposable according to the will of its subject, as it happens with the objects of any other right. If this assumption is valid, then persons may exercise not only in a positive way the right to life by defending it against potential threats, but they may also exercise this right negatively by terminating their life. The wording of article 1 of the ECHR, declaring a "right to life", seems to confirm that position.[2]

Still, a difficulty occurs here because a logical condition for the existence of any fundamental right is life as a fact. A living human being is the only meaningful subject of law, a "person" in a literal legal sense. The absence of life is not compatible with that legal quality; it corresponds to the end of all rights. If this is correct, then the logical presupposition of all rights cannot be considered an object of a particular right. Life, thus, can be nothing but a mere fact, understood as the necessary basis for building the normative world of rights.[3] This conclusion seems to ensure logical consistency for any modern legal system; we are obliged to accept life as a mere fact, just for placing ourselves within the legal system as subjects of rights, which means that we deny the possibility to "negotiate" freely the fact that we are living beings.[4]

Legally, that conclusion derives from another basic assumption which denies the possibility for a person to dispose freely of their human dignity; no act of self-humiliation or abandonment of fundamental rights could be considered a legitimate right-option because it would be inconsistent with human dignity. An intentional transformation of a person's status from a legal "subject" into a mere legal "object" (like a slave in the past, or a deceased human) would be absurd in modern legal systems.

Nevertheless, the above argumentation is valid under one logical condition; that the special quality of life does not entail the transformation of its subject into a mere object while still living! Because, if the existence of a person's life does not guarantee human dignity, his/her decision to die should be considered as nothing but the ultimate confirmation of biological autonomy, that is, of the normative maxim demanding control of our biological state in terms of freedom. This extreme possibility seems sufficient to justify death as a legitimate option for a person and is also entirely consistent with the rationale of a legal system.[5]

---

[2] Same wording, see in art. 2 para 1 of the EU Charter of F. Rights, and 6 para 1 of the UN Covenant of Civil and Political Rights.

[3] Cf. the classic Kantian position against suicide, in Kant (1790/1963), pp. 147 seq.

[4] In the same line, cf. Hufen (2001), p. 855. Yet, the German Ethics Committee (Nationaler Ethikrat) does not consider life as an obligation of persons. Cf. Nationaler Ethikrat (2006), p. 53.

[5] Kass (2002, pp. 251 seq.) has expressed an opinion against the enlargement of autonomy and personal choices, particularly regarding the choice of death. In his view, there is social pressure towards certain vulnerable groups (elderly, patients in a critical state) to accept death, which eventually restricts their self-determination. Yet, any personal choice may be subjected to social pressure within a certain community; it is our ability to encounter such a situation that confirms, in fact, our autonomy.

However, to confirm that a person has been transformed into a mere object while still living, we need evidence beyond that person's subjective thoughts. The elements supporting that evidence are critical when addressing end-of-life decisions under the law, especially when we appraise the role of third persons involved, like the attending physicians and the relatives of the person concerned.

5.3 Letting Die and Killing

"Assisted suicide" describes a situation where a person in terminal stage, unable to act physically, but having full ability to form and express free will, asks for assistance in order to terminate his/her life. The attending physician or other healthcare professional or a family member are those expected to provide assistance (access to a lethal substance, usually) to that person, which is the one who performs the critical act of suicide.

"Euthanasia" is a similar situation, with the difference that the patient asks the attending physician to perform the critical act of "killing" because physically or psychologically is unable to perform suicide.

"Withdrawal of life support" means that either the person concerned or (if unable to express will) third persons decide on the critical act of "letting die" when vital functions (including feeding and hydrating) are supported artificially. In the latter case, third persons are fully responsible for the judgment that the life of the person concerned is not any longer worthy.[6]

In ethical terms, all the above situations are characterized by a conflict between, on the one hand, the will of the patient or (if unable to express will) of their legal representatives, intending to terminate life and, on the other, the primary medical duty to maintain human life.[7] The legal question is whether the patient's informed consent as a necessary prerequisite for any medical act explicitly recognized in the Oviedo Convention and the EU Charter of Fundamental Rights, should be considered legally valid when its objective is to terminate life.

For answering that question, we need to refer also to the "right to life" of the ECHR. We have seen that, literally interpreted, this article does not exclude the option of death. Nevertheless, the European Court of Human Rights denied that interpretation, judging that the meaning of this right refers only to legitimate actions of life protection against external threats and does not cover free disposal of life by its subject.[8] Still, the Court acknowledges certain discretion for member states of the Council of Europe to decide on whether the option of death may be legitimate at a

[6] See, in the U.S. case-law, In re Quinlan, 355 A 2D 647 (N.J. 1976), In re Conroy, 486 A. 2D 1209 (N. J. 1985). Also, Cruzan v. Director, Missouri Department of Health 497 U.S. 261 (1990), even if, in that case, the S. Court concluded that no sufficient evidence confirmed the original will of the patient.

[7] See on that conflict, Kouwenhoven et al. (2019).

[8] See, ECtHR, Pretty v. U.K. (2002).

national level;[9] this means that, although the Convention does not recognize a "right to death", it does not prohibit it. Therefore, national laws in Europe adopting any of the above sorts of end-of-life-decisions are still in conformity with the ECHR; this is, for instance, the case of France or Spain, regarding withdrawal of life support, the case of Belgium, the Netherlands, and Luxemburg regarding euthanasia,[10] and that of Switzerland, regarding assisted suicide;[11] although the Strasbourg Court refuses the recognition of end-of-life-decisions as a legitimate interpretation of article 2 ECHR, such decisions are accepted at the national level with no consistency problems with that article, meaning that the European law remains open to end-of-life-decisions, at least implicitly. In other words, if a national legislator in Europe takes the initiative to legalize one or more end-of-life-decisions, the Strasbourg Court would not contest that initiative on the ground of article 2 ECHR if addressing a relevant case.

Given that, a further question is raised: Should we consider, contrary to the Court's current attitude, that a national legislator in Europe is not only allowed but moreover obliged to legalize end-of-life-decisions due to general principles recognized by the European law? A positive answer to that question would mean that, if such national legislation still is missing in a certain state, a citizen could refer to those general principles in order to oblige the state to take legislative measures for complying with them, arguing on the ground of the overall framework of the European legal order. In the following, we will discuss that issue.

5.3.1 The Minimal Quality of Life

Supporters of end-of-life decisions defend them as an ultimate expression of individual autonomy, reflecting *par excellence* the very essence of human dignity. Following that approach, choosing the time and conditions of death freely represents one of the most personal, private choices, of crucial significance for the individual autonomy and dignity: This is the absolute relief from a life's drama, sealed by pain and desperation!

It is difficult to imagine what social interest or social value could be considered more important than that ultimate personal freedom. No matter which end-of-life decision we refer to, this is grounded on a genuine expression of the individual free will. Is there any difference from suicide, then? In the view of the person concerned, suicide is also a conscious act of relief from an unaffordable life. Still, there is, here, a difference.

---

[9] See, e.g., ECtHR, Haas v. Switzerland (2011), Koch v. Germany (2012), Lambert and Others v. France (2015).

[10] Regarding the development of legislation in the Netherlands and Australia towards the acceptance of euthanasia, cf. Chesterman (1998), pp. 375 seq. In Australia, see in relevance the Rights of the Terminally Ill Act (1995).

[11] See, also, the more recent laws in Italy, in Ciliberti et al. (2018), and in Spain, in Sanz et al. (2021).

A characteristic element of all end-of-life decisions related to health is an objective situation of natural incapacity of the person concerned that excludes almost every act of self-determination, including the freedoms of movement, expression, and private life. This makes a difference with the wish of suicide expressed by persons biologically capable of exercising their self-determination but disappointed due to their personal relationships, social environment, etc. Suicide, in such cases, does not involve necessarily biological incapacity leading to limited self-determination. In end-of-life decisions, that element is significant; even if the patient has no doubts about the worth of life in general, the fact that he/she experiences a fall of vital functions yields misgivings about the worth of living in such particular conditions. That difference may reveal a particular meaning for the legal concept of a person's "life," as mentioned, for instance, in article 2 ECHR, or, to be more precise, about a potential legal significance of life's "quality."[12]

As we have seen, in modern legal systems, a human person's life is always associated with the feature of autonomy; the person's life is legally essential only because it supports autonomy. The law, therefore, does not refer to "life" as a mere biological fact, a simple component of the natural world, as it happens with the other species' life; human life is not considered a sum of biochemical reactions, but a unique biological condition that enables persons to act autonomously. If that condition is not confirmed in certain circumstances, "life" may be preserved as a natural fact, but its connection with the legal property of "person" does no longer exist. This is the minimal "quality" of life that may be confirmed objectively, and it is always required for attributing the supreme legal protection to humans, as "persons".[13]

Hence, as a natural fact, persons' lives are nothing but the "shell" of autonomy. However, this does not mean that a "meaningless" life is plainly identifiable. Two conditions are necessary to demonstrate that; the first is objective, and the second is depending on the person's particular view.

i) The minimal "quality" of human life refers exclusively to the person's condition of health, that is, to the conservation of biological functions in a situation permitting exercise of self-determination acts, no matter if social or psychological reasons eventually impede these acts. Therefore, the capacity of self-determination, considered in purely biological terms, is necessary to define that minimal "quality" of life.
ii) Still, the above condition is not sufficient; because no one but the person his/herself is in place to appraise genuinely if and to what extent the fall of vital functions allows self-determination. This additional "subjective" condition is the logical consequence of the life/autonomy essential bond. Indeed, the core of

[12] On that concept, see from an ethical point of view, Paterson (2017), chapt. 5, and arguments in Keown (2018), pp. 42 seq.

[13] See the distinction between "natural" and "human" definitions of life in Dworkin (1993), pp. 68 seq. The first term refers to the "sanctity" of life as a natural phenomenon, while the second refers to its social significance. For a psychological approach of the "quality of life," cf. Boddington and Podpadec (1992), pp. 273 seq., particularly pp. 280–281.

autonomy is shaped by our ability to develop our own lifestyle based on purely personal choices. Thus, judgments by other persons suggesting a specific health condition as necessary for enjoying a certain lifestyle are irrelevant, even if these judgments may refer to objective facts. For example, even a painful life in terminal situations may be meaningful and worthy for a patient, if that person believes so, choosing that particular way of self-determination and denying to ending that life. Otherwise, we would reject the essential subjective aspect of autonomy; to replace the will of the patient concerned by the will of another person, suggesting that the life of the first is no longer worthy and must be terminated, is nothing but a direct violation of human dignity, even if the motives could be compassion or love for that patient.[14]

5.3.2 Quality of Life and Human Dignity

Following the above argumentation, if in terminal situations with no therapeutic hope the patient is convinced that he/she cannot exercise acts of self-determination anymore, the intentional termination of life is ethically justified. In such cases, if life is maintained, it will be dissociated from autonomy, which means that the person concerned will be transformed into a mere object determined by others, no matter if continuing to be a living human being. The same argument is also valid for a legal approach; the end of life in such situations is embedded in the principle of human dignity, which cannot accept that transformation for law subjects.

Nevertheless, even if all therapeutic hopes have been scientifically abolished, and the patient is aware of that, a counter-argument still might refuse the intentional termination of life. We cannot exclude future progress in medicine or even a sudden, unforeseen reaction of the person's organism, leading to recovery or even to complete therapy. This possibility makes illegitimate any intentional termination of life just because our knowledge is practically limited, and we cannot logically exclude even a miracle!

That objection would be reasonable only if we undermined the crucial element of human dignity. Indeed "forced" maintenance of a person's life, in the name of uncertain hopes, that would be both unjustified scientifically and contrary to the person's own will, ultimately would be equivalent to an instrumentalization of the latter. Despite pain and suffering that make the patient's biological functions unaffordable, maintaining these functions with no rational prospect of recovery does not differ from that person's submission to involuntary experimentation. This treatment equates a human to a mere laboratory object. Indeed, in that case, a human being

[14]See the opinion of Robertson (1975), pp. 370 seq, objecting to judgments of others than the person concerned, about the quality of life, even if that person has no experience of self-determination, like for instance newborns suffering from serious diseases. In contrast, it is worth noting that, for some, the choice of death may be considered as a "duty," not as a mere "right," a position that undermines the aforementioned subjective condition on the determination of life's "quality." Cf. Hardwig (1997), pp. 339 seq., particularly 343 seq.

would be considered "available" just for testing the unconfirmed expectations of others. Even if the law accepts for persons unable to consent to undergo clinical experimentation, when a direct benefit for their health is reasonably expected, in the case we examine, the ability of free will does exist, and yet it plays no role whatsoever. It is precisely this fact that violates human dignity.

To conclude on the legitimacy of end-of-life decisions: Human life as a natural phenomenon becomes legally significant only through its subject's biological autonomy, not separately from that element. In that meaning, biology is submitted to our freedom. Therefore, article 2 of the ECHR is consistent with a "right to death," the other dimension of the explicitly recognized "right to life," because it cannot be interpreted but in conformity with the human dignity principle as declared in the European legal instruments. It is the human dignity principle that, in fact, imposes the life/autonomy strict tie on any legal approach relevant to the handling of human life.[15] For that reason, our initial question regarding a possible obligation of national legislators in Europe to legalize end-of-life-decisions is valid and should be answered affirmatively. Indeed, the central role of the human dignity principle in the European legal instruments produces certain effects here, even if we admit the position of the ECtHR against the recognition of a "right to death"; it is sufficient that the Court accepts a margin of discretion enabling national recognition of that right, for confirming that legal obligation on the basis of the European law's general principles.

5.3.3 The Genuine Character of End-of-Life Decisions

A person asking for the termination of his/her life must decide freely. Nonetheless, freedom is under pressure in such situations, as the patient stands before a crucial dilemma between life and death, with no intermediate options. This condition seems to restrain biological autonomy from the mere fact that the person is forced to cope with that dilemma, no matter what the final decision would be.

Still, under these circumstances, free will must be guaranteed, not only for maintaining the person's autonomy but also for reasons related to those that are to be involved in the execution of life-ending; these persons need to be sure that their acts are lawful and they are not asked to perform a crime against human life. That purpose is served if appropriate guarantees on the genuine free will of the patient are in place. A specific legislative intervention establishing such guarantees is necessary here, for reasons of legal certainty, as indicated by the example of several national European laws, like in countries adopting euthanasia, or assisted suicide.

In general, a crucial guarantee is, for instance, the appropriate information provided to the patient regarding the disease's development, the possibilities of recovery, and those of palliative care regardless of therapeutic expectations. That information must be provided prior to the final decision on life-ending, leaving

[15] On the correlation between autonomy and end-of-life-decisions, see also Besirevic, *supra* ref. 1.

sufficient time for enabling the person to reflect rationally, appraise all crucial factors, and form a genuinely free will. This is a sort of informed consent since it concerns a medical act, the purpose of which is the patient's benefit, even if recovery may not be possible.

An additional guarantee for a genuine will could be the evaluation of the patient's mental capacity by an independent expert, who needs to confirm whether the patient in such conditions is in place to decide consciously or, on the contrary, whether his/her mental functions do not support rational decision-making.[16] However, that evaluation cannot be extended to judgments on end-of-life justification based upon a neutral observer's "objective" standards. This would jeopardize the patient's biological autonomy by challenging his/her free will with a substituted judgment; free will should be expressed even with decisions considered as "wrong" for a third observer, on condition that the capacity of logical reasoning is confirmed, which is the only matter that the expert needs to examine.[17]

5.3.4 The Role of Legal Representatives

Considering that, in principle, only the patient is competent to appraise the quality of his/her life (and thus to make decisions for its termination), we need to address the problem of end-of-life decision-making for persons with limited or no ability to form or express free will, due to their age or mental state.

The first question, here, is whether a legal representative has the authority to ask for the unable patient's termination of life, like in the case of a patient child or a psychiatric patient. Given that a legal representative should act in the best interest of that person, we cannot exclude the possibility of making end-of-life decisions as well. Assuming that such decisions are legally valid as medical acts (following the above-suggested interpretation of the ECHR's "right to life"), the best interest of a person living in a situation of a continuous artificial support of vital functions with no hope of recovery may be the withdrawal of this support. Such a person usually is unable to appraise the quality of his/her life; therefore, the subjective element of this judgment, as presented before, is lacking. What remains is an objective situation of human life with no expectation of self-determination, which justifies an end-of-life-decision to be taken by legal representatives. As obvious, a disagreement between them impedes that decision because it raises doubts about the patient's best interest, particularly regarding his/her quality of life.

A second question concerns the role of the patient's opinion if he/she has a limited ability to form and express will. According to the Oviedo Convention, the

[16] See in relevance, ECtHR, Haas v. Switzerland (*supra*).

[17] Highlighting the significance of genuine end-of-life-decisions, some opinions sometimes prefer "active" euthanasia over "letting die." Cf. Rachels (1975), pp. 227 seq., and Kuhse (1998), pp. 236 seq. *Contra*, Nesbitt (1995), pp. 231 seq. However, problems regarding the practical implementation of legal safeguards may occur. See in Belgium, Raus et al. (2021).

opinion of minors and adults unable to provide consent about medical acts should be taken into account by their legal representatives when deciding. This opinion has practical importance when the unable person refuses to undergo a specific medical act, although the legal representative has provided consent. Again, the legal standard to address such disagreements is the patient's best interest, in its objective meaning, which presumably is expressed by the legal representative. In end-of-life cases, though, the patient's limited ability to judge his/her own quality of life exists. This element is crucial, particularly when a patient experiences pain and suffering. As a minimal indication of self-determination, it should be binding for the legal representative's final decision, although it cannot be considered a substitute decision.[18]

5.3.5 Advance Directives and Living Wills

Still, in cases of patients completely unable to form or express their will, a problem that we need to examine is whether a genuine will relevant to the end of life, expressed by them in a previous time when they were able to make decisions for their own health, should play a certain role at the crucial time.[19]

Indeed, in article 9, the Oviedo Convention accepts that possibility. It states that previously expressed wishes of the person concerned should be "taken into account" in final decisions. This is the only institutional expression of "advance directives"[20] in European law.

Through these directives, a person makes known wishes for their medical treatment at a previous time, for the event to become unable to form or express genuine will in the future. Their content refers mostly to the denial of medical acts concerning the artificial support of vital functions (resuscitation, feeding, hydrating, etc.) to avoid prolonging a painful life; thus, they imply end-of-life decisions. Advance directives differ from the legal representation precisely to the point that it is the person concerned and not a third person that has expressed a valid will on medical treatment, even if not at the crucial time of decision-making.

Directives may either authorize a third person as a proxy for taking medical decisions when the patient becomes unable to decide, or determine wishes about specific medical acts ("living wills") directly.[21] Certainly, they may also contain both options. The advantage of directives authorizing a proxy is that informed consent of the latter is expressed at the crucial time of medical intervention. On the other hand, they do not comprise a genuine will of the person about specific

[18] Regarding euthanasia in minors, see Brouwer et al. (2018). On the case of dementia, where the adult patient experiences a gradual loss of the consent capacity, see Cipriani and Di Fiorino (2019).

[19] See the case-law in the U.S. mentioned above (ref. 6), and the famous case Schiavo (T. M. Schindler Schiavo incap. ex rel. v. M. Schiavo, Dist. Court of Florida, 25.3.2005).

[20] Cf. Capron (2001) and Mappes (1998).

[21] Cf. Brock (1991), pp. 353–354. See the DNR orders' example, in Tomlinson and Brody (1988).

interventions, which is the advantage of living wills. Nonetheless, living wills cannot capture the possibility of mind-changing, given that the patient still may be able to form will (and thus, change wishes) although unable to express that. This possibility may occur under the direct experience of pathology at a crucial time, which is very different from the "sober" condition of the person when deciding and formulating a living will.[22]

As mentioned, the Oviedo Convention accepts the advance directives and states that they need to be taken into account by the physician; this stipulation obliges the latter to justify specifically a medical decision declining from their content. An interesting question, here, is whether that discretion to deviate from advance directives meets a certain limit, in other words, whether under certain circumstances the content of advance directives is binding for the physician or the patient's legal representatives.[23]

Given that the patient's will regarding their biological autonomy is binding insofar as it does not violate human dignity, it is important to point out relevant consequences applicable in advance directives and living wills.

i) An advance directive authorizing a proxy to make medical decisions on behalf of the person concerned is binding only on the condition of acceptance by the latter. Indeed, the putative proxy may refuse to accept due to conscience or religious reasons. Here, the directive's binding nature refers only to third persons who need to act in compliance with the proxy's decisions, except those contravening either the patient's human dignity or their fundamental rights or other provisions of the law. Third persons cannot decline from the proxy's decisions, since in that case, they would directly violate the patient's human dignity, as expressed by the advance directive. Indeed, its content is the only guarantee for avoiding an absurd treatment of the patient, being a minimal indication of his/her autonomy. On the other hand, the proxy authorization is always exercised under a strict framework requiring respect and protection of the patient's human dignity, compliance with the law, and appropriate proxy information before any decision. In principle, the proxy needs to respect end-of-life wishes, if expressed by the patient either with an additional clause included in the content of advance directives or even with non-verbal signs at the crucial time, indicating unaffordable pain and suffering. This is consistent with what we have already suggested for minors and mentally ill persons, and, moreover, it involves an appraisal of the quality of life, as expressed authentically by the person concerned.
ii) The content of living wills is binding for both the attending doctor and the patient's relatives or close friends. Still, relevant wishes should not be in contrast with the protection of the latter's human dignity (allowing, for instance, experimental treatment with no expected benefit). The wishes oblige third persons

---

[22] Similar argumentation against the real value of advance directives for dementia patients see in Walsh (2020).

[23] See in relevance, Pascalev and Vidalis (2010).

insofar as they respect their fundamental rights, and also the physician, insofar as they comply with medical ethics and law.

In any of the above cases, the possibility of a wish-change by the patient at the crucial time (if still able to form a valid will), creates a major question, particularly when the original living will requires termination of life. The only solution here is to accept deviation from the living will if indications of such a change are uncontested, following specific standards that the national legislation should provide (i.e., confirmation by an independent organ). If the patient remains in a situation of permanent inability to form a genuine will (being, for instance, in a permanent vegetative state), there is no question to decline from the living will's content.[24]

In conclusion, the need for specific regulation of advance directives and living wills by national law is evident, as the general wording of article 9 of the Oviedo Convention does not ensure certainty of law. The content of that regulation also needs to comprise a requirement regarding the form of advance directives and living wills for validating their force (notary form or other), as suggested already by the model that several national European laws in relevance follow.

5.3.6 The Liability of the Attending Doctor: Futile Treatment

From the part of patients, end-of-life decisions may represent an authentic expression of biological autonomy. Still, we need to discuss also problems of the attending doctor's liability. So far, in most European national legislations, end-of-life decisions, if executed, involve doctors' criminal liability. Yet, even under strict criminal regulations, we must accept an exception, where liability for "letting die" decisions does not exist; this is the case of medical futility.[25]

As mentioned, in case of disease or other pathological situations, our biological autonomy is expressed through personal control of the treatment that we need to follow, ensured by the procedure of informed consent. The attending doctor is not entitled to act without this prerequisite (with the exception of emergencies), being obliged by the patient's free will to accept or not a specific intervention, and having full responsibility regarding the patient's previous and appropriate information.

In some instances, though, to continue treatment means no benefit for the patient, as no effects for improvement of health conditions are expected anymore. Such a conclusion may be indisputable if based upon strict scientific data, regardless of what other persons or the attending physician define as "benefit" for the concrete patient,[26]

[24] Cf. Dworkin (1993), p. 228.

[25] Regarding distinctions in medical futility, see Brock (2001), p. 234; Wicclair (2001), pp. 340 seq. Against the existence of objective standards to identify futility, see Lelie and Verweij (2003), pp. 23 seq., White (2020), pp. 181 seq (where alternative terms are mentioned also, such as non-beneficial' or 'potentially inappropriate' treatment', 'disputed treatment').

[26] See ECtHR, Afiri, and Biddarri v. France (2018). Against a strict medical determination of futility, see Close et al. (2018).

taking into account psychological or social beliefs that may affect the latter. Thus, when an oncologic patient experiences a terminal condition, it is difficult to consider as "benefit" the continuous administration of chemotherapy for prolonging temporarily vital functions with no hope of any recovery; on the contrary, this practice, if painful and stressful for the patient, could easily be defined as a sort of "harm" unacceptable for any person.[27] In similar cases, the patient's will for stopping medical intervention is not an expression of "treatment refusal" precisely because no treatment exists in any possible meaning whatsoever; the patient refuses only the artificial support of his/her vital functions, not an intervention expected to yield reasonable hopes for recovery or improvement of health.

Here, a counter-argument could suggest that conservation of life in any possible means is always a "benefit" for the person concerned, and in that sense, the physician has both an ethical and a legal duty to ensure this. This argument would undermine, though, the fact that the law does not protect human life as a merely natural condition, but only in its connection with autonomy. If medical intervention entails a violation of the autonomy components, like corporal integrity, or human dignity, with no reasonable hope for recovery or health improvement, it could be equated with experimentation on life by transforming a human being into its mere instrument; by definition, this is nothing but "harm" for the patient.[28]

Here, it is important to underline that the "benefit" standard is, in principle, irrelevant to economic considerations regarding the costs of a particular treatment or, in general, the allocation of sources and means for citizens' health in a certain society. In terms of legal liability, the attending physician must inform the patient or his/her legal representatives about the futility of a particular treatment, if this is confirmed by scientific evidence. This information has no economic dimension; it must be limited to the strict "beneficence" principle. Assuming that there is a treatment promising for recovery or health improvement but still extremely demanding in terms of costs, no futility exists; the doctor has the legal duty to provide information about that possibility also, even if the patient or relatives eventually decline from accepting, for economic reasons.[29]

5.3.7 The Liability of the Attending Doctor: Palliative Care

A cessation of futile treatment in terminal situations changes the scope of medical intervention if the patient experiences unaffordable pain and suffering. In these conditions, the attending doctor's duty needs to focus not anymore on therapeutic

[27] See in relevance, ECtHR, Gard, and Others v. U.K. (2017).

[28] It should be noted that a refusal of the attending physician to cease futile treatment on the ground of conscientious objection produces the same effect for the patient, even if the physician may be excused in terms of liability (when the law recognizes that objection). See the Spanish law in relevance, in Sanz et al. (2021).

[29] See on that dimension, in general, Angell (1993).

acts, which proved hopeless, but on acts of palliative care intending to relieve pain and suffering to the extent possible. Acts of palliative care again presuppose informed consent by the patient or, if unable, by their legal representatives.

With palliative care, both the patient and the doctor have accepted the possibility of death, although they do not intend for that purpose, namely, they do not act to hasten death, like in assisted suicide, withdrawal of artificial support, or euthanasia. The benefit for the patient consists exclusively of the relief from the painful implications of pathology, ensuring a peaceful and dignified end of life. This is also the standard for judging the attending doctor's liability in legal terms since no liability could be confirmed in relation to therapeutic medical acts, which have been proved futile.

In terminal situations, pain is usually addressed with the administration of opioids (like morphine, etc.) to the patient. A serious issue here is that a known implication of opioids involves the potential limitation of life expectancy in terminal patients. Thus, the attending physician faces a crucial ethical dilemma to choose for the patient between a longer life, which will remain painful, and a potentially shorter one, but peaceful and dignified until the end.[30]

This dilemma also presents a serious legal dimension. In countries prohibiting end-of-life decisions, the administration of such palliative treatment might be considered a decision on euthanasia, which involves the doctor's criminal, civil, and disciplinary liability. An interpretation of the ECHR's fundamental right to life that also covers end-of-life decisions would suggest, here, the priority of palliative care as the doctor's legal duty, even implicating death hastening. Otherwise, given that any treatment with therapeutic scope has proved futile, the only option for the doctor is to leave the patient with no care at all since palliative care has been excluded as well. This would be an illegitimate option, though, entailing harm to the patient, which is contrary to article 1 of both the Oviedo Convention and the EU Charter of Fundamental Rights; a temporary and painful prolongation of vital functions with no hope of therapy or pain relief cannot be considered in any meaning a "benefit" for a human being.

5.3.8 Premature and Anencephalic Newborns

Issues of life-support withdrawal are raised regarding newborns before completing the 22nd week of pregnancy (a viability time limit confirmed by current methods of artificial support) and also newborns with no brain development (usually, with only the brain stem existing).[31]

---

[30] Cf. Doerflinger and Gomez (2016). *Contra*, Sykes and Thorns (2003), pp. 312 seq. This dilemma reflects the double-effect philosophical doctrine on acts resulting in both positive and negative consequences simultaneously. Cf. Chesterman (1998), p. 370; Brock (2001), pp. 239–240.

[31] Cf. Nuffield Council on Bioethics (2006). Between 22nd and 26th week, there is a grey zone where viability remains uncertain. See also, Campbell (1994), pp. 243 seq.; Kluge (2001), pp. 246–247.

These cases need specific examination since, for different reasons, the organism's development during pregnancy remains incomplete, which means that the biological element of autonomy is missing. The newborn's viability which may last from a few hours to several weeks is dependent on the artificial support of vital functions in intensive care units. Similarly to the example of terminal patients, there is a question, here, on futile treatment and palliative care since, on the one hand, unsupported survival of the newborn is absolutely impossible and, on the other, the artificial conservation of the newborn's life involves pain and suffering for the latter. A crucial difference from terminal situations we discussed already is that the newborn not only lacks the experience of unsupported life (being completely unable to form and express wishes) but, moreover, it remains unclear whether it can be considered a "person" in the first place, since the biological element of autonomy does not exist.

This last remark seems to justify an end-of-life decision, that is, the withdrawal of artificial support. Indeed, a pediatric physician has a legal duty to provide medical care (including palliative care) on the condition that the human organism to be treated is identified legally as a "person"; otherwise, no patient/doctor relationship is established. In such cases, medical liability should be judged similarly to what is acceptable for a mature human fetus's medical care. As we will see in the human reproduction chapter, in legal terms, care for mature fetuses is based on the expectation of a future person's existence, namely a human life in its potential transformation into a complete person, viable or not. In front of that expectation, pediatric physicians are responsible for engaging all available means to achieve the newborn's autonomous viability, with no artificial support. If this scope is proved scientifically unfeasible, the attending doctor needs to confirm medical futility and provide palliative care to the extent possible, similarly to what is legally accepted for persons, even if the newborn is not viable.

5.4 Transplantations Involving Deceased Donors

For a short time, even after death, previous acts of biological autonomy by the person while living may have serious consequences. This is the case of organ and tissue removal from deceased donors for transplantation purposes.

As mentioned before, transplantations gain considerable importance for the protection of health nowadays. Nevertheless, the removal of tissue and organs from living donors implicates risks for their health and autonomy. Therefore, ensuring transplants' availability from deceased individuals represents a practice more acceptable in ethical terms. As such, it is legally prioritized, as indicated by article 19 para 1 of the Oviedo Convention and its Additional Protocol in relevance. Since deceased humans are not considered persons legally, no issues of autonomy or health protection occur, which means that, in principle, their body is available for transplantation purposes. Still, the law does not recognize this availability without conditions.

5.4.1 Confirming the Death

Here, the first problem is related to the confirmation of death as a biological fact. Diagnosis of death must be exact and timely; exact, for making sure that no question regarding the protection of the donor's autonomy exists, and timely, because if vital organs remained attached to the deceased body, they would become useless shortly. It is worth noting the possibility of prolonging the function of vital organs artificially in order to ensure their availability as transplants, thanks to our technology. However, this creates reasonable doubts about the exact moment of death, which was not the case in the past, before achieving the artificial prolongation of vital functions.

In any case, a general agreement in modern Medicine suggests that death as a biological fact is certain from the moment of cessation of the brain stem's function, entailing the termination of all brain activities, which is known as "brain death".[32] Even if other vital organs (like the heart, the kidneys, the liver, etc.) continue to function, the moment of brain death represents an irreversible condition for the life's end, that is, a moment from which no recovery is expected, according to scientific evidence. Although the European law does not refer explicitly to the crucial condition of "brain death",[33] this is accepted by several national legislations, and recognized not only as a medical but also as a legal standard for determining the point of death.

Even so, the diagnosis of brain death may create objections in certain cases on the part of relatives or friends of the deceased person, or even on the part of attending physicians, due to psychological, emotional, or philosophical personal attitudes. In ethical terms, such objections may be excused regardless of their scientific irrelevance. Indeed, here, we need to bear in mind that the transplantation procedures usually involve deceased potential donors of a relatively young age. The event of death is unexpected for these persons, and if occurring (due to a street accident, etc.), relatives and friends are not prepared to accept it. For a short time before death, this sudden dramatic event seems to be moderated under the hopes of rescue that appropriate treatment in the intensive care unit creates. In such circumstances, it is extremely tough for anyone to accept any scientific evidence of brain death as an irreversible point for life's ending, particularly if knowing that other vital organs continue to function regularly. Only if all vital organs cease to function, the fact of death is certain for the common understanding; at that later moment, though, organs become useless for transplantations.

Similarly, attending doctors may be reluctant to withdraw the artificial support of a person's biological functions after confirming brain death, not because they may

[32] For some, cessation of the brain hemispheres' function is sufficient evidence for brain death. See, in relevance, McMahan (2001), particularly p. 257, Rhodes (2001), pp. 330–331. For an "intermediate" position, according to which patients in a permanent vegetative state are not dead but they cannot be considered persons either (therefore euthanasia or withdrawal of life support are justified), see Culver and Gert (1982), pp. 310 seq. On brain death in general, see Wijdicks (2001).

[33] The Transplantation Protocol of the Oviedo Convention does not make reference to the determination of death, leaving that issue to the discretion of national law.

feel compassion for the deceased person's family members or friends, but mostly for personal reasons of moral, religious, or psychological nature. We cannot exclude that even a medical expert may believe in the "miracle" of recovery after experiencing a hard but hopeless struggle against death, no matter if scientific evidence suggests the opposite.

In legal terms, these objections are associated with rules protecting the fundamental rights of those invoking them, such as the freedom of thought, or religious freedom, when the objection is grounded on relevant beliefs. This indicates a condition of conflicting interests, that is, the social interest of health protection by promoting transplantations (which presupposes organ availability after brain death for those needing that kind of treatment[34]), and the individual interest represented by fundamental rights of relatives, friends, and doctors involved, which may object to the standard of brain death and the collection of vital organs.

The European legislator acknowledges that particular conflict situation, embracing alternative solutions to resolve it at the level of national law. A general remark on the issue is that the attending doctor is recognized as the first responsible for making feasible the collection of organs and the transplantation procedure. National laws in Europe do not exclude criminal and civil sanctions in relevance if doctors fail to meet this responsibility. Therefore, the European law indirectly recognizes the brain death standard, not leaving a margin for third persons to object on the ground of their emotions and beliefs, or for the doctor to block necessary acts of organ removal by invoking personal reasons, even if this doctor certainly holds a right to abstain from their performance for such reasons. Thus, according to the current European law, the social interest that transplantations represent for protecting health prevails over legitimate individual objections, at least as a general principle.

5.4.2 Consenting to the Organ Removal

Removing organs from deceased individuals raises the most crucial question we have already mentioned, relevant to the extent of biological autonomy. Who should be entitled to consent to organs collection when the individual concerned, while living, had expressed neither a will nor a refusal of organ donation?

In terms of legal certainty, this question is critical since no answer is provided by the Oviedo Convention and its Protocol on Transplantations. Currently, we can distinguish two trends occurring in national legislations. The majority of them have adopted the solution of the so-called "presumed consent" of the deceased individual, according to which if no explicit refusal for the organ donation had been expressed by the latter while living, we can conclude that he/she consented

[34] Cf. Rhodes (2001), pp. 334 seq. Access must be ensured, even for those who provoked damages to their vital organs, due to irresponsible behavior (*liver damage* caused by excess *alcohol* intake); otherwise, the social right to Health would be unreasonably restricted. Indeed, we are responsible for most of the pathologies that we experience, since we expose ourselves deliberately to harmful environmental conditions. See, also, Franklin (2003), p. 54.

tacitly to the removal of organs after death. In contrast, only a few national legislations require the explicit consent of close family members about organ donation. The European law accepts both as equally possible options to be followed by national law.

Obviously, the first option of "presumed consent" facilitates the collection of organs. In that sense, it promotes health as a social right ensuring the existence of biological material available for transplantations with no need for an "intermediate" approval by anyone. Nevertheless, this advantage could be challenged, given that organ removal means an intervention in the dead body's integrity. Even if the Oviedo Convention underlines the need for respecting the human body, such an intervention may infringe strong emotions of other persons, particularly the emotions of close family members and close friends. For these persons, conserving the dead body's integrity may be important, either for "keeping alive" the image of their beloved person until the moment of burial, or because they consider themselves "guardians" of its supposed tacit refusal to donate organs after death. In both cases, that attitude is strongly related to their own personality, expressing a dimension of their respective fundamental right that needs to be taken into account.

There is no question that paying respect to deceased individuals illustrates a value not only for our civilization but also for civilizations of the past. In legal terms, though, that value is not absolute; as an "object" of law, the dead individual is not considered an agent of human dignity and human rights; therefore, its body's value is not a value *per se* like that of living persons. Nevertheless, we cannot accept unlimited possibilities for using the dead body, that is, the free instrumentalization of it for serving various purposes. This is because, in modern societies, respect for the human body is justified in relation to active rights of living persons, like family members or friends, particularly to their right to personality that we mentioned before.[35] From that aspect, no use of the dead body is considered legitimate if it undermines that last right.

The emotions of persons closely related to the deceased individual belong to the inherent core of their personality, in the sense that the individual's existence, while living, influenced these persons' life (private or public) significantly. In family life, this fact seems legally evident considering the exceptional worth that the law attributes to family relations; on the ground of these mutual emotions flourishing between family members, it is based on the early socialization of any person, which develops the essential elements of personality. It would be unreasonable, then, to believe that such strong personal ties (existing, besides, between close friends and colleagues too) disappear after a person's death.

The right to personality of living persons is precisely what justifies a requirement to ask for their consent to organ removal by deceased individuals.[36] Strongly

[35] This is the right at stake, here. In contrast, economic rights are highly controversial, despite ethical opinions that accept the economic motivation in transplantations (which is excluded legally, according to art. 21 of the Oviedo Convention). Cf. Blumstein (1996). On motivation in transplantations, see also Siegel (2000), pp. 947 seq.

[36] See about that, ECtHR, Petrova v. Latvia (2014), Elberte v. Latvia (2015).

recognized by the European law, that right may be invoked for preventing a relevant intervention in the dead body, even in countries having adopted a "presumed consent" system. From that point of view, the system of "presumed consent" cannot be grounded on arguments implying that the dead body "belongs" by definition to the society or the state, for serving the social right to Health. That system is legitimate on the condition that the rights of other persons are respected; it plays then mostly an instructive role as part of a public strategy to promote the idea of organ donation, and should not be considered a coercive measure of organ collection, which would be diametrically opposite to that consent-based idea.

5.5 Further Studying...

ECtHR: Pretty v. the United Kingdom, 29 April 2002 (Chamber judgment)

"The applicant was dying of motor neurone disease, a degenerative disease affecting the muscles for which there is no cure. Given that the final stages of the disease are distressing and undignified, she wished to be able to control how and when she died. Because of her disease, the applicant could not commit suicide alone and wanted her husband to help her. But, although it was not a crime in English law to commit suicide, assisting a suicide was. As the authorities refused her request, the applicant complained that her husband had not been guaranteed freedom from prosecution if he helped her die.

The Court held that there had been no violation of Article 2 (right to life) of the Convention, finding that the right to life could not, without a distortion of language, be interpreted as conferring the diametrically opposite right, namely a right to die.

The Court also held that there had been no violation of Article 3 (prohibition of inhuman or degrading treatment) of the Convention. Even if it could not but be sympathetic to the applicant's apprehension that without the possibility of ending her life she faced the prospect of a distressing death, nonetheless, the positive obligation on the part of the State which had been invoked would require that the State sanction actions intended to terminate life, an obligation that could not be derived from Article 3.

The Court lastly held that there had been no violation of Articles 8 (right to respect for private life), 9 (freedom of conscience) and 14 (prohibition of discrimination) of the Convention".[37]

Question

Comment on the following paragraphs, under the light of the ECHR and the Oviedo Convention:

[37] ECtHR, Press Unit.

38. The text of Article 2 expressly regulates the deliberate or intended use of lethal force by State agents. However, it has been interpreted as covering not only intentional killing but also the situations where it is permitted to "use force" which may result, as an unintended outcome, in the deprivation of life (ibid., p. 46, § 148). Furthermore, the Court has held that the first sentence of Article 2 § 1 enjoins the State not only to refrain from the intentional and unlawful taking of life, but also to take appropriate steps to safeguard the lives of those within its jurisdiction (see L.C.B. v. the United Kingdom, judgment of 9 June 1998, Reports of Judgments and Decisions 1998-III, p. 1403, § 36). This obligation extends beyond a primary duty to secure the right to life by putting in place effective criminal-law provisions to deter the commission of offences against the person backed up by law-enforcement machinery for the prevention, suppression and sanctioning of breaches of such provisions; it may also imply in certain well-defined circumstances a positive obligation on the authorities to take preventive operational measures to protect an individual whose life is at risk from the criminal acts of another individual (see Osman v. the United Kingdom, judgment of 28 October 1998, Reports 1998-VIII, p. 3159, § 115, and Kılıç v. Turkey, no. 22492/93, §§ 62 and 76, ECtHR 2000-III). More recently, in Keenan, Article 2 was found to apply to the situation of a mentally ill prisoner who disclosed signs of being a suicide risk (see Keenan, cited above, § 91).

39. The consistent emphasis in all the cases before the Court has been the obligation of the State to protect life. The Court is not persuaded that "the right to life" guaranteed in Article 2 can be interpreted as involving a negative aspect. While, for example in the context of Article 11 of the Convention, the freedom of association has been found to involve not only a right to join an association but a corresponding right not to be forced to join an association, the Court observes that the notion of a freedom implies some measure of choice as to its exercise (see Young, James and Webster v. the United Kingdom, judgment of 13 August 1981, Series A no. 44, pp. 21-22, § 52, and Sigurđur A. Sigurjónsson v. Iceland, judgment of 30 June 1993, Series A no. 264, pp. 15-16, § 35). Article 2 of the Convention is phrased in different terms. It is unconcerned with issues to do with the quality of living or what a person chooses to do with his or her life. To the extent that these aspects are recognised as so fundamental to the human condition that they require protection from State interference, they may be reflected in the rights guaranteed by other Articles of the Convention, or in other international human rights instruments. Article 2 cannot, without a distortion of language, be interpreted as conferring the diametrically opposite right, namely a right to die; nor can it create a right to self-determination in the sense of conferring on an individual the entitlement to choose death rather than life.

40. The Court accordingly finds that no right to die, whether at the hands of a third person or with the assistance of a public authority, can be derived from Article 2 of the Convention. It is confirmed in this view by the recent Recommendation 1418 (1999) of the Parliamentary Assembly of the Council of Europe (...).

41. The applicant has argued that a failure to acknowledge a right to die under the Convention would place those countries which do permit assisted suicide in breach of the Convention. It is not for the Court in this case to attempt to assess whether or not the state of law in any other country fails to protect the right to life. As it recognised in Keenan, the measures which may reasonably be taken to protect a prisoner from self-harm will be subject to the restraints imposed by other provisions of the Convention, such as Articles 5 and 8, as well as more general principles of personal autonomy (see Keenan, cited above, § 92).

Similarly, the extent to which a State permits, or seeks to regulate, the possibility for the infliction of harm on individuals at liberty, by their own or another's hand, may raise conflicting considerations of personal freedom and the public interest that can only be resolved on examination of the concrete circumstances of the case (see, mutatis mutandis, Laskey, Jaggard and Brown v. the United Kingdom, judgment of 19 February 1997, Reports 1997-I). However, even if circumstances prevailing in a particular country which permitted assisted suicide were found not to infringe Article 2 of the Convention, that would not assist the applicant in this case, where the very different proposition – that the United Kingdom would be in breach of its obligations under Article 2 if it did not allow assisted suicide – has not been established.

ECtHR: Haas v. Switzerland, 20 January 2011 (Chamber judgment)

"This case raised the issue of whether, by virtue of the right to respect for private life, the State should have ensured that a sick person wishing to commit suicide could obtain a lethal substance (sodium pentobarbital) without a prescription, by way of derogation from the law, so as to be able to end his/her life without pain and with no risk of failure. The applicant, who had been suffering from a serious bipolar affective disorder for around twenty years and considered that, as a result, he could no longer live in a dignified manner, argued that his right to end his life in a safe and dignified manner had been violated in Switzerland as a result of the conditions that had to be met – and which he had not met – in order to be able to obtain the substance in question.

The Court held that there had been no violation of Article 8 (right to respect for private life) of the Convention, finding that, even assuming that States had a positive obligation to take measures to facilitate suicide in dignity, the Swiss authorities had not breached that obligation in the applicant's case.

The Court noted in particular that the member States of the Council of Europe were far from having reached a consensus as regards the right of an individual to choose how and when to end his life. Although assistance in suicide had been decriminalized (at least partly) in certain member States, the vast majority of them appeared to attach more weight to the protection of the individual's life than to his right to end it. The Court concluded that States had a wide margin of appreciation in such matters."[38]

Question

Comment on the following paragraphs, under the light of the ECHR and the Oviedo Convention:

[38] ECtHR, Press Unit.

51. In the light of this case-law, the Court considers that an individual's right to decide by what means and at what point his or her life will end, provided he or she is capable of freely reaching a decision on this question and acting in consequence, is one of the aspects of the right to respect for private life within the meaning of Article 8 of the Convention.

52. In the Court's opinion, however, the instant case is to be distinguished from the above-cited Pretty case. Like the Federal Court, it considers that it is appropriate to state at the outset that the instant case does not concern the freedom to die and possible immunity for a person providing assistance with a suicide. The subject of dispute in this case is whether, under Article 8 of the Convention, the State must ensure that the applicant can obtain a lethal substance, sodium pentobarbital, without a medical prescription, by way of derogation from the legislation, in order to commit suicide painlessly and without risk of failure. In other words, unlike the Pretty case, the Court observes that the applicant alleges not only that his life is difficult and painful, but also that, if he does not obtain the substance in question, the act of suicide itself would be stripped of dignity. In addition, and again in contrast to the Pretty case, the applicant cannot in fact be considered infirm, in that he is not at the terminal stage of an incurable degenerative disease which would prevent him from taking his own life (see, conversely, Pretty, cited above, § 9).

...

56. With regard to the balancing of the competing interests in this case, the Court is sympathetic to the applicant's wish to commit suicide in a safe and dignified manner and without unnecessary pain and suffering, particularly given the high number of suicide attempts that are unsuccessful and which frequently have serious consequences for the individuals concerned and for their families. However, it is of the opinion that the regulations put in place by the Swiss authorities, namely the requirement to obtain a medical prescription, pursue, inter alia, the legitimate aims of protecting everybody from hasty decisions and preventing abuse, and, in particular, ensuring that a patient lacking discernment does not obtain a lethal dose of sodium pentobarbital (see, mutatis mutandis, with regard to restrictions on abortion, Tysiąc v. Poland, no. 5410/03, § 116, ECtHR 2007-I).

ECtHR: Lambert and Others v. France, 5 June 2015 (Grand Chamber)

"The applicants are the parents, a half-brother and a sister of Vincent Lambert who sustained a head injury in a road-traffic accident in 2008 as a result of which he is tetraplegic. They complained in particular about the judgment delivered on 24 June 2014 by the French *Conseil d'État* which, relying on, among other things, a medical report drawn up by a panel of three doctors, declared lawful the decision taken on 11 January 2014, by the doctor treating Vincent Lambert, to discontinue his artificial nutrition and hydration. The applicants submitted in particular that withdrawing his artificial hydration and nutrition would be contrary to the State's obligations under Article 2 (right to life) of the European Convention on Human Rights.

The Court held that there would be no violation of Article 2 (right to life) of the European Convention on Human Rights in the event of implementation of the *Conseil d'État* judgment of 24 June 2014.

(continued)

It observed in particular that there was no consensus among the Council of Europe member States in favour of permitting the withdrawal of life-sustaining treatment. In that sphere, which concerned the end of life, States must be afforded a margin of appreciation. The Court considered that the provisions of the Act of 22 April 2005, as interpreted by the *Conseil d'Etat*, constituted a legal framework which was sufficiently clear to regulate with precision the decisions taken by doctors in situations such as that in the present case."[39]

Question

Comment on the following paragraphs, under the light of the ECHR and the Oviedo Convention:

158. In its judgment of 24 June 2014, the Conseil d'État detailed the factors to be taken into account by the doctor in assessing whether the criteria for unreasonable obstinacy were met, while making clear that each situation had to be considered on its own merits. These were: the medical factors (which had to cover a sufficiently long period, be assessed collectively and relate in particular to the patient's current condition, the change in that condition, his or her degree of suffering and the clinical prognosis) and the non-medical factors, namely the patient's wishes, however expressed, to which the doctor had to "attach particular importance", and the views of the person of trust, the family or those close to the patient.

159. The Court notes that the Conseil d'État established two important safeguards in that judgment. Firstly, it stated that "the sole fact that a person is in an irreversible state of unconsciousness or, a fortiori, has lost his or her autonomy irreversibly and is thus dependent on such a form of nutrition and hydration, does not by itself amount to a situation in which the continuation of treatment would appear unjustified on grounds of unreasonable obstinacy". Secondly, it stressed that where a patient's wishes were not known, they could not be assumed to consist in a refusal to be kept alive (see paragraph 48 above).

160. On the basis of this analysis, the Court cannot subscribe to the applicants' arguments. It considers that the provisions of the Leonetti Act, as interpreted by the Conseil d'État, constitute a legal framework which is sufficiently clear, for the purposes of Article 2 of the Convention, to regulate with precision the decisions taken by doctors in situations such as that in the present case. The Court therefore concludes that the State put in place a regulatory framework apt to ensure the protection of patients' lives (see paragraph 140 above).

...

[39] ECtHR, Press Unit.

177. The applicants submitted, relying on Article 8 of the Convention, that the Conseil d'État should not have taken into consideration Vincent Lambert's spoken remarks, which they considered to be too general.

178. The Court points out first of all that it is the patient who is the principal party in the decision-making process and whose consent must remain at its heart; this is true even where the patient is unable to express his or her wishes. The Council of Europe's "Guide on the decision-making process regarding medical treatment in end-of-life situations" recommends that the patient should be involved in the decision-making process by means of any previously expressed wishes, which may have been confided orally to a family member or close friend (see paragraph 63 above).

179. The Court also observes that, according to the comparative-law materials available to it, in the absence of advance directives or of a "living will", a number of countries require that efforts be made to ascertain the patient's presumed wishes, by a variety of means (statements of the legal representative or the family, other factors testifying to the patient's personality and beliefs, and so forth).

180. Lastly, the Court points out that in its judgment in Pretty (cited above, § 63), it recognised the right of each individual to decline to consent to treatment which might have the effect of prolonging his or her life. Accordingly, it takes the view that the Conseil d'État was entitled to consider that the testimony submitted to it was sufficiently precise to establish what Vincent Lambert's wishes had been with regard to the withdrawal or continuation of his treatment.

ECtHR: Petrova v. Latvia, 24 June 2014

"Having sustained life-threatening injuries in a car accident, the applicant's son was taken to hospital, where he died. Shortly afterwards, a laparotomy was performed on his body, in the course of which his kidneys and spleen were removed for organ-transplantation purposes. The applicant alleged that the removal of her son's organs had been carried out without her or her son's prior consent and that, in any event, no attempt had been made to establish her views.

The Court held that there had been a violation of Article 8 (right to respect for private and family life) of the Convention. It found that the Latvian law in the area of organ transplantation as applied at the time of the death of the applicant's son had not been sufficiently clear and had resulted in circumstances whereby the applicant, as the closest relative to her son, had certain rights with regard to removal of his organs, but was not informed – let alone provided with any explanation – as to how and when these rights could have been exercised".[40]

[40]ECT HR, Press Unit.

Question

Comment on the following paragraphs, under the light of the ECHR and the Oviedo Convention:

54. The Court notes at the outset that an application cannot in principle be brought in the name of a deceased person, because he or she cannot be considered to be a "person" ("personne physique" in French) for the purposes of Article 34 of the Convention (see Dvořáček and Dvořáčková v. Slovakia, no. 30754/04, § 41, 28 July 2009, and Aizpurua Ortiz and Others v. Spain, no. 42430/05, § 30, 2 February 2010).

...

56. In the present case the rights of the deceased, Mr Petrov., and his mother, the applicant in the present case, are closely related. The domestic law at the material time explicitly provided that the right to express one's wishes in relation to removal of organs or body tissue after death pertained not only to the person concerned but also to his or her closest relatives, including parents (see paragraphs 36 and 37 above). The Court considers, however, that there is no need to examine the issue of transferability of rights in more detail in the present case since the applicant complains of a violation of her own rights in connection with the removal of her son's organs after his death. Contrary to what has been argued by the Government, the Court finds that on the application form the applicant expressly indicated her wish to complain in her name and she maintained that position in her observations on the admissibility and merits of the case.

...

87. As to the alleged interference, turning to the circumstances of the present case, the Court notes that following a car accident the applicant's son sustained life-threatening injuries of which, after an attempt to save his life had been made, he had died. Immediately after his death, his kidneys and spleen had been removed for organ transplantation purposes. The applicant, who was one of his closest relatives, was not informed of this and could not therefore exercise certain rights allegedly established under domestic law – to express consent or refusal in relation to the removal of her son's organs.

...

90. As to whether the interference was "in accordance with the law", the Court observes that Latvian law at the material time explicitly provided for the right on the part of not only the person concerned but also his or her closest relatives, including parents, to express their wishes in relation to removal of organs after that person's death (see paragraphs 36 and 37 above). The parties did not contest this. However, their views differed in so far as the exercise of this right was concerned. The applicant's view was that the domestic authorities had not fulfilled their duty to provide conditions whereby her wishes in relation to the removal of the applicant's son's organs for transplantation purposes could be expressed. According to the Government, the "presumed consent system" in Latvia was an active process and the persons involved were expected to take positive steps if they wished to veto any organ removal. It is the Court's view that these issues appertain to the quality of

domestic law, in particular, whether the domestic legislation was formulated with sufficient precision or afforded adequate legal protection against arbitrariness in the absence of the relevant administrative regulation.

Provide legally appropriate solutions

1. The hospitalized patient, Arthur B. suffering from cancer at the terminal stage, experiences unaffordable pain from the treatment and agony for his health condition for the next days and even hours. Arthur B. feels that he needs to make a critical end-of-life decision as soon as possible. What are the legal options for the attending physician under the current European Biolaw,

- if Arthur B. asks the physician to interrupt the treatment provided?
- if the physician believes that this treatment has become futile for the condition of Arthur B. and no other therapeutic option exists?
- if the relatives of Arthur B. believe that the treatment is futile and disagree on that with the attending physician?
- if Arthur B. asks the attending physician, instead of treatment, to administer a lethal drug?
- if Arthur B. asks the attending physician for a lethal drug to commit suicide?

2. Harold A., a physician, is aware of a "Do-Not-Resuscitate" order issued by his patient Clara B., now unable to express her will. Both the husband and son of Clara B. disagree with that order and tell the doctor to proceed immediately to the resuscitation.

Under the light of articles 6 and 9 of the Oviedo Convention, what should the physician do?

References

Angell M (1993) The doctor as double agent (Kennedy Inst. of Ethics J. 3: 279-286). In: Mappes ThA, Degrazia D (eds) Biomedical ethics, 5th edn. McGraw-Hill, New York 2001, pp 132–136

Besirevic V (2016) Mission (im) possible: defending the right to die. In: Himma K, Spaić B (eds) Fundamental rights: justification and interpretation. Eleven Publishing International, The Hague, pp 149–177

Blumstein JF (1996) Legalizing payment for transplantable cadaveric organs (Thomasma, D.C., Kushner, T. H., Birth to Death: Science and Bioethics, Cambridge U.P., pp. 119-132). In: Kuhse H, Singer P (eds) Bioethics. An anthology. Blackwell, Oxford 1999, pp 390–398

Boddington P, Podpadec T (1992) Measuring quality of life in theory and in practice (Bioethics 6: 201-217). In: Kuhse H, Singer P (eds) Bioethics. An anthology. Blackwell, Oxford 1999, pp 273–282

Brock DW (1991) Surrogate decision making for incompetent adults: an ethical framework (The Mount Sinai J. of Medicine 58: 388-392). In: Mappes ThA, Degrazia D (eds) Biomedical ethics, 5th edn. McGraw-Hill, New York 2001, pp 350–356

Brock DW (2001) Medical decisions at the end of life. In: Kuhse H, Singer P (eds) A companion to bioethics. Blackwell, Oxford, pp 231–241

Brouwer M, Kaczor C, Battin MP, Maeckelberghe E, Lantos JD, Verhagen E (2018) Should pediatric euthanasia be legalized? Pediatrics 141. https://doi.org/10.1542/peds.2017-1343

Campbell N (1994) When care cannot cure: medical problems in seriously ill babies (F. K. Beller, R. F. Weir/ eds, The Beginning of Human Life, Kluwer, pp. 327-344). In: Kuhse H, Singer P (eds) Bioethics. An anthology. Blackwell, Oxford 1999, pp 243–254

Capron AM (2001) Advance directives. In: Kuhse H, Singer P (eds) A companion to bioethics. Blackwell, Oxford, pp 261–271

Chesterman S (1998) Last rights: euthanasia, the sanctity of life, and the law in the Netherlands and the Northern Territory of Australia. Int Comp Law Q 47:362–393

Ciliberti R, Gorini I, Gazzaniga V, De Stefano F, Gulino M (2018) The Italian law on informed consent and advance directives: new rules of conduct for the autonomy of doctors and patients in end-of-life care. J Crit Care 48:178–182

Cipriani G, Di Fiorino M (2019) Euthanasia and other end of life in patients suffering from dementia. Leg Med 40:54–59

Close E, Willmott L, White BP (2018) Charlie Gard: in defence of the law. J Med Ethics 44:476–480

Culver CM, Gert B (1982) The definition and criterion of death (Philosophy in Medicine, Oxford U. P.). In: Mappes ThA, Degrazia D (eds) Biomedical ethics, 5th edn. McGraw-Hill, New York 2001, pp 310–316

Doerflinger RM, Gomez CF (2016) Killing the pain not the patient: palliative care v. assisted suicide. In: United States Conference of Catholic Bishops, accessed May (vol 19). http://www.nccbuscc.org/prolife/programs/rlp//98rlpdoe.htm

Dworkin R (1993) Life's Dominion. An argument about abortion and euthanasia. Harper Collins, London

Franklin P (2003) The recipient's perspective. In: Morris P (ed) Ethical eye: transplants. Council of Europe Publ., Strasbourg, pp 51–62

Hardwig J (1997) Is there a duty to die? (Hastings Center Report 27: 34-42). In: Kuhse H, Singer P (eds) Bioethics. An anthology. Blackwell, Oxford 1999, pp 339–348

Hufen F (2001) In dubio pro dignitate. Selbstbestimmung und Grundrechtsschutz am Ende des Lebens in Aktive und Passive Sterbehilfe, Wilhelm Fink., pp 85-113 Neue Juristische Wochenschrift: 849–857

Kant I (1790/1963) Lectures on ethics. Harper & Row, New York

Kass LR (2002) Life, liberty and the defense of dignity. The challenge for bioethics. Encounter Books, S. Fransisco

Keown J (2018) Euthanasia, ethics and public policy: an argument against legalization. Cambridge University Press, Cambridge

Kluge E-HW (2001) Severely disabled newborns. In: Kuhse H, Singer P (eds) A companion to bioethics. Blackwell, Oxford, pp 242–249

Kouwenhoven PS, Van Thiel GJ, Van Der Heide A, Rietjens JA, Van Delden JJ (2019) Developments in euthanasia practice in the Netherlands: balancing professional responsibility and the patient's autonomy. Eur J Gen Pract 25:44–48

Kuhse H (1998) Why killing is not always worse – and sometimes better – than letting die (Cambridge Quarterly of Healthcare Ethics 7: 371-374). In: Kuhse H, Singer P (eds) Bioethics. An anthology. Blackwell, Oxford 1999, pp 236–239

Lelie A, Verweij M (2003) Futility without a dichotomy: towards an ideal physician-patient relationship. Bioethics 17:21–31

Mappes TA (1998) Some reflections on advance directives (American Philosophical Association Newsletters 98: 106-111). In: Mappes ThA, Degrazia D (eds) Biomedical ethics, 5th edn. McGraw-Hill, New York 2001, pp 356–363

McMahan J (2001) Brain death, cortical death, and persistent vegetative state. In: Kuhse H, Singer P (eds) A companion to bioethics. Blackwell, Oxford, pp 250–260

Nationaler Ethikrat (2006) Self-determination, and care at the end of life. Opinion, Berlin

Nesbitt W (1995) Is killing no worse than letting die? (J. of Applied Philosophy 12: 101-105). In: Kuhse H, Singer P (eds) Bioethics. An anthology. Blackwell, Oxford 1999, pp 231–235
Nuffield Council on Bioethics (2006) Critical care decisions in fetal and neonatal medicine: ethical issues. London
Pascalev A, Vidalis T (2010) "Vague Oviedo": autonomy, culture and the case of previously competent patients. Bioethics 24:145–152
Paterson C (2017) Assisted suicide and euthanasia: a natural law ethics approach. Routledge, London
Rachels J (1975) Active and passive euthanasia (N. England J. of Medicine: 78-80). In: Kuhse H, Singer P (eds) Bioethics. An anthology. Blackwell, Oxford 1999, pp 227–230
Raus K, Vanderhaegen, Sterckx S (2021) Euthanasia in Belgium: shortcomings of the law and its application and of the monitoring of practice. J Med Philos 46:80–107
Rhodes R (2001) Organ transplantation. In: Kuhse H, Singer P (eds) A companion to bioethics. Blackwell, Oxford, pp 329–340
Robertson JA (1975) Involuntary euthanasia of defective newborns, (Stanford L.R.: 213-214, 251-261). In: Mappes ThA, Degrazia D (eds) Biomedical ethics, 5th edn. McGraw-Hill, New York 2001, pp 369–375
Sanz TRV, Pastor PP, Moreno-Milán B, Hanlon LFM, Herreros B (2021) Spanish regulation of euthanasia and physician-assisted suicide. J Med Ethics 47. https://doi.org/10.1136/medethics-2021-107523
Siegel LR (2000) Re-engineering the laws of organ transplantation. Emory Law J 49:917–954
Sykes N, Thorns A (2003) The use of opioids and sedatives at the end of life. Lancet, Oncology 4: 312–318
Tomlinson T, Brody H (1988) Ethics and communication in do-not-resuscitate orders (N. England J. of Medicine 318: 43-46). In: Mappes ThA, Degrazia D (eds) Biomedical ethics, 5th edn. McGraw-Hill, New York 2001, pp 335–340
Walsh E (2020) Cognitive transformation, dementia, and the moral weight of advance directives. Am J Bioethics 20:54–64
White BP, Willmott L, Close E (2020) Futile, non-beneficial, potentially inappropriate or 'disputed' treatment. In: Emmerich N, Mallia P, Gordijn B, Pistoia F (eds) Contemporary European perspectives on the ethics of end of life care. Springer, Cham, pp 181–198
Wicclair MR (2001) Medical futility: a conceptual and ethical analysis (adapted from Ethics and the elderly, Oxford U.P., 1993). In: Mappes ThA, Degrazia D (eds) Biomedical ethics, 5th edn. McGraw-Hill, New York, pp 340–344
Wijdicks EFM (ed) (2001) Brain death. Lippincott Williams & Wilkins, Philadelphia

Data

6

6.1 General

The discovery of the structure of DNA in the middle of the twentieth century represents a crucial landmark in the history of Life Sciences and Genetics in particular. It was a decisive step for understanding life as a natural phenomenon at the molecular level, and developing a technology that enables us to intervene in any living being's vital functions.

In humans, especially, the knowledge of genetic information that constitutes the basic features of a person's organism has impressive consequences. That knowledge may be of interest both for the individuals' welfare and for their relationships with others. A wide possibility of access to genetic information already portrays a future with big changes in personal and even social life.[1]

Information about the organism's constitution comprises not only genetic data but also phenotypic characteristics. These may be of interest when they reveal their subject's identity, as it happens with fingerprints or the eye iris morphology. Nowadays, a rapidly developing technology permits the use of such biometric data to a substantial extent regarding contacts with the state (i.e., for forensic purposes) or in private relationships and transactions. The data protection law governs the management of biometric data.[2] Questions in relevance are similar to those emerging with the management of genetic data, concerning either the collection of biological samples (from which we can obtain data) or any further processing of samples and data. Moreover, the management of genetic data, in particular, is proved crucial for

[1] This development involves extensive uncertainty regarding social implications, which is accepted by scientists and the industry though. On the other hand, the public interest for applications is constantly increasing, making necessary a process of deliberation on multiple levels. See, in relevance, Gottweis (2005), pp. 175 seq.

[2] Cf. Consultative Committee of the Convention for the Protection of Individuals with regard to Automatic Processing of Personal Data (2005).

T. Vidalis, *The Emergence of Biolaw*, Springer Textbooks in Law,
https://doi.org/10.1007/978-3-031-02359-0_6

the status of a person's biological autonomy. In the following, we will focus on that topic, as the most interesting for Biolaw.

A primary normative condition for any management of genetic information is its direct control by the data subject.[3] Personal genetic data constitute a significant part of two substantial dimensions of our personality—biological autonomy, and informational self-determination.[4] To understand what exactly means this direct control over genetic information, from a legal point of view, we need first to explain some essential elements of Molecular Genetics and the respective technology based upon this, the modern Biotechnology.

6.2 Revealing Information: Molecular Genetics and Biotechnology

Molecular Genetics is the study of the structure and functions of genetic material, namely the DNA of natural species—microorganisms, plants, animals, including humans. A significant part of genetic material, particularly through its components, the genes, codifies the basic information for every living organism's development. This phenomenon of genetic codification is the result of a long-lasting evolution of species. It represents a process of infinite "tests" by Nature for the selection and survival of different living beings within a continuously changing through long time periods natural environment. This process will stand as long as life exists.

Codified information in different genomes is transmitted from generation to generation through the natural process of reproduction. Insofar as we understand the mechanisms through which this information is expressed, we are in place to know and explain the previous, the present, and, to a significant degree, the future condition of a living organism. This information includes the location of concrete genes in the whole genome, the nature of the information they codify, their mutual influence and the influence they receive by environmental factors, and, lastly, the ways they affect concrete phenotypic characteristics. Based on that knowledge, we can also perform tests involving modifications on the DNA, with the aim to alter genetic characteristics that influence specific phenotypes. By adding or removing parts of DNA, usually corresponding to concrete genes (a technique already in use from the early 70s), we became able to attribute to certain plants and animals properties existing in other species, even distant ones in biological terms. These interventions of "genetic engineering" have changed the prospects of modern Biotechnology completely.

In general, Biotechnology intends to offer products and services by exploiting biological processes, that is, functions of living organisms, like, for instance, the process of reproduction. However, until recently, biotechnological interventions did not focus directly on these functions' molecular basis, particularly their genetic

[3] See, for example, Clayton et al. (2019), p. 6.

[4] Cf. Simitis (1984), pp. 398 seq.

composition. Traditional plant or animal breeding was developed through crosses between closely related species, with the intention to create a new variety of plant or a new animal race having desired properties. In that sense, it was a very restrictive methodology since no hope of success existed, except when the breeding experimentation was performed with species presenting only trivial genetic differentiations.

The discovery of DNA as the molecular mechanism of any manifestation of life was the basis for developing advanced biotechnological applications, practically unlimited. Indeed, the nature of life at the molecular level is the same, regardless of the species concerned. When comparing species, what is different is the number of cells and genes encompassing the nuclear DNA, that is, the source of genetic information. Slight differences also occur in individual genomes that explain the different phenotypic characteristics for each individual belonging to the same species.

This common "quantitative" molecular basis of any existing form of life allowed experimentations performed at that fundamental level. Tests of inclusion or removal of genes performed on biologically simple microorganisms guided scientists to repeat them with more complex organisms. The result was to achieve the mixing of genomes deriving from different species and the creation of new transgenic species, the appearance of which in Nature would probably be impossible (at least, completely unforeseen) following merely the process of Evolution and natural selection. These new organisms are expected to meet crucial human needs, particularly in nutrition and health, with the development of better food products and more efficient pharmaceuticals.

Moreover, from this "invasion" of technology to the common to all species molecular basis of life, human life itself is not excluded.[5] The more we collect information referring to the human genome's composition, the closer potential interventions for changing that composition arise. Current methods of gene therapy already indicate that the human genome may be the objective of genetic engineering, which involves tremendous possibilities but also unknown risks for the future of our species;[6] questions related to some kind of "designed" human offspring and the prospects of the so-called "transhumanism" emerge, all relevant to the limits of biological autonomy, as we will see in Chap. 8 concerning human enhancement.

[5] Particularly due to the mapping of the human genome. See Murphy (2001). Nevertheless, our knowledge about the precise functions and interactions of human genes remains still very limited.

[6] See the human genome's characterization as a "common heritage of humanity", also associated with the dignity and diversity of individual humans, in art. 1 of the UNESCO's Human Genome Declaration.

6.3 Biolaw and the Management of Human Genetic Information

The management of human genetic information, that is, the collection and processing of genetic data for various purposes falls within the object of Biolaw.

Genetic information is legally significant if connected with concrete conditions related to the physiology of the human organism. A certain sequence of nucleotides in the human DNA, forming or not genes, is of legal interest only if it has a confirmed link to a specific organism's phenotypic characteristic. Thus, parts of the human genome the functions of which remain still unknown ("junk" DNA), are not significant from a legal point of view. Having concluded the mapping of the human genome, modern Genetics offered an important tool for determining more rapidly and more accurately in the future the concrete functions of genes and their correspondence to human phenotypic characteristics. Following this scientific process, it is evident that human genetic information will become significant for the law to a much wider extent.

We can analyze the legal consequences from the management of genetic data in three different fields; in relation to persons, in relation to a human population, and in relation to human embryos or fetuses. That distinction is based on the difference between the corresponding legal interests at stake in each case. In this chapter, we will study the two first topics, and we will examine the legal issues that emerge in the practice of biobanking, that is, the storage of genetic data in massive collections, mostly for research purposes. The legal issues concerning the processing of genetic data obtained by human embryos and fetuses will be examined later in Chap. 7 dedicated to human reproduction.

6.3.1 Genetic Data of Persons

Nowadays, the management of genetic information obtained by a person's genome becomes essential not only for scientific purposes but for various public and private relationships also. The knowledge of genetic data explains phenotypic characteristics with a genetic cause; this causal link leads to identify a person's genetic predisposition for the manifestation of serious diseases, disorders, or other pathologies, like the chromosomal ones (Down syndrome, etc.), thalassemia, the Huntington's chorea, a variety of cancers, cardiovascular diseases, etc. Besides, genetic data deriving from non-identified biological samples may lead with almost absolute certainty to their identification, that is, to revealing their subject since the genetic profile of each human individual is unique.

These two remarks illustrate the connection between genetic information and the very essence of personhood, an intimate area of our biological formation which, if disclosed to others, may influence important aspects of our personal and social life profoundly. More specifically, a genetic predisposition for the manifestation of serious pathologies may change our plans for the future as well as the way third persons are approaching us for concluding various relationships. Moreover, being located to the biological nature of their subject, genetic data reveal certain limits to that person's autonomy, which is of interest, particularly for the others.[7] Identifying a person's genetic profile is sufficient for confirming kinship with other persons, unknown previously, a possibility with severe consequences for the private and family life of all involved. This is the case, for instance, of DNA testing upon judicial order for confirming paternity.[8] Also, DNA testing supports forensic investigation purposing to identify a suspect for the commitment of crimes, which entails the mandatory submission of that person to criminal procedures.[9]

In private relationships, disclosure of genetic information has been associated with certain risks for its subject. The genetic profile of that person may affect his/her opportunities to find employment or continue to be employed, be insured, or attend education, influencing, thus, the overall personal autonomy. Indeed, suppose that employers, insurers, or educational institutes (when granting scholarships) have access to that genetic information; in that case, they could require interested persons to meet standards of "genetic suitability" for excluding the possibility of genetically caused pathologies in order to ensure a "safe" agreement with these persons. The obvious question is whether the collection and processing of such sensitive data may be justified in societies that, in principle, have excluded any discrimination between persons on the basis of their natural characteristics.[10]

6.3.1.1 Genetic Data as Sensitive Personal Data

Genetic data belong to the category of sensitive personal data due to their direct connection with the essence of an individual's personhood.[11] In the EU law, the General Data Protection Regulation (GDPR) has adopted a special status of protection for sensitive data (*"special categories of personal data"*—art. 9), similar to the relevant provisions of the previous legislation on data protection. Moreover, the GDPR dedicates a special article for the collection and processing of genetic data.[12]

In principle, sensitive data are protected under stricter terms regarding their collection and processing than those required for simple data.[13] The basic rule,

---

[7] Cf. Ram (2015), esp. 899 seq.

[8] Cf. ECtHR, Mikulic v. Croatia (2002). On the illegal practice of paternity tests without consent, see Clayton et al. (2019), pp. 30 seq. Regarding the identification of children's families in war conditions, see Hocking and Harvey-Blankenship (2003), pp. 81 seq.

[9] On the extended practice of genetic data processing in that area, and the relevant ethical and legal problems (including also the question of the forensic authorities' legitimate access to private collections of data—Direct-To-Consumer etc.), see Scudder et al. (2019).

[10] On other areas where genetic data may be considered important, see Clayton et al. (2019), pp. 21 seq.

[11] Cf. Lehtonen (2017), pp. 103 seq.

[12] For some, although related to the biological condition of a person, genetic data should be distinguished from health data, as they present special characteristics. See on "genetic exceptionalism", Clayton et al. (2019), pp. 7 seq, and criticism in Martani et al. (2019).

[13] Article 9 of the GDPR is also relevant to health data in general, which can be subjected to unauthorized processing, not necessarily after a breach of medical secrecy from the attending

here, is that any collection and processing of sensitive data is prohibited unless specific terms are met concerning the scope intended.

6.3.1.2 Informed Consent

A general requirement for the lawful collection and processing of genetic personal data is that of their subject's informed consent.[14] Similarly to what we have already mentioned for medical acts, consent must be the product of free and genuine will, as risks for the overall individual autonomy are significant here.

"Free" is the expression of will, not only when the person is not forced by external factors, but also on the condition that previous appropriate information has been provided to that person about the scopes, the methodology, and potential implications of the genetic data collection and processing. It is important to underline that information is crucial, particularly in cases where the knowledge and experience of an average individual are not sufficient for supporting genuine consent, which is most common with the management of genetic data.[15] Informed consent is also a requirement for the lawful collection and processing of non-sensitive data. Yet, in the sensitive data, informed consent must be obtained according to certain formalities that ensure its genuine nature.

A collection of genetic data may presuppose the extraction of a biological sample from the person concerned (saliva, hair, sperm, blood, or other tissue, following the scope of the relevant genetic testing). Informed consent, therefore, refers, firstly, to that intervention in the person's body. In the EU countries, the latter needs to comply with the general provisions of the EU Charter of Fundamental Rights regarding corporal integrity protection (art. 3 para 1). For countries having adopted the Oviedo Convention, the extraction of samples also falls within the objective of its specific provision regarding "interventions for health." This is because a minimal influence on the individual's biological condition always occurs, even if the purpose of the samples' removal may not be a clinical one. Withdrawal of consent is possible, but only before the act of extraction.

The use of DNA for forensic purposes also presupposes informed consent for the extraction of biological samples by the person involved, even if allegedly suspicious of crime commitment. Forced removal of samples, here, contravenes article 3 ECHR on torture prohibition and article 3 para 1 of the EU Charter on corporal integrity mentioned above, as well as its article 4 relevant to tortures and degrading

doctor. See ECtHR, Panteleyenko v. Ucraine (2006), L.L. v. France (2006), Avilkina and Others v. Russia (2013), L.H. v. Latvia (2014).

[14] At the level of the EU Charter of F. Rights, see art. 8 para 2. Also, see art. 5b of the UNESCO's Human Genome Declaration, accepting a right-not-to-know in relevance (art. 5c).

[15] Regarding the "direct-to-consumer" (DTC) tests available to the public, notably via e-commerce applications, and the problems of their reliability also related to the absence of genetic counselling and appropriate procedures of informed consent, see Borry (2014) and Hendricks-Sturrup and Lu (2019). On the EU regulation of DTC genetic testing (Directives 98/79, 2005/29, Regulation 746/2017), see in detail, Kalokairinou (2018).

treatment.[16] It is a different issue, whether a person's denial to undergo that intervention may be considered evidence by judicial authorities to support the allegation on suspicion of that person. Similarly, we must accept that no forced removal of samples for genetic testing is allowed, even after a judicial order on paternity proof, no matter if that ultimately may affect the evidence procedure.

In the field of genetic testing for health purposes, the Oviedo Convention sets a general rule (article 12), according to which:

> Tests which are predictive of genetic diseases or which serve either to identify the subject as a carrier of a gene responsible for a disease or to detect a genetic predisposition or susceptibility to a disease may be performed only for health purposes or for scientific research linked to health purposes, and subject to appropriate genetic counseling

This provision underlines the importance of genetic counseling, that is, specific information on health matters that needs to be provided to the person concerned before obtaining consent for the collection and processing of genetic data.[17]

Regarding further processing of genetic data, the subject's consent needs to cover the means and purpose of that new processing,[18] and the relevant information must provide details on their protection from risks of illegal use.

6.3.1.3 Persons Unable to Provide Consent

Consent cannot be obtained if the subject concerned is a minor or a mentally ill person. Regarding collecting samples in these cases, legal representatives provide valid informed consent, similar to what we mentioned regarding medical acts.

Further genetic processing is regulated by the relevant provision of the GDPR, according to which processing even without consent is legitimate for ensuring a vital interest of the subject (article 9 para 2 c). Practically, that means that the consent of legal representatives for collecting samples is sufficient for covering any further processing of deriving genetic data, on the condition that this is necessary for the unable person's health.

6.3.1.4 Processing Without Consent

In exceptional cases, the law accepts data processing without the consent prerequisite, either of the person concerned or of their legal representatives.

Processing of genetic data for health or research purposes illustrate, here, characteristic examples. According to article 9 para 2 of the GDPR, processing without consent is allowed for preventive, diagnostic, or therapeutic medical acts, as well as

---

[16] See also art 7 of the UN Covenant of Civil and Political Rights.

[17] Regarding the legal regulation of genetic testing for health purposes, mainly focusing on personalized medicine and big data processing, see Sándor (2018) and Price and Cohen (2019).

[18] This is important in the case of "secondary findings", that is, investigation of further health-related genetic data beyond the initial scope of testing. See Isidor et al. (2019), arguing that this may modify the notion of illness towards a molecular condition, and the person as "patient-in-waiting". See also the remark on "genetic screening", *infra*.

for the management of national systems of health and social services, for the protection of public health; this includes serious cross-border threats (pandemics, etc.), and quality measures when developing medicinal products. Processing for research purposes falls within the objective of these exceptional provisions, on the condition that specific measures of confidentiality are in place (like anonymization of data or encryption techniques) by national legislation.[19]

It must be pointed out that in all the above cases, processing without consent is lawful only as an ultimate means of access to the relevant data, a condition indicated by the term "necessary" that the law uses; if the processing is not "necessary," the informed consent either of the person concerned or (when unable) of legal representatives stands as the primary legal requirement in this respect. Regarding, in particular, cases of processing for preventive, diagnostic, or therapeutic purposes, an additional guarantee that the law sets out, refers to the processor, which needs to be a professional, subject to a relevant institutional duty of secrecy, according to relevant legislation of the EU or member-states. Medical doctors and other professionals (nurses, physiotherapists, psychologists, etc.) represent here the most common examples.

A different case concerns the processing of sensitive data without the subject's consent for defending a right before the courts (i.e., for confirming kinship or proving innocence before a civil or a criminal court). In such cases, the GDPR allows the disclosure of data, including genetic data. Is that applicable even when the processor of genetic data is a professional, subject to professional secrecy, especially a medical doctor? To answer presupposes a specific balancing between the duty of secrecy and the right to be defended through genetic data disclosure. For instance, genetic data disclosure may demonstrate the innocence of a person considered suspect; in such a case, the fundamental right to personal freedom, which is involved in criminal cases, should prevail over medical secrecy if that data is not matching genetic information from biological samples collected in the site of a criminal act. This means that the court may enable a medical doctor to disclose data as an exception to the secrecy duty. Nevertheless, this is not an obligation for the doctor; indeed, although the law accepts legitimate exceptions from the secrecy duty, the duty still stands as an ethical rule; thus, the doctor needs to decide freely on data disclosure, based on ethical considerations.

It is worth noting that the problem of medical secrecy that we already discussed regarding the HIV issue, emerges again in genetic data. This is because a person's genetic profile is partially common with relatives' profiles due to heredity. The extent of the common part between related genomes varies according to the degree of kinship. Therefore, genetic disorders or predispositions for the manifestation of serious diseases in the person concerned (if revealed through genetic testing) may be detected to close relatives (children, parents, siblings), which need to undergo similar testing to confirm that or not.

[19] See also, art. 7 of the UNESCO's Human Genome Declaration.

Similarly to what we have accepted for HIV, a relevant exception to the medical secrecy duty is also legitimate for protecting these persons' right to Health.[20] The sensitive nature of genetic data should not be considered more significant, also given that disclosure of information will be limited within the private space of family relationships, not involving a risk of uncontrolled data flow.

6.3.1.5 Consent Under Pressure

Previously we analyzed issues related to the person's consent as the crucial term that ensures the protection of genetic data or other sensitive data. As we have seen, this term is not always required when other legal interests, considered more important by the law, are to be served. Nonetheless, consent represents a real guarantee referring to the essence of our biological autonomy.

What happens when consent is provided under pressure? Occasionally, to exercise a specific right, a person may be forced to disclose genetic information or even undergo genetic testing in order to reveal genetic data that others consider necessary, notably when such data is associated with the subject's future health. This information should not be overestimated; other health data may be more important for a particular human organism's future. However, genetic data, as permanent characteristics of human biology, attract the state's or other persons' interest when it comes to concluding relationships with the person concerned, and may determine the nature of these relationships. We will examine variations of that issue and the solutions provided by the law, both in compulsory and optional relationships of the person with the state and in relationships of private nature with other persons.

6.3.1.6 Compulsory Relationships with the State

Here, the question is whether a state authority may demand disclosure of already existing or even collection of new genetic data, in the framework of its subject's public obligations, like the military duty or the imprisonment obligation following judicial sentence.

Since relationships of that kind are not dependent on the free will and consent of the person involved, we understand that autonomy in its biological dimension is at serious risk. Indeed, if state authorities, acting in their sovereign power, were entitled to access to genetic or other information of biological nature, the autonomy of that person would be inevitably undermined since "consent" to that access would be forced. Mainly when that person is subject to compulsory confinement, or lives under conditions of hierarchical submission backed by sanctions, any consent related to that sensitive area of personhood cannot be considered "free"; any potential denial to allow access, either will lead to the undesired treatment of that person in the future, or it will enforce the sanctions in relevance.

---

[20] Cf. Resnik (2003), pp. 250 seq. On the conflict between the right-not-to-know and the disclosure of genetic data, see Wilson (2005). In favor of other persons' interests against the right-not-to-know, see Skene (1998).

Therefore, insofar as the guarantee that the consent reflects does not work, the answer to the initial question is negative. Hierarchical orders in such conditions (prisons, army, including their hospitals or clinics) or even recommendations aiming at obtaining genetic information and, moreover, the collection of biological samples, are contrary to the provisions on personal integrity and human dignity of the ECHR, and the EU Charter of Fundamental Rights, even if the purpose might be an important research project.

A further question is whether the collection and processing of genetic data may be possible without consent; in other words, if the law's exceptional provisions that we have mentioned before apply here. For answering that, we need to make a distinction: i) if no biological sample already exists, the lack of consent impedes any original collection; otherwise, the problem of personal integrity violation occurs again; ii) in contrast if biological samples do exist, any further processing with no need of consent, under the exceptional conditions prescribed by the GDPR, is possible, regardless of the special nature of the person's placement.

6.3.1.7 Optional Relationships with the State

Any person may conclude relationships with the state by exercising a relevant right. Thus, by exercising the freedom to Work, anyone may conclude an employment contract, or by exercising the freedom to Education, a person has access to educational institutions or public scholarships. Also, along with their family members, a public servant or employee participate in the social security system, meaning further relationships with the state, generated by the initial optional relationship in employment.

In such intentional relationships, the issue of genetic data management reveals a different dimension. Could be legally accepted a potential requirement from the part of the public authority on genetic data disclosure for concluding or maintaining such relationships with a particular person? In such cases, the latter needs to balance the relative importance of his/her genetic data; to exercise the rights to Work or Education, that person is obliged to abandon, at least temporarily, the protection of these sensitive data.

In principle, various legitimate relationships that we conclude voluntarily involve a previous balancing of rights. Yet, this is justified only when balancing is not creating circumstances of irreversible damage to personality, an inalienable part of which constitutes the status of personal data. If that infringement results from the person's own will, it illustrates evidence of pressure excluding freedom of that will, which cannot be considered a genuine act of self-determination. Therefore, since forced disclosure of genetic data in order to make possible certain relationships undermines the confidential nature of data, we conclude that a public authority is not entitled to ask for such information even when it comes to optional relationships with individuals in employment, insurance, education, etc.

6.3.1.8 Private Relationships: Employment and Insurance Contracts

The third aspect of that topic concerns private relationships between individuals. Again, in such relationships, the problem is whether a person needs necessarily to disclose genetic information or to undergo genetic testing for concluding a relevant contract.

The most discussed cases in relevance are those of employment and private insurance contracts. In these cases, employers or insurers may ask for genetic data disclosure (already existing, or obtained after genetic testing for the first time) as a necessary term for contracting or keeping a contract in force with the person interested.

It is essential to notice, here, a difference with the case of optional relationships between the state and individuals; in private relationships, there is not a general guarantee for respecting fundamental rights, including those related to the core of personhood, on the condition that human dignity remains unaffected.[21] This means that contracting persons, acting in the area of self-determination, may accept temporary limitations to their fundamental rights (like the data protection right), on the apparent condition that these limitations are the result of free will and not of forced situations. Forced situations do exist though in the field of working relationships usually, given the dominant position that the employer holds.[22] In such cases, disclosure of genetic information on the part of the employee (or putative employee) should be excluded as a non-legitimate term, even if the person interested consents to that; since consent is not provided in conditions of real freedom of will, it must be considered "forced," resulting in violation of the employee's biological autonomy and human dignity.[23]

On the other hand, occasionally, disclosure of specific genetic data may intend to protect the employee or third persons. When the working environment (for instance, a toxic environment in the chemical industry) favors the possibility for the manifestation of a genetically based disease, a predisposition of which exists on the individual's genome, relevant genetic data must be disclosed for the protection of that individual. The same necessity occurs when the potential manifestation of serious genetic diseases may be dangerous for third persons; this is the case, for instance, of the Huntington's neurological disease in pilots, being responsible for the air flight passengers' safety. In both examples, either the interest of the individual health of the person concerned or that of public safety are considered more significant if Compared to the previously discussed situation, where only the economic

[21] Otherwise, though, if we accept the "Drittwirkung" doctrine; according to that, fundamental rights are legally binding in private relationships also. See Engle (2009).

[22] See, in detail, Malauzat (2000), pp. 317 seq. The collection of employees' medical data is justified if these data are relevant to the concrete work position. See, in France, Cour de Cassation (Soc.) 17.10.1973 and, in Germany, BAG, NZA 1985, 57. The German administrative case-law accepts the collection of medical data (including genetic data), indicating only potential implications for the employee's health in the future. See BVerwG, ZBR 2002, 184, 185, VGH Munchen, NJW 1989, 790, 791. Nevertheless, this position is disputable under the view of the fundamental freedom to Work. See an administrative court's judgment that annulled a public school's rejection to hire a teacher due to her family medical record demonstrating the Huntington disease in family members (VG Darmstadt 470/2004).

[23] See European Group on Ethics (2003).

interest of the employer is at stake. In any case, the principle of proportionality (and, in GDPR terms, the principle of "data minimization"[24]) excludes any performance of "genetic screening",[25] that is, testing on the entire individual genome, that may reveal data irrelevant to the purpose intended but still of sensitive nature.[26]

In insurance relationships, the legal rationale is similar.[27] In private insurance, particularly, there is the same unequal placement of the person to be insured (or already insured) when they are not covered by public insurance. Here, we have objective conditions that favor a "forced" will of this person, as it happens in employment relationships. Thus, following the above analysis, the disclosure of genetic data is not legitimate. In contrast, if the person interested is already covered by the public insurance system and wishes supplementary benefits provided by a private insurance contract, the two parts cannot be considered unequal in contracting; therefore, a consent to disclose genetic information, if required by the insurer, is legally valid, since the contracting circumstances do not force the will of the data subject.[28]

6.3.2 Genetic Information of Populations

Genetic information has legal significance, even when its subjects' identity is unknown if the population's characteristics to which these subjects belong are identifiable.

6.3.2.1 Purposes of Data Processing

Genetic testing in populations has, in general, two purposes; either to diagnose genetic predisposition for the manifestation of certain diseases in a specific community of people or to correlate the genetic profile of a particular population with geographical or historical data in searching for its origin.

Epidemiological genetic testing for medical purposes intends to detect specific genetic causes, like gene mutations, correlated with clinical symptoms of

[24] Art. 5 para 1 c GDPR.

[25] See EGE (2003).

[26] See in the U.S. law, the relevance of GINA legislation with the employment matter, in Areheart and Roberts (2018).

[27] See, in detail, Malauzat (2000), pp. 310 seq.

[28] According to the EGE, any disclosure of genetic data to insurers should be prohibited. See EGE (2003). For a comparison of genetic data regulation between EU member states and the U.S., see Simon (2003), pp. 59 seq. The study identifies three relevant trends in legislation (permissive, moderate, and prohibitive, which is the current trend of international regulation and the GINA legislation of 2008 in the U.S.). Regarding concerns expressed by the insurance industry about that regulation, and the gradual understanding of its significance concerning the insurers' risks, see Van Hoyweghen et al. (2005), pp. 79 seq., esp. pp. 83 seq. A detailed presentation of the data protection legislation in the U.S. (GINA, HIPAA Privacy Rule on Health Data), see in Clayton et al. (2019), pp. 10 seq.

multifactorial diseases that appear at statistically considerable rates in concrete populations. Thus, it is feasible to design preventive measures for avoiding exposure of these populations' members to environmental factors that favor the manifestation of disease.

Regarding the historical, ethnological, or anthropological purposes of genetic testing, scientists intend to combine respective social data with the genetic characteristics of populations in order to reveal the concrete course of heredity during long periods and, through that element, the origins, the immigration flows, the contacts with other populations, etc. of the population examined.

6.3.2.2 Genetic Discrimination

As we have mentioned, genetic testing in populations uses individuals' data, the identity of which is unknown; thus, in principle, no issue of data protection emerges. For reaching results of interest is sufficient to know the particular identity of the population to which data subjects belong. Nevertheless, the fact that the population's identity is known creates legal problems of a different sort, concerning the possibility of unaccepted discrimination and the application of equal treatment.

More specifically, if a genetic predisposition for the manifestation of a severe disease would be confirmed for the population of a particular city or village or that of immigrants of a particular origin, this could lead to practices of harmful discrimination imposed on these populations' members in work, insurance or other social relationships. Opposite practices of positive discrimination in social life could also occur if no such genetic predisposition is confirmed for a certain population, which means that its members' treatment would be "privileged" Compared to others. In fact, here, we encounter an indirect disclosure of sensitive genetic data, even if the identities of the population's members remain unknown, and these persons had not undergone any specific genetic testing; it is sufficient that they belong to the concrete population.

To avoid genetic discrimination, we can refer to the general provisions of the ECHR and those of national Constitutions that prohibit any sort of discrimination based on the natural or cultural properties of individuals. An explicit prohibition is also provided by the Oviedo Convention (art. 11), and by the EU Charter of Fundamental Rights (art. 21 para 1).[29]

For genetic testing in populations, the informed consent of biological samples donors stands as a necessary condition since the collection of these samples presupposes a medical act of intervention in their organism. Information needs to refer to the research purposes that can be generic, according to the GDPR. If these samples are collected anonymously, their further processing does not require specific consent or confidentiality measures as it happens with identified samples and data. Thus, even if the data processing does not refer to particular individuals, these are involved only voluntarily by expressing explicit consent to the biological material

[29] Also, art. 2, 6 of the UNESCO's Human Genome Declaration.

collection. This excludes any mandatory genetic testing, as being contrary to the individual's biological autonomy, even on the grounds of public health protection.

The fact that we address the indirect disclosure of individual genetic data, if knowing the identity of a certain population, raises a question of limitation to the right "not-to-know." Indeed, even if someone does not want to know whether he/she is subject to particular genetic predispositions of serious diseases, it is sufficient to detect that predisposition at the population level, for disclosing the information to any member of it automatically, despite the individual right not-to-know. This is a condition of conflict between that right and the opposite right to information of other population members regarding conditions affecting their health. To resolve that conflict makes necessary to establish guarantees of secrecy for the anonymous data of population tests, as well. These guarantees would aim to avoid an extensive disclosure of the data processing results, establishing limited access only to scientists involved and persons wanting to be informed for protecting their health. Dissemination of results through scientific publications is acceptable only to the extent that no severe risks for the right not-to-know of the tested population members or the imposition of negative discrimination exist.

Finally, it is needless to say that the concrete population, as an organized collective entity with its legal representatives (i.e., municipal authorities, etc.), is not entitled to decide on behalf of its individual members a certain performance of genetic testing. However, it may encourage this by providing relevant information or even consulting services to its members.

6.4 Biobanking

The management of genetic data is often organized in special facilities, the biobanks. By this term, we define large collections of biological samples, genetic data, medical data (i.e., medical records on pathologies and therapies), biometric data, and data referring to individual lifestyles (concerning professional life, sportive activities, smoking, etc.).[30] All these personal data derives from a significant number of individuals, usually representing the general population, and they are stored as pseudonymized or anonymous (non-identifiable) information. In that sense, biobanks should be distinguished from collections of biological material storage (i.e., blood, tissue, organs, gametes, embryos, etc.) and simple registries of genetic or other personal data.

Biobanks support, mostly, the study of multifactorial genetic diseases or pathologies by enabling scientists to perform combined processing of data deriving from various origins, biological, medical, or social, a work which is not feasible with the processing of isolated genetic data. Like cardiovascular diseases, neurological diseases, or cancer, widespread serious diseases may be investigated in-depth, thanks to this combined data processing. The research results may support

[30] On biobanks definition, see also, Hallinan (2021), pp. 26 seq.

preventive measures for the protection of individual and public health as valuable elements of appropriate consulting about environmental factors that trigger a pathological genetic predisposition. They are also valuable for the development of new pharmaceuticals, the design of which is based on genetic data, in order to achieve the best efficiency and to avoid implications (pharmacogenetics, pharmacogenomics, precision medicine).[31]

The legal issues on biobanking are basically related to guarantees of data protection.[32] These guarantees are of crucial importance since the collection of multiple different categories of data is associated with a large number of individual donors. The enormous potential in combining this massive data may lead to accurate knowledge regarding the genetic profile and, subsequently, the foreseen future development of health, either of individuals or population groups. Possible dissemination of that knowledge may influence personal lives significantly by restricting the status of biological autonomy.

So far, no specific international or EU legislation exists on biobanking regulation.[33] In contrast, countries that have developed biobanking activities, usually with a pivotal role of the state, have enacted such laws since the early 2000s (Iceland, Estonia, Sweden, Norway, etc.).[34] Nonetheless, the GDPR covers in detail the data protection issues in relevance across the EU.[35] More specifically:

Collections in biobanks may be composed of anonymous biological specimens (cell lines, blood, etc.) or genetic, medical, lifestyle, etc. data, not raising data protection issues. They may also contain pseudonymized specimens or data, which, in contrast, need to be treated under the light of the GDPR. Pseudonymized data, in particular, are considered personal data if the data controller (or authorized person) has access to the identification link.[36] In that case, all provisions of the GDPR regarding the responsibility of the data controller and the data processors or other authorized persons (i.e., administrative personnel, etc.) of the biobank are applicable, since all these people perform data processing having access to the data identities. Regarding anonymous data deriving from identifiable populations, the biobank is responsible for guaranteeing confidential processing to ensure that no discrimination threats occur, following the previous analysis we made.

The method of data collection is an issue, as the practice has shown. The reliability of research is dependent directly on the amount of data that a biobank collection comprises. Insofar as the amount of data reaches the entire population of a

---

[31] Cf. Van Delden et al. (2004), pp. 303 seq., referred to the relevant ethical issues.

[32] Although a complex range of rights of "custodians" and sample donors is also discussed. See in relevance, Verlinden et al. (2016).

[33] On incomplete international regulation in relevance, see Hallinan (2021), pp. 67 seq. For non-legal regulation, see the WMA's Taipei Declaration (2016), in Chassang and Rial-Sebbag (2018).

[34] Cf. Nationaler Ethikrat (2004), pp. 35 seq. Against the necessity of specific legislation, see Goebel and Scheller (2019), referring to the German law.

[35] A critical approach, see in *Hallinan* (2021), pp. 192 seq.

[36] See, in relevance, Elger and Caplan (2006), pp. 663–664.

certain country, we can expect more accurate research results. Therefore, the more extensive possible participation of the population is of crucial importance. The relevant practice includes a) either original collections of data formed by volunteer donors, after previous information about the purposes of the collection, following the model of informed consent («opt-in»), b) or collections of already existing data without consent, on condition that the data subjects are entitled to withdraw their "presumed" will afterward («opt-out»).

That latter method became subject to criticism, mostly because it looks like "imposing" participation, not involving a necessary timeframe for consideration on the part of putative donors. Even if previous detailed information of the general public has been scheduled about the biobanking scopes and functions, there is no question here of individual "previous consent"; this practice may be considered an "approval" for the use of individual data, namely a sort of "presumed consent." The question, thus, is whether this method is accepted as a valid expression of "informed consent."

Certainly, any further use of already collected samples and relevant data that have been processed for specific purposes requires the "renewal" of the donor's consent after information about the new purposes of use. This, of course, refers to identifiable samples and data since further use of anonymous material and data is not considered as processing of personal data.[37] Still, regarding identifiable data (including those pseudonymized), a problem in terms of research promotion occurs because it is practically impossible to predict all potential research use of genetic data in order to obtain specific consent for each and every one beforehand. Moreover, it would be impossible to search and call past donors just for asking them to provide new consent, whenever required, in future research activities.[38]

A suitable solution to that problem could be accepting a sort of "generic" consent that could cover larger fields of potential future research, but without granting absolute discretion with no description of such fields. This solution ensures minimal control by the donor over the samples and genetic data, in the sense that any research considered in the donor's view as ethically unacceptable (i.e., for military purposes, etc.) would be excluded by definition. With recital 33 of the GDPR, that solution of generic consent has been adopted legally in the EU.[39]

However, the GDPR moves one step beyond, for the sake of research facilitation, by accepting in parallel the principle of presumed consent under strict conditions. Thus, according to article 9 para 2 j, the use of sensitive data (including genetic data) without the consent of the individual concerned is allowed, if adequate specific guarantees for data protection and confidential processing have been adopted by law

[37] On the problem of anonymous data re-identification, see Townend (2018), p. 661.

[38] On problems related to consent, see Nationaler Ethikrat (2004), pp. 48 seq., particularly pp. 56–59. The practice of engaging citizens in research policies, as active agents encouraged by regulatory institutions, also illustrates an approach aiming to overcome informed consent issues, by suggesting an "active opt-out" system based on "informed trust". Cf. Arnason (2009), pp. 136–138.

[39] See Townend (2018), p. 660.

at the national level (like, for instance, requirements for certified and classified personnel, implementation of encryption methods, firewalls, etc.).

This important exception from the principle of informed consent does not mean that donors of data lose any control over this; they always remain subjects of the specific rights stated by the GDPR, including the right to object (art. 21) and the right to "erasure" (art. 17). Still, even under these guarantees, the issue remains delicate since the central element of the subject's control, the previous informed consent, does not exist. Therefore, research agents in general, and biobanks in particular, need to provide appropriate information highlighting these rights, if their sample or data collections comprise content deriving from other facilities (hospitals, clinics, insurance agents, workplaces, etc.) with the intention to make it available for research purposes without the consent of the subjects concerned.

In conclusion, the current legal framework in Europe has facilitated biobanking research activities substantially by adopting a) the concept of generic consent, which may cover multiple research uses for the future, and b) even research with no consent, if specific national measures guaranteeing confidential storage and processing are in place, which is applicable in already existing data collections.[40]

6.5 Further Studying...

ECtHR: Avilkina and Others v. Russia, 7 October 2013

"A deputy prosecutor required medical institutions to report all refusals of a blood transfusion by Jehovah's Witnesses. As a result, he was informed of the second applicant's chemotherapy in a public hospital following a non-blood management treatment plan and of the fourth applicant's refusal of the use of foreign blood for surgical treatment. The Court observed that the applicants were not suspects or accused in any criminal proceedings and the prosecutor was merely conducting an investigation into the activities of a religious organisation in response to complaints received by his office. There had consequently been no pressing social need to request the disclosure of the confidential medical information concerning the applicants. In fact, other options had been available to the prosecutor to follow up on the complaints, such as seeking the applicants' consent to disclosure or questioning them about the matter. The Court found a violation of Article 8 of the Convention".[41]

[40] On soft-law and guidelines that facilitate the implementation of GDPR in research activities, see Staunton et al. (2019).

[41] See ECtHR, Thematic Report. Health-related issues in the case-law of the European Court of Human Rights, Strasbourg 2015: https://www.ECtHR.coe.int/Documents/Research_report_health.pdf.

Question

Comment on the following paragraphs, under the light of the ECHR, the GDPR, and the Oviedo Convention:

45. The Court reiterates that the protection of personal data, including medical information, is of fundamental importance to a person's enjoyment of the right to respect for his or her private and family life guaranteed by Article 8 of the Convention. Respecting the confidentiality of health data is a vital principle in the legal systems of all the Contracting Parties to the Convention. The disclosure of such data may seriously affect a person's private and family life, as well as their social and employment situation, by exposing them to opprobrium and the risk of ostracism (see Z v. Finland, 25 February 1997, §§ 95-96, Reports 1997-I). Respecting the confidentiality of health data is crucial not only for the protection of a patient's privacy but also for the maintenance of that person's confidence in the medical profession and in the health services in general. Without such protection, those in need of medical assistance may be deterred from seeking appropriate treatment, thereby endangering their own health (see Z., cited above, § 95, and Biriuk v. Lithuania, no. 23373/03, § 43, 25 November 2008). Nevertheless, the interests of a patient and the community as a whole in protecting the confidentiality of medical data may be outweighed by the interest of investigating and prosecuting crime and in the publicity of court proceedings, where such interests are shown to be of even greater importance (see, Z., cited above, § 97).

46. The Court further reiterates that in cases concerning the disclosure of personal data, it has recognised that a margin of appreciation should be left to the competent national authorities in striking a fair balance between the relevant conflicting public and private interests. However, this margin goes hand in hand with European supervision (see Funke v. France, judgment of 25 February 1993, Series A no. 256-A, p. 24, § 55) and the scope of this margin depends on such factors as the nature and seriousness of the interests at stake and the gravity of the interference (see Z, cited above, § 99).

47. Turning to the circumstances of the present case, the Court observes that the applicants were not suspects or accused in any criminal investigation. The prosecutor merely conducted an inquiry into the activities of the applicants' religious organisation in response to complaints received by his office. The medical facilities where the applicants underwent treatment did not report any instances of alleged criminal behaviour to the prosecutor's office. In particular, it was open to the medical professionals providing treatment to the second applicant, who was two years old at the time, to apply or to ask the prosecutor to apply for judicial authorisation for a blood transfusion if they believed her to be in a life-threatening situation. Likewise, there is nothing in the materials before the Court to suggest that the doctors who reported the fourth applicant's case to the District Prosecutor opined that her refusal of a blood transfusion was not an expression of her true will but rather the product of pressure exerted on her by other adherents of her religious beliefs (see, mutatis mutandis, Jehovah's Witnesses of Moscow, cited above, §§ 137-38). In such circumstances, the Court does not discern any pressing social need for requesting the disclosure of the confidential medical information concerning the applicants. It therefore considers that the means employed by the prosecutor in conducting the inquiry need not have been so oppressive for the applicants. 48. In this connection the Court does not lose sight of the fact that there were options, other than ordering the disclosure of confidential medical information, available to the prosecutor to follow up on the complaints lodged with his office. In particular, he could have tried to obtain the applicants' consent for the disclosure and/or questioned them in relation to the matter (see paragraphs 23-24 above). Nevertheless, the prosecutor chose to order the disclosure of the confidential medical information without giving the applicants any notice or an opportunity to object or to agree.

ECtHR:* Mikulić *v. Croatia, 7 February 2002

"M. Mikulić is a Croatian national born out of wedlock in 1996. On 30 January 1997, she and her mother filed a paternity suit against H.P. before Zagreb Municipal Court. A number of hearings scheduled by the court were adjourned because H.P. failed to appear. H.P. also failed to abide by court orders to attend appointments to undergo DNA tests to establish paternity, which were scheduled six times. After three-and-a-half years, the court concluded that H.P. was the applicant's father. It based its conclusion on the testimony of the applicant's mother and on the fact that H.P. had been avoiding DNA tests.

The Court noted that no measures existed under Croatian law to compel H.P. to comply with the court order for DNA tests to be carried out. Nor was there any direct provision governing the consequences of such non-compliance.

The Court considered that people in the applicant's situation had a vital interest, protected by the Convention, in receiving the information necessary to uncover the truth about an important aspect of their personal identity. On the other hand, it had to be borne in mind that the protection of third persons might preclude their being compelled to make themselves available for medical tests of any kind, including DNA tests......The Court considered, however, that under such a system the interests of the individual seeking the establishment of paternity must be secured when paternity could not be established by means of DNA testing. The lack of any procedural measure to compel the alleged father to comply with the court order was only in conformity with the principle of proportionality if it provided alternative means enabling an independent authority to determine the paternity claim speedily. No such procedure was available to the applicant in the present case.

Furthermore, in determining an application to have paternity established, the courts were required to have regard to the basic principle of the child's interests. The Court found that the procedure available did not strike a fair balance between the right of the applicant to have her uncertainty as to her personal identity eliminated without unnecessary delay and that of her supposed father not to undergo DNA tests. Accordingly, the inefficiency of the courts had left the applicant in a state of prolonged uncertainty as to her personal identity. There had, consequently, been a violation of Article 8"[42]

[42] ECtHR, Press release.

Question

Comment on the following paragraphs, under the light of the ECHR, the GDPR, and the Oviedo Convention:

56. The applicant argued in effect not that the State should refrain from acting but rather that it should take steps to ensure adequate measures, in the context of a paternity dispute, to efficiently resolve her uncertainty as to her personal identity. Thus, the applicant complained in substance not of something that the State did, but of its lack of action.

57. The Court reiterates that while the essential object of Article 8 is to protect the individual against arbitrary interference by the public authorities, it does not merely compel the State to abstain from such interference: in addition to this negative undertaking, there may be positive obligations inherent in an effective respect for private or family life. These obligations may involve the adoption of measures designed to secure respect for private life even in the sphere of the relations of individuals between themselves (see X and Y v. the Netherlands, judgment of 26 March 1985, Series A no. 91, p. 11, § 23, and Botta v. Italy, judgment of 24 February 1998, Reports 1998-I, p. 422, § 33).

58. However, the boundaries between the State's positive and negative obligations under Article 8 do not lend themselves to precise definition. The applicable principles are nonetheless similar. In determining whether or not such an obligation exists, regard must be had to the fair balance which has to be struck between the general interest and the interests of the individual; and in both contexts the State enjoys a certain margin of appreciation (see, for instance, Keegan, cited above, p. 19, § 49, and M.B. v. the United Kingdom, no. 22920/93, Commission decision of 6 April 1994, Decisions and Reports 77-A, p. 116).

59. The Court reiterates that its task is not to substitute itself for the competent Croatian authorities in determining the most appropriate methods for establishing paternity through judicial proceedings in Croatia, but rather to review under the Convention the decisions that those authorities have taken in the exercise of their power of appreciation. The Court will therefore examine whether Croatia, in handling the applicant's paternity claim, has been in breach of its positive obligation under Article 8 of the Convention (see, for instance, Hokkanen v. Finland, judgment of 23 September 1994, Series A no. 299-A, p. 20, § 55; and, mutatis mutandis, Handyside v. the United Kingdom, judgment of 7 December 1976, Series A no. 24, p. 23, § 49).

60. In the present case the only avenue by which the applicant may establish whether or not H.P. is her biological father is through judicial proceedings before a civil court, since H.P. denies paternity.

61. The Court notes in this connection that no measures exist under domestic law to compel H.P. to comply with the first-instance court's order that DNA tests be carried out. Nor is there any direct provision governing the consequences of such non-compliance. It is true, however, that in civil proceedings, pursuant to section 8 of the Civil Procedure Act, the courts must give judgment according to their own discretion after assessing the evidence presented individually and as a whole. The courts are, in this respect, free to reach conclusions taking into consideration the fact that a party has been obstructing the establishment of certain facts.

62. After three and a half years, during which time H.P. failed to appear at six appointments for DNA testing, the first-instance court concluded that H.P. was indeed the applicant's father. It based its conclusion on the testimony of the applicant's mother and on the fact that H.P. had been avoiding DNA tests. The appellate court, on the other hand, found this evidence insufficient for establishing his paternity. In this connection, the Court observes that a procedural provision of a general character, giving discretionary power to courts to assess evidence, is not in itself a sufficient and adequate means for establishing paternity in cases where the putative father is avoiding the court's order that DNA tests be carried out.

63. In addition, the first-instance court has been ineffective in resolving the question of paternity through the assessment of other relevant evidence. The Government argued that this was due to H.P.'s refusal to cooperate in the proceedings. It appears, however, that the court has been unable to find adequate procedural means to prevent H.P. from impeding the proceedings.

64. In the Court's opinion, persons in the applicant's situation have a vital interest, protected by the Convention, in receiving the information necessary to uncover the truth about an important aspect of their personal identity. On the other hand, it must be borne in mind that the protection of third persons may preclude their being compelled to make themselves available for medical testing of any kind, including DNA testing.

The States parties to the Convention have different solutions to the problem that arises when a putative father refuses to comply with court orders to submit to the tests which are necessary to establish the facts. In some States the courts may fine or imprison the person in question. In others, non-compliance with a court order may create a presumption of paternity or constitute contempt of court, which may entail criminal prosecution.

A system like the Croatian one, which has no means of compelling the alleged father to comply with a court order for DNA tests to be carried out, can in principle be considered to be compatible with the obligations deriving from Article 8, taking into account the State's margin of appreciation. The Court considers, however, that under such a system the interests of the individual seeking the establishment of paternity must be secured when paternity cannot be established by means of DNA testing. The lack of any procedural measure to compel the alleged father to comply with the court order is only in conformity with the principle of proportionality if it provides alternative means enabling an independent authority to determine the paternity claim speedily. No such procedure was available to the applicant in the present case (see, mutatis mutandis, Gaskin, cited above, p. 20, § 49).

65. Furthermore, in determining an application to have paternity established, the courts are required to have regard to the basic principle of the child's interests. The Court finds that the procedure available does not strike a fair balance between the right of the applicant to have her uncertainty as to her personal identity eliminated without unnecessary delay and that of her supposed father not to undergo DNA tests, and considers that the protection of the interests involved is not proportionate.

66. Accordingly, the inefficiency of the courts has left the applicant in a state of prolonged uncertainty as to her personal identity. The Croatian authorities have therefore failed to secure to the applicant the "respect" for her private life to which she is entitled under the Convention.

There has, consequently, been a violation of Article 8 of the Convention.

ECtHR: Biriuk v. Lithuania, 25 February 2009

"The applicants sued a newspaper for breach of privacy after it published a frontpage article quoting hospital staff as saying they were HIV positive. The article went on to give other details about their private life. Although the domestic courts found in the applicants' favour, they were unable to award more than the statutory maximum of EUR 2,900. The Court considered it crucial for domestic law to safeguard patient confidentiality and discourage any disclosures of personal data, especially bearing in mind the negative impact of such disclosures on the willingness of people at risk to take voluntary tests for HIV and seek treatment. In such cases of outrageous abuse of press freedom, the severe statutory limitations on judicial discretion in redressing the damage suffered thereby deterring recurrences had failed to provide the applicants with the protection of privacy they could have legitimately expected. There had thus been a violation of Article 8. On the other hand, the Court has also acknowledged that the interests of a patient and the community as a whole in protecting the confidentiality of medical data may, in certain situations, be outweighed by the interests of investigating crime or of holding court proceedings in public."[43]

Question

Comment on the following paragraphs, under the light of the ECHR, the GDPR, and the Oviedo Convention:

41. The Court notes at the outset that publication of the article about the applicant's state of health, doctors having confirmed that she was HIV-positive, as well as references to her sexual life (paragraph 6 above) were of a purely private nature and therefore fell within the protection of Article 8 (see, for example, the aforementioned Dudgeon v. the United Kingdom judgment, § 41). The Court takes particular note of the fact that the applicant lived not in a city but in a village, which increased the impact of the publication on the possibility that her illness would be known by her neighbours and her immediate family, thereby causing public humiliation and exclusion from village social life. Moreover, the applicant was entitled to respect for her privacy regarding her sexual life, whatever her neighbours' concerns. In this respect the Court sees no reason to depart from the conclusion of the national courts, which acknowledged that there had been interference with the applicant's right to privacy.

42. The Court will next examine whether there existed a public interest justifying the publication of this kind of information about the applicant. However, the Court sees no such legitimate interest and agrees with the finding of the Vilnius City Third District Court, which found that the main purpose of the publication was to increase the respondent newspaper's sales. In the Court's view, the publication of the article in question, the purpose

[43] Ibid.

of which was apparently to satisfy the prurient curiosity of a particular readership and boost the defendant's commercial interests, cannot be deemed to contribute to any debate of general interest to society (see, among many authorities, Prisma Presse v. France (dec.), nos. 66910/01 and 71612/01, 1 July 2003).

43. As for the reasoning of the Supreme Court in the present case, the Court does not accept that the purported concerns of the local population for their safety were legitimate, either socially or scientifically, thereby justifying a publication about the applicant's state of health and her life style. In the Court's view the opposite holds true: respecting the confidentiality of health data is crucial not only for the protection of a patient's privacy but also for the maintenance of that person's confidence in the medical profession and in the health services in general. Without such protection, those in need of medical assistance may be deterred from seeking appropriate treatment, thereby endangering their own health and, in the case of transmissible diseases, that of the community (see Z v. Finland, cited above, § 95). In this respect the Court attaches particular significance to the fact that, according to the newspaper, the information about the applicant's illness had been confirmed by the medical staff of Pasvalys hospital (see paragraph 6 above). However, it is of special importance that domestic law provides appropriate safeguards to discourage any such disclosures and the further publication of personal data.

44. The Court takes into account that the national law at the material time did contain norms protecting the confidentiality of information about the state of health of a person. It has regard to the existence of the judicial guidelines to be followed if the right to privacy of a person has been breached (see paragraphs 12-19 above). The Court also notes that the domestic courts indeed awarded the applicant compensation for non-pecuniary damage. However the principal issue is whether the award of LTL 10,000 was proportionate to the damage she sustained and whether the State, in adopting Article 54 § 1 of the Law on the Provision of Information to the Public, which limited the amount of such compensation payable by the mass media, fulfilled its positive obligation under Article 8 of the Convention.

45. The Court agrees with the Government that a State enjoys a certain margin of appreciation in deciding what "respect" for private life requires in particular circumstances (cf. Stubbings and Others v. the United Kingdom, 22 October 1996, §§ 62-63, Reports 1996-IV; X and Y v. the Netherlands, 26 March 1985, § 24, Series A no. 91). The Court also acknowledges that certain financial standards based on the economic situation of the State are to be taken into account when determining the measures required for the better implementation of the foregoing obligation. The Court likewise takes note of the fact that the Member States of the Council of Europe may regulate questions of compensation for non-pecuniary damage differently, as well as the fact that the imposition of financial limits is not in itself incompatible with a State's positive obligation under Article 8 of the Convention. However, such limits must not be such as to deprive the individual of his or her privacy and thereby empty the right of its effective content.

46. The Court recognises that the imposition of heavy sanctions on press transgressions could have a chilling effect on the exercise of the essential guarantees of journalistic freedom of expression under Article 10 of the Convention (see, among many authorities, Cumpănă and Mazăre v. Romania [GC], no. 33348/96, §§ 113-114, ECT HR 2004-XI). However, in a case of an outrageous abuse of press freedom, as in the present application, the Court finds that the severe legislative limitations on judicial discretion in redressing the damage suffered by the victim and sufficiently deterring the recurrence of such abuses, failed to provide the applicant with the protection she could have legitimately expected under Article 8 of the Convention. This view is confirmed by the fact that the impugned ceiling on judicial awards

of compensation contained in Article 54 § 1 of the Law on the Provision of Information to the Public was repealed by the new Civil Code soon after the events in the present case (see paragraph 32 above).

47. In the light of the foregoing considerations, the Court rejects the Government's objection as to the applicant's victim status and concludes that the State failed to secure the applicant's right to respect for her private life.

There has therefore been a violation of Article 8 of the Convention.

Provide legally appropriate solutions

1. A public biobank operating in the field of precision medicine announces its interest to collect biological material (blood or urine samples) and health data from the general population. Also, the biobank concludes agreements with public hospitals, private clinics, and private insurance companies for the transfer of their health data collections.

Based on the current European legislation, comment on the legal validity of the above,

- if the identities of the biological samples' donors and the data subjects are known to the biobank.
- if the data transferred from hospitals, clinics, and insurance companies are pseudonymized
- if these collections also include statistical data from epidemiological research
- would you change your answers if the biobank determined a relevant fee for those wishing to contribute with original material and health data and for hospitals, clinics, and insurance companies for the transfer of their collections?

2. According to a new national law of an EU member state, the system of public insurance includes compulsory genetic testing of all insured citizens to determine the individual risk related to genetic diseases and organize the national health system accordingly. Also, for citizens wishing to work in the public sector, the law excludes hiring those identified as carriers of serious genetic disorders to ensure safety throughout the individual working career.
Based on the current European legislation, comment on the legal validity of these stipulations.

References

Areheart BA, Roberts JL (2018) GINA, big data, and the future of employee privacy. Yale Law J 128:710–783

Arnason V (2009) Scientific citizenship, benefit, and protection in population-based research. In: Solbakk JH, Holm S, Hofmann B (eds) The ethics of research biobanking. Springer, Boston, pp 131–141

Borry P (2014) Genetic tests via the internet: limitations and risks. In: Glasa J, Glasova H (eds) Genetic testing for health purposes in Central and Eastern Europe. Institute of Medical Ethics and Bioethics n.f., Bratislava, pp 10–15

Chassang G, Rial-Sebbag E (2018) Research biobanks and health databases: the WMA Declaration of Taipei, added value to European Legislation (soft and hard law). Eur J Health Law 25:501–516

Clayton EW, Evans BJ, Hazel JW, Rothstein MA (2019) The law of genetic privacy: applications, implications, and limitations. J Law Biosci 6:1–36

Consultative Committee of the Convention for the Protection of Individuals with regard to Automatic Processing of Personal Data (T-PD) (2005) Progress Report on the Application of the Principles of Convention 108 to the Collection and Processing of Biometric Data, Council of Europe, Strasbourg

Elger BS, Caplan AL (2006) Consent and anonymization in research involving biobanks. EMBO Rep 7:661–666

Engle E (2009) Third party effect of fundamental rights (Drittwirkung). Hanse Law Rev 5:165–173

European Group on Ethics (EGE) (2003) Ethical Aspects of Genetic Testing in the Workplace, Opinion No 18

Goebel JW, Scheller J (2019) Do we need a biobank law? J Lab Med 43:311–315

Gottweis H (2005) Governing genomics in the 21st century: between risk and uncertainty. New Genet Soc 24:175–194

Hallinan D (2021) Protecting genetic privacy in biobanking through data protection law. Oxford University Press, Oxford

Hendricks-Sturrup RM, Lu CY (2019) Direct-to-consumer genetic testing data privacy: key concerns and recommendations based on consumer perspectives. J Pers Med 9:25. https://doi.org/10.3390/jpm9020025

Hocking BA, Harvey-Blankenship M (2003) Tracing the family accounts? Law, biology, and family identity. Journal International de Bioethique 14:79–92

Isidor B, Julia S, Saugier-Veber P, Weil-Dubuc PL, Bézieau S, Bieth E, Bonnefont JP, Munnich A, Bourdeaut F, Bourgain C, Chassaing N, Corradini N, Haye D, Plaisancie J, Dupin-Deguine D, Calvas P, Mignot C, Cogné B, Manouvrier S, Pasquier L, Héron D, Boycott KM, Turrini M, Vears DF, Nizon M, Vincent M (2019) Searching for secondary findings: considering actionability and preserving the right not to know. Eur J Hum Genet 27:1481–1484

Kalokairinou L (2018) Ethical and legal aspects of direct-to-consumer genetic testing in Europe, Diss. KU Leuven, Biomedical Sciences Group, Leuven

Lehtonen LA (2017) Genetic information and the data protection directive of the European Union. In: Beyleveld D, Townend D, Rouille-Mirza S, Wright J (eds) The data protection directive and medical research across Europe. Ashgate, Hants, England, pp 103–112

Malauzat M-I (2000) Le droit façe aux pouvoirs des données génétiques. Presses Univ. D' Aix-Marseille, Aix-en-Provence

Martani A, Geneviève LD, Pauli-Magnus C, McLennan S, Elger BS (2019) Regulating the secondary use of data for research: arguments against genetic exceptionalism. Front Genet 10: 1254. https://doi.org/10.3389/fgene.2019.01254

Murphy TF (2001) Mapping the human genome. In: Kuhse H, Singer P (eds) A companion to bioethics. Blackwell, Oxford, pp 198–205

Nationaler Ethikrat (2004) Biobanks for Research, Berlin

Price WN, Cohen IG (2019) Privacy in the age of medical big data. Nat Med 25:37–43

Ram N (2015) DNA by the entirety. Columbia Law Rev 115:873–940

Resnik DB (2003) Genetic testing, and primary care: a new ethic for a new setting. New Genet Soc 22:245–256

Sándor J (2018) Genetic testing between private and public interests: some legal and ethical reflections. Front Public Health 6: file:///C:/Users/USER/Downloads/fpubh-06-00008.pdf

Scudder N, McNevin D, Kelty SF, Funk C, Walsh SJ, Robertson J (2019) Policy and regulatory implications of the new frontier of forensic genomics: direct-to-consumer genetic data and genealogy records. Curr Issues Crim Justice 31:194–216

Simitis S (1984) Die informationelle Selbstbestimmung – Grundbedingung einer verfassungskonformen Informationsordnung. Neue Juristische Wochenschrift 37:398–405

Simon J (2003) Genetic testing and insurance: an international comparison. Journal International de Bioéthique 14:59–78

Skene L (1998) Patients' rights, or family responsibilities? Two approaches to genetic testing. Med Law Rev 6:1–41

Staunton C, Slokenberga S, Mascalzoni D (2019) The GDPR and the research exemption: considerations on the necessary safeguards for research biobanks. Eur J Hum Genet 27:1159–1167

Townend D (2018) Conclusion: harmonization in genomic and health data sharing for research: an impossible dream? Hum Genet 137:657–664

Van Delden J, Bolt I, Kalis A, Derijks J, Leufkens H (2004) Tailor-made pharmacotherapy: future developments and ethical challenges in the field of pharmacogenomics. Bioethics 18:303–321

Van Hoyweghen I, Hortsman K, Schepers R (2005) "Genetics is not the issue": insurers on genetics and life insurance. New Genet Soc 24:79–98

Verlinden M, Nys H, Ectors N, Huys I (2016) Qualitative study on custodianship of human biological material and data stored in biobanks. BMC Med Ethics 17:1–10

Wilson J (2005) To know or not to know? Genetic ignorance, autonomy, and paternalism. Bioethics 19:492–504

7 Reproduction

7.1 General

The most outstanding enlargement of our biological autonomy, thanks to modern technology, is, with no doubt, related to human reproduction. Since the early 1960s, technology has offered the option to control fertility and births through the development and marketing of chemical contraceptives, which revealed a new dimension of personal freedom. Until then, reproduction was considered a dimension of human nature "out" of autonomy. To have children was a matter related exclusively to sexual intercourse and the natural result of heterosexual relationships. Dominant social perceptions confirmed and promoted this reality, regardless of national, religious, social, or cultural differences, on the one hand, by exaggerating marriage as a social institution, on the other, by determining its purpose in reference to procreation and the establishment of families.

In these conditions, everyone was submitted to a given social norm, guiding directly from marriage to childbirth, with practically no lifestyle alternatives. This pattern was disputed by social movements after World War II, but mostly it was deeply changed thanks to new technological applications during the next decades. These were methods permitting a clear dissociation between sexual life and reproduction, and also permitting the close control of each step of the latter. Control was proved crucial since the process of reproduction passed from the area of the "unlocked" human nature to that of our free will and biological autonomy.

This new era of freedom was expressed initially through its negative dimension; free sexual life with no fertility implications. The "negative" freedom of reproduction is realized either as prevention, thanks to new, safe, and comfortable means of contraception, or as the artificial termination of pregnancy, the abortion, under conditions ensuring the protection of the womens' health.

Since the late 1970s, the freedom of reproduction is fully conceptualized, from the moment that its positive dimension appears through impressive technological innovations. With the use of in vitro fertilization (IVF) and embryo transfer to the woman's womb, human reproduction became entirely part of the person's biological

T. Vidalis, *The Emergence of Biolaw*, Springer Textbooks in Law,
https://doi.org/10.1007/978-3-031-02359-0_7

autonomy. Ever since, any person may have children, even if not able to, through sexual intercourse (due to pathological reproductive organs or fertility problems), or even if traditional artificial support (i.e., artificial insemination) is proved ineffective.[1]

Nevertheless, the fact that reproduction became part of our biological autonomy, although impressive, creates many ethical and legal issues, which is a reasonable result if we consider the entailed significant changes that occurred in personal and social relationships. Until the IVF medical revolution, our systems of values were formed and tested, ignoring the possible consequences of that innovation. We need then to investigate to what extent these values, as reflected in the current European law, encompass that "enlarged" freedom of reproduction.[2]

7.2 Negative Reproduction

For a long time, biological autonomy in reproduction meant the freedom not-to-have-a child. This is expressed by various options to prevent conception and pregnancy or to interrupt pregnancy. The first issue in relevance is related to the legal ground of that negative freedom. Free control of our biological ability to reproduce is a necessary aspect of human dignity, expressed by a specific right that belongs to our privacy.

In the European legal framework, control of reproduction reflects an aspect of the fundamental right to private life, referring also to corporal integrity and, thus, to the principle of human dignity. For the woman especially, this means her freedom to accept or not the development of another human life into her body, which involves serious implications for her biological functions. Nevertheless, the relevant article 8 of the ECHR mentions the possibility of limitations when the right to private life conflicts with the protection of health, public morality, or others' rights. Is the negative right to reproduce also subject to such limitations?

7.2.1 Mandatory Sterilization

The above question is raised, in particular, with the practice of sterilization without the consent of the person concerned. For several decades, mostly from the last quarter of the nineteenth century until World War II, several states, democratic or not, had adopted public policies on eugenics. These policies included mandatory sterilization of certain population groups, mostly those experiencing serious mental diseases, with the intention to prevent hereditary transmission of such diseases to the

[1] On the origins of the right to reproduce, see Ramm (1989), pp. 862–863. See, also, Buchanan et al. (2001), pp. 206 seq.

[2] On the characteristic convergence between different cultural and religious approaches, in favor of the right to reproduce, see Obermeyer (1995).

offspring. It was an era favoring racist and discriminative attitudes. Besides, methods of contraception and interruption of pregnancy were not popular, being considered immoral or unsafe.[3]

Mandatory sterilization was based, then, on reasons of an alleged public interest. Could that justification or another referring, for instance, to a supposed inability of these persons to undertake parental responsibilities, be legally sound in our days as well? The answer is negative. Indeed, personal control of our body is entirely independent of expressions of our personality that may affect social interests. A social interest to avoid the transmission of hereditary diseases is limited when the biological autonomy of persons is threatened; otherwise, these persons would be treated as mere means for serving the common good, contrary to the protection of their human dignity. Even the counterargument that justifies mandatory sterilization as a method for protecting unable persons, either from health implications (due to pregnancy) or from the burden resulting from potential parental responsibilities, is not convincing. Nowadays, alternative ways of protection are available without violating biological autonomy, like, for instance, the involvement of family members in decisions about the health protection of the unable person or the future care of children. But, mostly, this counter argumentation is unacceptable because it implies differentiations in the autonomy status between groups of persons, suggesting some of them a complete deprivation of biological autonomy in the name of their "protection." This profoundly paternalistic approach ultimately is incompatible with the grounds of modern law.

Also, it is worth noting a confusion that may occur between the right to reproduce and the right to create a family and undertake parental responsibilities. The first refers only to the freedom to exercise or not a simple biological ability, while the latter is associated with social interests and, as such, it may be subjected to certain limitations in the best interest of future children. Nevertheless, these limitations may be imposed only after the fact of reproduction and not beforehand (by "preventing" reproduction or, in other words, by denying the right to reproduce). The right to create a family is equally fundamental in the sense that all persons, regardless of their health status, are considered subjects of it. This, necessarily presupposes their status as subjects of the right to reproduce.

In conclusion, the example of mandatory sterilization indicates the reasons for which the "social" limitations to the right to private life, as provided by the ECHR, need to be considered irrelevant regarding the right to reproduce; embedded in the core of our biological autonomy, this right (particularly in its negative dimension) enjoys absolute legal protection incompatible with reservations of any kind.

[3]Cf. Muller-Hill (1994), pp. 182 seq.

7.2.2 Prevention

7.2.2.1 Voluntary Sterilization

The next issue that we need to address when we discuss reproduction as a dimension of biological autonomy refers to voluntary sterilization; is it a legitimate option for someone to accept this radical intervention in his/her body in order to enjoy a sexual activity free of any reproductive implication? For women, in particular, that question is critical, given that the unintentional conception and pregnancy means a biological burden, possibly entailing unpleasant decisions, that is, a necessary abortion.

The act of voluntary sterilization may be considered a person's consent to abandon the right to reproduce. Nonetheless, the issue is more complicated here. The close relationship between sexual freedom and the right to reproduce and the latter's relevance with the right to create a family produce a significant influence on the exercise of biological autonomy. Indeed, the exercise of sexual freedom may result in undesired changes in personal biology (due to potential conception and pregnancy), and subsequently to further consequences for the overall personal autonomy involving the creation of a family. Are these serious implications a sufficient reason for the person to abolish an expression of biological autonomy like the right to reproduce?

We need to notice here that, in general, the human dignity principle is not compatible with a waiver of fundamental rights by their subject. A "voluntary" loss of the freedom of speech, the right to private life, or even the "habeas corpus" would be meaningless as expressions of autonomy. On the contrary, they should be understood only as acts of compulsory submission to others' will, which may exercise pressure restraining the autonomy of the person concerned; they reflect acts of forced consent to the waiver of rights, entailing directly the violation of human dignity.

However, when considering voluntary sterilization, it is difficult to indicate a third person as the one who exercises pressure by "forcing" the subject to abandon his/her right to reproduce. This is a crucial difference from the forced expressions of the person's will mentioned above. Here, no external pressure is involved, as it is the interested person the one who decides to transform elements of their biology for meeting their overall autonomy potential; this represents a genuine, free will that results not in restrictions, but in the enlargement of self-determination, contrary to what is happening with the examples of rights' "forced" abolition.[4] In that sense, the act of voluntary sterilization does not contravene human dignity. Therefore, it is acceptable within the framework of the European Biolaw, as a conventional medical act, subjected to article 5 and in full compliance with article 1 of the Oviedo Convention.

[4]The absence of free will, after information, is crucial here. See the ECtHR case-law in relevance, e.g., Csoma v. Romania (2013), V. C. v. Slovakia (2011), N. B. v. Slovakia (2012), I.G., M.K., and R.H. v. Slovakia (2012).

7.2.2.2 Contraception

Contraception represents the negative freedom to reproduce in its simplest expression.[5] Here, the methods, chemical or mechanical, vary and may be applied safely in men and women, ensuring almost absolute efficiency.[6] These methods obstruct the fusion of gametes, that is, the oocytes' fertilization separating the sexual act from the possibility of procreation. In terms of biological autonomy, that illustrates our ability to exercise sexual freedom without the risk to reproduce, or in other words, it establishes our discretion in procreation.[7]

At the legal level, sexual freedom concerns our body's control, namely our corporal integrity, and represents a dimension of our privacy, as protected by article 8 of the ECHR. As with any right relevant to the body's control, sexual freedom is also not subjected to legal reservations. Indeed, limitations, here, can be justified only when risks for infringement of other persons' rights occur, for instance, regarding corporal integrity, personal freedom, etc.. of the sexual partner.[8]

Therefore, a relevant question is whether contraceptive methods may jeopardize such rights of the sexual partner, particularly his/her positive rights to reproduce and create a family. Here, the answer is negative either when the partner is aware of the use of contraceptives or even when he/she ignores that use. The opposite solution would result in the complete loss of the negative rights to reproduce and create a family of the partner desiring the contraception. Indeed, this one would be obliged to exercise sexual freedom with the risk of unintentional procreation and, subsequently, of an involuntary future lifestyle. In contrast, his/her sexual partner takes no risk of similar burdens and loss of rights since he/she reserves the option to reproduce at any other time. However, it is worth noting that a disagreement between spouses regarding contraception may influence the status of marriage seriously, justifying a relevant cause for divorce.

7.2.3 Beyond Prevention: The Case of Abortion

After the oocyte's fertilization resulting in the creation of the first human cell and the beginning of embryonic development, the control of reproduction is concentrated on the voluntary artificial interruption of the embryo implantation, or the interruption of pregnancy.

[5] See the first recognition of the right in the U.S. case-law, in Griswold v. Connecticut, 381 U.S. 479 (1965).

[6] Cf. Baylor College of Medicine (2000), with further references.

[7] Certainly, this is possible without using methods of contraception, but only on the condition that sexual contacts are excluded during the female fertility cycle. However, this confirms that Nature imposes an involuntary restriction on our sexual freedom; this restriction is removed by the use of contraception, which in that sense "liberates" the sexual life.

[8] And not a specific "value" of gametes, as such; human sperm and oocytes are not considered as agents of value, even from those defending an "absolute" value of the unborn human life. See Marquis (1989), pp. 46 seq., particularly pp. 55–56.

Abortion differs from the natural loss of the unborn human life since it represents an intentional act, an expression of the woman's biological autonomy. As such, it also differs from any act of forced interruption of pregnancy, which violates biological autonomy and, subsequently, the woman's human dignity.[9]

The European legislation does not include specific stipulations on abortion, yet we can trace legal guidance on the topic considering the ECHR's and the Oviedo Convention's general provisions. In most national legislations, abortion is regulated in the context of criminal law, which indicates a "disapproval" of that practice as a matter of principle on the part of legislators. This attitude is related directly to the recognition of special value to the unborn human life.

For the law, the embryo *in utero* and the fetus are forms of human life that deserve special protection since they prepare the human person's existence; both may be considered "potential" persons. Nevertheless, we cannot equate the unborn human life to the person's legal status; insofar as an embryo or a fetus remain biologically dependent on the woman's organism, they are deprived of the necessary biological presupposition for developing autonomy. Indeed, that biological dependence is crucial, in normative terms also, since it makes the unborn human life not only part of the woman's organism but, moreover, part of her own personality. This condition differs from the social dependence that a newborn, a child, or even an adult experiences for surviving in a particular family or community environment. The difference is embedded in two facts; first, a logical fact that excludes the coexistence of two persons (the pregnant woman and the embryo or fetus) in the same organism; second, a fact related to the very nature of biological dependence, that excludes any possibility of self-determination. Even if, in a futuristic scenario, the development of unborn human life could be accomplished in an artificial womb (with no involvement of the woman's organism),[10] a dependence would persist until the complete biological disengagement of the newborn, which ultimately activates the biological possibility for developing autonomy.[11]

[9]Although social attitudes and official narratives on abortion continue to underestimate the women's views in relevance. Cf. Priaulx (2017), pp. 73 seq.

[10]On efforts intending to the development of "artificial womb," see Simonstein (2005). It is worth noting that a similar absolute dependence occurs, sometimes, even after premature newborns' birth when their survival is impossible without artificial support of vital functions; this support imitates, in fact, conditions of life support before birth. This proves that birth is only a conventional threshold for the beginning of persons (legally practical, though) since, in exceptional situations, it cannot ensure the elements needed for the development of personal autonomy (see Chap. 1).

[11]That condition differs from other situations of biological coexistence between humans (like that of Siamese twins), where we can observe a mutual manifestation of autonomy by both individuals, characterized by free and voluntary self-limitation of the one part's personal expressions for the sake of the other part's free exercising of rights. Indeed, the Siamese twins live in enduring conditions of potentially conflicting interests, resolved by themselves, as autonomous persons; from a legal point of view, this situation of biological coexistence cannot be considered a sort of ontological "dependence" like the one we observe between the pregnant woman and the embryo or fetus. See, in relevance, Cowley (2003), pp. 69 seq.

That condition of ontological dependence produces specific legal consequences. Thus, even if abortion is in principle disapproved by the legislator, on exceptional occasions it is accepted in every modern legal system.[12] This is because, despite the value acknowledged to the unborn human life, if the latter's existence is not conciliated with certain interests of the pregnant woman, these interests are considered to have a legitimate priority.

At the level of the European law, we can argue that, at least, protection of the woman's life represents such a legitimate interest prevailing over the life of the embryo or fetus, on the ground of article 2 of the ECHR referring to the right to life.[13] Indeed, this right is attributed only to "persons," not to embryos or fetuses, for defending life against any threat, including threats from a pathological process of reproduction. We can, thus, conclude that the ECHR justifies abortion if the life of the pregnant woman is in danger.[14]

Nevertheless, the European law does not provide ground for supporting with certainty other interests that may prevail over the unborn life to justify abortion, like, for instance, the woman's health (physical or mental) and her social relationships (in cases where pregnancy is the result of rape or incest). Moreover, the European law does not cover the case of free abortion,[15] as a woman's right for defending her lifestyle, or that of negative eugenics, for avoiding the birth of children with serious pathologies. All these cases occur in most national European legislations, though, as legitimate reasons that justify abortion, corresponding to the woman's fundamental rights and prevailing over the value of the embryo or fetus.[16] The Strasbourg Court is not expected to oppose these national legal provisions, following its policy to consider a "margin of appreciation"recognized for national legislators, particularly in the area of private and family life. Therefore, it is unlikely for claims against abortion on that ground to be accepted by the ECtHR case-law.

Still, the essential issue remains, and we need to cope with it at the level of human rights as protected by the existing European legal instruments. In the following, we will develop an argumentation that scrutinizes all potential reasons for abortion, as a matter of principle for any right-based legal system, including the European one. A practical result of such scrutiny would be the precise determination of that "margin of appreciation"and a possible justification of additional reasons for terminating

[12] On the conflict emerging between the pregnant woman's interests and the value of the embryo or fetus, see, e.g., ECtHR, A, B and C v. Ireland (2010) and Purdy (1990), pp. 71 seq.

[13] Although the Strasbourg Court addresses the protection of unborn life in terms of ethics. See Open Door and Dublin Well Woman v. Ireland (1992).

[14] See, e.g., ECtHR, Boso v. Italy (2002). See case-law in relevance in Fenwick (2012), pp. 249–276.

[15] Against free abortion, see ECtHR, A, B, and C v. Ireland (2010).

[16] Affecting also the issue of medical liability: In cases where physicians may invoke conscientious objection to perform abortion (for religious reasons, etc.), they still need to ensure that the legal right of the woman to terminate her pregnancy will be fulfilled otherwise (by recommending another physician, etc..). On that problem, see Nelson (2018), referring to the U.S. law.

a pregnancy, besides protecting the woman's life, on the ground of the European legal instruments.

7.2.3.1 The Value of an Embryo or Fetus In Utero

To address that issue, we need first to examine the value of an embryo or fetus *in utero*. The fact that these organisms, although human, are considered objects of law, being biologically dependent on the woman's organism, is not sufficient to specify their value. For that reason, we need an additional, positive determination, which derives from a more comprehensive meaning of human value.

Here, a problem arises since, for the law, the human value is attributed to persons only, and therefore it seems incompatible with any law object. However, this correlation presupposes the real prospect of persons' existence, that is, the human species'existence as such, as an apparent interest of any human society and its legal system. Under this view, the known concept of human value, pertinent to each and every human person after birth, justifying the principle of human dignity and, hence, the distinction between subjects and objects of law, is grounded on an "objective" dimension of that value. This reflects a primary interest of the entire society, as a guarantee for the perpetual existence of the whole human species, and it is not attributed to particular individuals.[17] The deriving value of the unborn human life at any stage of development after the embryo implantation into the woman's body is grounded precisely on that objective meaning of the human value. Indeed, pregnancy illustrates the "potential" for the human species' reproduction, which is getting more critical following the embryonic organism's development. Accordingly, we must accept that the weight of the unborn life's legal value increases, also, to the point of that life's conversion into a human person (a subject of law), when the fact of birth puts an end to its biological dependence.

That legal justification of the unborn life's value is radically different from metaphysical justifications that approach the human embryo or fetus (from the very moment of conception or at a later moment) as a real person, a subject of "rights." It is also different from other justifications, inspired by a particular biological or psychological rationale. These last tend to correlate an embryo or fetus's value with the development of vital organs or functions (like the central neural system, the feeling of pain, etc..). Both justifications are incompatible with the logic of modern law, no matter how we can judge them in terms of ethics.

Metaphysical approaches, usually related to religious beliefs, tend to enlarge the concepts of "person"and "rights" because they fail to associate these concepts with the ability of autonomy, which is substantial in modern law. In these approaches, both concepts reflect an existentialistic relationship-with-others, an element distant from the legal approach of autonomy as a field of self-determination where, in principle, any intervention of other individuals is excluded. No matter if our autonomy is subjected to limitations, its starting point is a free- from- the- presence-

[17] See the ECtHR's case Vo v. France (*supra*), according to which human dignity is not associated by definition with a person's quality.

of- others, space of personal activity; based upon this elementary self-determination, a person decides freely to accept restrictions when exercising rights, for the sake of other persons, equally free. If, contrary to that view, the others would determine "inherently" a person's freedom, then autonomy would be deprived of any meaning as such; it would be dependent on limitless burdens, to the extent that self-determination would disappear ultimately, along with the substantial distinction between subjects and objects of law.

On the other hand, those who underline the importance of biological or psychological factors for determining the value of the unborn human life fail to explain why some specific and not other facts are crucial indeed (for instance, why that value does not emerge from the moment of conception, or that of implantation or perhaps when the full development of vital organs is concluded). This is a logical error since we cannot deduce normative consequences with direct reference to pure facts.[18] For instance, we can ask why we do not recognize equal value to the unborn human life and vertebrate animals that also develop a central neural system and feel pain. For such approaches rejecting the necessary recognition of autonomy as a characteristic value pertaining to humans only, questions like this remain unexplained.

7.2.3.2 The Relative Value of the Unborn Human Life

Bearing in mind the special status that the law recognizes to the unborn human life, some national laws permit free abortion explicitly, that is, a woman's right to interrupt pregnancy with no specific reason, within certain time limits. These laws evaluate the negative right to reproduce of pregnant women as prevailing over the value of the unborn life.[19]

This balancing of values is justified by the fact that a woman cannot control with precision the reproductive process, for preventing an unintended pregnancy some moments before or after the conception. The first indications of pregnancy appear relatively late, and only at that late time, a woman becomes able to exercise as a biologically autonomous person the right to reproduce. Given that, the option of free abortion functions as a guarantee for the full respect of that right, securing the possibility of negative exercise by the woman concerned.

Would it be possible to justify free abortion, even if a woman might be aware of the fact of pregnancy and accepted it initially but changed her mind afterward? Here, the answer is related to the concrete legal status of the unborn life, which, as mentioned, is not stable but becomes more significant following the stages of embryonic development.

The value of an embryo *in vivo* is undoubtedly at a lower level when the process of pregnancy has not started yet, that is, between the moment of conception and the moment of the implantation in the woman's womb, a time frame that lasts more or

[18] See the classic argument in Hume (1739/1984), p. 521.

[19] After conception, the first trimester is considered an appropriate time frame within which free abortion is justified, following most national laws, like in France, the U.K., and the U.S.

less 14 days.[20] Within that time, implantation as a necessary biological condition for embryonic development still is missing; therefore, the "potential" for a future person does not exist.[21] The value of the early not-implanted embryo, here, cannot be more significant than the value of a biological product of the woman's organism, like, for instance, the value of an oocyte. As far as no potential for embryonic development exists, we must accept that the biological contribution of the woman's partner to the creation of that embryo remains irrelevant in terms of values. At that level, thus, the negative right to reproduce for women through free abortion cannot be contested.[22]

In contrast, after the embryo implantation, the prospect of a future person's development is visible, making the value of the unborn life more significant. Free abortion in these circumstances still may be acceptable as a means of the (negative) right to reproduce, but the embryo's increased value also needs to be taken into account, which makes sense at the European legal level.

Under the view of the European law, we can accept that, insofar as the ECHR recognizes explicitly the right to create a family (article 12), and implicitly the right to reproduce, it covers the option of free abortion also, at least within the timeframe after conception and before implantation, for securing negative reproduction. Given that no other means for preventing implantation exists, free abortion, here, is the only option for serving that purpose; in that sense, it cannot be considered an issue included within the national "margin of appreciation", and influenced by particular ethical or religious beliefs of a certain national community. This means that national laws prohibiting free abortion at that very early stage are not found in compliance with the ECHR and need to be revised.

The same cannot be suggested, though, concerning free abortion after implantation, which practically matters much more. Given the continuously increasing value of the developing human life, the option of free abortion certainly is subjected to concrete balancing between that value and the negative right to reproduce of the woman concerned. This work falls with no doubt within the "margin of appreciation"of national legislations, as national ethical and religious attitudes play a significant role in evaluating the conflicting interests at stake.

Some national laws add other reasons that justify abortion, even when the embryonic development is at an advanced level, meaning that the potential of the new person's emergence is strong; therefore, the value of the unborn life is

[20] Until the beginning of the third week after conception, there is not yet an embryo individualization, which means that it is possible the early embryo's splitting into more embryos, able to be developed separately as complete human organisms (this is the case of monozygotic twins). See Nationaler Ethikrat (2003), p. 14.

[21] It is worth noting that opinions determining the beginning of person from the moment of conception influence both the current policies on abortion and IVF, as shows the example in the U.S. with the anti-abortion movement. See Boys and Harris (2019).

[22] That primordial condition of the embryonic life excludes, in biological terms, any considerations defending a certain "future" human life that is to be lost if free abortion is accepted. See in relevance, Stretton (2004), pp. 144 seq.

considered higher. We will examine three such reasons that refer to protecting the woman's health, her social status, and also the "eugenic"reason.

Compared to the protection of the woman's life, which is clearly covered as a legitimate reason for abortion by article 2 of the ECHR, the protection of health has not have an obvious justification at the same level. Certainly, individual health represents a significant dimension of the fundamental right to corporal integrity, which is explicitly recognized by the European legal instruments. Nonetheless, not any health condition prevails over the unborn life's value when they are found to conflict. Indeed, a temporary manifestation of a disease or a less serious accident, although they affect the woman's individual health as such, cannot be considered sufficient reasons for observing abortion.

On the other hand, a problem of violation of the right to corporal integrity and, potentially, of the woman's overall autonomy occurs, if her condition of physical or mental health may be damaged permanently by the maintenance of pregnancy. In such circumstances, what is at stake is the woman's human dignity because the continuation of pregnancy represents a *de facto* instrumentalization of her body, just for ensuring the birth of a new person. This option is, thus, unacceptable at the level both of the ECHR and the EU Charter of Fundamental Rights, regardless of national attitudes concerning the value of the unborn life; therefore, it cannot be included within the "margin of appreciation" justifying differences between national legislations in relevance. In that case, also, we need to adopt the same conclusion as in the case of the woman's life's protection, accepting that the European law covers the protection of health as a legitimate reason for abortion.[23]

In contrast, the protection of the woman's social status, when pregnancy results from unwanted sexual intercourse, cannot be excluded from the "margin of appreciation." Here, we refer to a time beyond the limit for a legitimate free abortion, when the pregnant woman usually has the opportunity to reflect on her future social relationships and terminate a pregnancy. Beyond that time limit, the embryonic development and the corresponding value of the unborn life cannot be judged by definition as inferior compared to that woman's interest. Given that an opportunity to terminate the pregnancy is already secured (through the option of free abortion), for protecting the legitimate interests of the woman, we cannot generalize that reason by excluding it from national ethical or social attitudes; it is a matter of national law if, following these attitudes, it will accept abortion for "social" reasons.

Finally, the justification of abortion as a means of negative eugenics, when the future newborn is expected to present serious pathologies, falls equally within the margin of national legislative discretion.[24] A relevant option for the woman may represent her autonomy, particularly her right to create a family and enjoy a family life according to her preferences. Still, let's suppose that indications regarding

---

[23] See ECtHR, Tysiac v. Poland (2007). Still, a widely discussed decision of the U.S. Supreme Court [(Gonzales v. Carhart, 550 U.S. 124 (2007)] undermined that reason in certain circumstances (partial-birth abortion). See criticism in Annas (2007), pp. 2201–2207.

[24] On the risk of eugenics, see Kass (1973), passim.

pathological characteristics of the unborn life occur after the time limit of free abortion; in that case, the fetus's increased value becomes significant in balancing interests to different degrees, following the ethical and religious attitudes of each particular national community.

We must also take into account that the correlation between certain pathological findings with the expected influence on the future newborn's welfare is highly uncertain and involves ethical considerations, also given the progress of modern Medicine that nowadays addresses several pathologies after birth successfully. Down syndrome or cystic fibrosis are undoubtedly tough situations, but the welfare of many children born with these has been impressively improved thanks to currently available treatments.

In general, situations subjected to continuous medical appraisal and research do not correspond to a unique and absolute prediction for the child's welfare. In legal terms, numerous variants of these situations make it impossible to establish a rigid reason for accepting termination of pregnancy[25] that might overcome the need for a national legislative appraisal. In other words, although termination of pregnancy for eugenic purposes may be seen as an aspect of the woman's rights to create a family and to family life, we do not have a certain standard for determining the generally acceptable welfare of the future child at the European level, which would prevail in all cases over the unborn life's value. If, on the other hand, we considered the child's welfare a matter of the woman's personal assessment (when exercising her autonomy), we would take the risk to justify eugenic abortions even for trivial pathologies, undermining the fetus's value altogether.

A relevant matter is that of cases known as "wrongful birth" and "wrongful life." Here, the problem is whether the European law can accept arguments characterizing the lives of persons suffering from severe pathologies as "damage." This question is relevant to the legitimate option of terminating a pregnancy for eugenic purposes that accept some national legislations. Suppose an attending doctor failed to detect severe pathologies in the fetus or to inform parents about these pathologies when eugenic abortion was still legally possible. In that case, some argue that besides compensation due to the parents for malpractice or lack of information, additional compensation is due to the newborn as well, for the "damaging" life he/she will be obliged to experience.

Legally, this argumentation is misleading, though, mostly because it would be logically absurd to accept compensation as equivalent to the "absence of life"; the latter cannot be restored as a practical situation through compensation; it represents merely a definite situation with no return. It is entirely different from the necessary support that persons living in such conditions need to receive in order to cope with suffering and pain throughout their life. That support falls mostly within the duties of public authorities providing the means for healthcare in the context of the social right to Health.

[25] See, in relevance ECtHR, R.R. v. Poland (2011), held that prenatal tests access does not necessarily mean access to abortion.

7.2.3.3 The Role of the Woman's Husband or Companion

An issue arising with regard to the woman's decision to terminate a pregnancy is whether her husband or companion are justified to intervene since they have contributed biologically to the embryo conception. To rephrase in legal terms that question, can we consider their rights to reproduce, create a family, or enjoy family life, as entitling these persons to also decide on abortion, having a veto when they disagree with the pregnant woman's view?

Regarding that potential conflict of interests, we need to pay attention, particularly to the fact that the woman is bearing the unborn life in her body. This means that if we prioritize her husband or companion's will, we automatically accept her treatment as a mere means of reproduction, contrary to her human dignity; that is why the woman's will is always the decisive one, following article 1 of both the ECHR and the EU Charter of Fundamental Rights.

This does not mean, however, that the other part remains unprotected. Indeed, a compulsory restriction of the husband's or companion's rights due to the woman's decision to continue or not a pregnancy may result in a permanent change in their family life and ultimately in their entire lifestyle. For that reason, it is up to the free will of the man involved if he will comply with the woman's view, or he will prefer to abstain from that situation by ending the common life with her. A relevant cause of divorce for the married man should be acknowledged, here, as the only solution that fits both parties' autonomy, without compromising their entire life through very sensitive decisions on family matters.

7.2.3.4 The Role of the Minor Woman's Parents

When the pregnant woman is a minor and subjected to parental care, do her parents have a decisive role in the termination of pregnancy for any of the legitimate reasons mentioned previously?

Parental care is recognized by the ECHR, either in the context of family life of article 8 or following the specific reference of article 2 para 2 of its 1st Additional Protocol that ensures the parents' right to educate children according to their philosophical or religious beliefs. The question, thus, is whether the abortion decision falls within the content of these provisions, being an expression of parental care, which needs to be exercised for the best interest of the minor concerned.

Regarding an abortion that becomes necessary for protecting the life or health of the minor woman, the answer here looks plain, as the latter's best interest obviously contains protection of life and health by any necessary means. Even if parents disagree with abortion as a matter of free choice, on such occasions, they are obliged to accept it as a medical intervention if no other alternative exists. In terms of medical liability, a physician who proceeds to termination of pregnancy with the sole consent of the minor pregnant for protecting her life or health, even acting without the consent of her parents, cannot be found liable if the abortion has become an urgent medical act, according to article 8 of the Oviedo Convention. Moreover, a disagreement between parents on the necessity of abortion does not affect the doctor's final decision in such conditions.

Beyond these issues, the most crucial question is about abortion as a free choice of the minor woman despite her parents' disagreement and vice-versa when parents favor abortion and the minor wishes to continue the pregnancy. These questions reveal the significance of abortion for the woman's biological autonomy characteristically.

Are decisions of free choice related to reproduction, like that of abortion, part of parental care? A possible objection could argue that termination of pregnancy (as any other intervention in a person's body) belongs to the intimate core of personal autonomy. This excludes any intervention of thirds, including parents, in order to secure a minimal space of freedom for facilitating the gradual process of the minor's socialization as an autonomous personality.

Nevertheless, even if it is true that minors need to enjoy a minimal space of autonomy, this does not encompass necessarily everything related to biological autonomy and matters of health. Usually, minors lack the knowledge for protecting their own health, which explains why relevant decisions are considered part of parental care and subject to parental will. Decisions on reproduction and particularly on abortion concern the condition of health and may be seen, in principle, as belonging to the parents' role. Nevertheless, two notes in relevance seem crucial.

(a) Parents have no role in imposing an abortion when there is no risk for the minor woman's life or health, even if they consider that the child's birth will jeopardize her future self-determination. If we admit the opposite, we undermine the developing emotional bond between the girl and the unborn life, which plays a crucial role in self-determination. In that case, an average adult's experience is not sufficient as a basis for decision instead of that personal bond. Forced termination of pregnancy, when no risk for health occurs, is much more likely to jeopardize the girl's current and future autonomy by disappointing her regarding the ability to control her body and privacy, which ultimately entails the violation of her human dignity.

(b) Also, parents have no role to object to the minor woman's decision to perform a free abortion if sufficient medical guarantees for the protection of her health exist. Forced maintenance of pregnancy involving the unintentional birth of a child, again would mean a permanent change for her future autonomy, and moreover, it might put at risk even her corporal or mental health as potential implications. For that reason, this also represents a violation of the young woman's human dignity.

In conclusion, when abortion is a matter of free choice, not related to protecting the pregnant woman's life or health, it belongs to the minimal biological autonomy acknowledged for minors as an expression of personal control over their corporal integrity. The experience of pregnancy and prospective birth of a child for self-determination is so intensive that any thirds' authorization to take decisions in relevance would necessarily entail a direct violation of the woman's human dignity, even if these are the parents, as subjects of family rights recognized by the ECHR.

7.3 The Positive Dimension of the Right to Reproduce: Assisted Reproduction

Biological autonomy in human reproduction was initially acknowledged with the options of contraception and abortion. However, the emergence of assisted reproduction has multiplied, to an unexpected degree, autonomous decisions in that field, establishing the positive dimension of the right to reproduce. This involves not only new methods that enable persons with fertility problems to have children but also new prospects for the development of modern biomedical research.

Assisted reproduction includes known from the past techniques, like artificial insemination, but mostly refers to the method of in vitro fertilization (IVF) and embryo transfer. This method is applied to humans since the late 70s and already encompasses several new techniques in a continuous development promising more efficiency. The possibility to split the reproduction process into a phase out of the human body (*in vitro*) and a phase into the body (*in vivo, in utero*) raises the most ethical and legal questions eventually.

In Europe, these questions are encountered by national laws. During the last three decades, the legislative activity is very intense in that field, responding to the necessity for a delicate regulation of values and rights in relevance; this process is often affected by deep ethical and religious convictions about the nature of human life, which differ widely in the European national societies. This is the reason for which only very minimal legal provisions exist at the level of the European Biolaw; a prospect for common legislation in assisted reproduction, either at the level of the EU law or that of the Council of Europe, looks at present less possible.[26]

For its part, the Strasbourg Court confirms that regulatory trend, through the known approach recognizing a "margin of appreciation"for national legislators, admitting the vital ethical significance of the relevant issues inevitably sealed by national attitudes.[27] Still, some general conclusions are deduced from the ECtHR case-law, which contributes to a common European normative stance, along with the very few provisions that we can find in the European legal instruments.

7.3.1 General Terms for the Methods of Assisted Reproduction

Overall, assisted reproduction is considered by the European law a legitimate option, both for creating a family and for performing biomedical research.

[26] Cf. Vidalis (2007), pp. 12 seq.

[27] See e.g., Evans v. U.K. (2007), D. And Others v. Belgium (2014), Parrillo v. Italy (2015). Consistency in national legislation is nevertheless a standard of the ECtHR's view. See ECtHR, Costa, and Pavan v. Italy (2012), where a factual inconsistency between the legal acceptance of abortion, on the one hand, and the prohibition of the PGD method for selecting embryos in assisted reproduction, on the other, led to the conclusion that a violation of private and family life of the applicants by the Italian state occurred, despite the margin of appreciation approach.

On the one hand, the right to create a family is not limited in terms of methods, according to articles 12 of the ECHR and 9 of the EU Charter of Rights; this right does not refer to natural sexual intercourse as the exclusive means for procreating. On the other, by regulating the selection of embryos, article 14 of the Oviedo Convention makes an explicit reference to the assisted reproduction methods, implying that these are acknowledged as accepted methods in this article's particular context.

However, this "programmatic" acceptance of assisted reproduction by the European law does not oblige the national legal systems in Europe to introduce permissive rules in relevance. Fertility methods certainly represent the positive dimension of the right to reproduce and, subsequently, to create a family according to the legal provisions mentioned above, but they cannot be considered exclusive means for that. Indeed, the right to create a family may be exercised by everyone, even by persons unable to reproduce naturally, through other means, like the minors' adoption and fostering. Therefore, no direct link exists between fertility methods and the ability to exercise positively that fundamental right.

This means that the European states always hold the option to accept or not through their national legal systems the assisted reproduction as a lawful practice producing valid effects on family relationships, or they may accept or prohibit specific fertility methods, following different ethical or religious considerations. On the other hand, national restricting policies need to be expressed explicitly by legislation or national courts; if no national regulation on assisted reproduction exists, persons seeking access to fertility methods are protected directly by the above general permissive attitude of the European law.

Beyond this programmatic reception by the European law,[28] assisted reproduction constitutes, in any case, a set of specific medical acts. Therefore, the Oviedo Convention, pertinent to all medical acts, is fully applicable here. Notably, we need to highlight article 5 on informed consent procedures for the persons wishing to undergo fertility treatment, and article 10 on medical confidentiality, along with the medical data detailed protection provided mostly by other instruments of the European law, like the GDPR. In conclusion, a national legal system, either adopting explicitly or even remaining with no specific regulation on assisted reproduction, needs to comply with these provisions, as referring to a particular category of medical acts.

7.3.2 Specific Methods

There is a vast area of regulation within the acknowledged "margin of appreciation"at the level of national legal systems. This includes:

[28] Suggesting also continuous updating of national laws in relevance, following the progress of science. See ECtHR, S.H. and Others v. Austria (2011).

7.3.2.1 Reservations on Access to Fertility Methods

The national law may establish restrictions by excluding from access persons under a certain age (minors), which are unable to provide valid consent. Also, it may exclude persons beyond a limit of age (elder persons) mostly for ensuring the creation of a balanced family, in terms of the age difference, for the sake of the future children best interests.

The latter restriction could be found contrary to the right to create a family and equal treatment of these persons; this could emerge, at least in countries that favor access to fertility methods. Indeed, the best interests of "future" children have no explicit protection at the level of the European law, which means that no conflict of interests can be indicated here for justifying restrictions on creating a family.

Another general restriction is accepting access only for persons unable to reproduce naturally, in other words, by defining assisted reproduction as a therapeutic practice and not as a free choice for everyone.[29]

A third restriction is the determination of a limited number of embryos to be transferred into the woman's womb in order to ensure the protection of her health and to avoid the practice of embryo reduction, which is ethically disputable.[30]

A fourth reservation requires a certain formality to the informed consent procedure, stricter than that of conventional medical acts, given the delicate nature of access to fertility treatment, in terms of personal and family engagements. That reservation, usually, is expressed by formal procedures of information and written consent by the individuals concerned and influences accordingly the liability of expert doctors and clinics.[31]

7.3.2.2 Selection of Methods

National legislations may not accept all the available assisted reproduction methods, opting to exclude some, following ethical, social, or religious considerations.

Only IVF with married couples' gametes is generally accepted by current national laws, mainly when these couples are unable to reproduce naturally due to fertility problems.

Fertilization with donors, that is, with the use of sperm or oocytes obtained from third persons, has no such general recognition, although it may be the only solution for a couple and even for unmarried, homosexual, or divorced persons to reproduce

[29] However, this is not limited to fertility problems but is applicable also to avoid infections if relevant tests are available in IVF procedures. Regarding the HIV example, see Sonnenberg-Schwan et al. (2005), pp. 603 seq. A specific problem arises when the exercise of the right to reproduce is impeded by certain social conditions (e.g., imprisonment). See ECtHR, Dickson v. U.K. (2007), in favor of access to artificial insemination of a prisoner.

[30] Cf. Comite Consultatif National d' Éthique (1991).

[31] A significant question concerns the use of cryopreserved embryos after divorce, separation, or disagreement of putative parents. The Strasbourg Court focuses on maintaining both gamete donors' consent, which excludes any future use if one of the donors withdrew consent, even if the other has no alternative to reproduce. See ECtHR, Evans v. U.K. (*supra*) See in the U.S., also, Robertson (2001), particularly pp. 1037 seq.

and create a family. Following the European law, it remains important that these persons still have the option to conclude a marriage, if unmarried or divorced, or proceed to adoption or fostering procedures, in order to fulfill that wish; this option deprives them of an argument directly related to the article 12 of the ECHR.

Fertilization *post mortem*, that is, with the use of gametes from a deceased person, after his/her previous explicit consent, while still living, should be addressed in the same line as well.[32]

With surrogacy, a method that involves a third person (the surrogate mother) for addressing problems in embryo gestation of the intentional mother, different legal problems occur. Surrogacy is based on a previous contract that recognizes respective rights and obligations between the intentional parental couple and the surrogate mother. Following such contracts, usually approved officially by judicial decision, the surrogate mother agrees on the embryo transfer and gestation in her body until the newborn's birth, with the obligation to deliver the child to the intentional parents after birth.[33] From their side, the intentional parents agree to support medically, provide the needed care, compensate the surrogate for relevant costs throughout pregnancy, and, in certain cases, pay a special lawful reward (commercial surrogacy).[34]

The big majority of national legislations in Europe prohibit that method due to ethical concerns related in particular to the protection of the surrogate from risks of exploitation and family implications that may occur after the child's birth.[35] For that reason, the couples interested often seek access to surrogacy in countries that have legalized the method, in Europe or elsewhere.[36] After having children, parents claim for these a legitimate status as citizens of their country of origin, although the method remains illicit there. In two famous cases,[37] the ECtHR addressed that problem after the denial of relevant parental claims by the administration and courts in France, a country that prohibits surrogacy, as contrary to the "ordre publique" clause of its national legislation. The Strasbourg Court overrode this by acknowledging a prevailing fundamental interest of the children in their own identity, which also

[32] See a relevant survey in the U.S., showing the interest for that method, in Trawick et al. (2020).

[33] Although the legal paternity of the child may need an official act to be concluded, as provided in the U.K. current law. Against that requirement, see Horsey (2018), pp. 63 seq, arguing for the recognition of the intentional parents' legal paternity by default (like in the system adopted in Greece).

[34] For some, commercial surrogacy is preferable, as it may ensure the protection of surrogates from uncontrolled exploitation practices due to their vulnerable situation. See Wilkinson (2003), pp. 184 seq. *Contra*, see Horsey (2018), p. 7, and Hodson et al. (2019), according to which legalizing commercial surrogacy does not prevent international exploitation of persons.

[35] See, for example, the case in Germany, and a criticism, in Dücker and Hörnle (2018).

[36] See Piersanti et al. (2021). On issues emerging with transnational surrogacy arrangements and the need for international regulation in relevance, see Sandor (2018), pp. 35–55, Rotabi et al. (2017).

[37] ECtHR, Mennesson, and Others v. France and Labassee v. France (2014).

includes a right to become a citizen of their parents' country of origin.[38] This judgment makes a significant difference regarding the extent of the national "margin of appreciation," as it produces legal effects for all European legal systems regarding surrogacy.[39] Thus, even if, at the national level, surrogacy remains an illicit method, and this falls undoubtedly within the "margin of appreciation," an indirect acceptance of the method, through the recognition of the children civil status, seems to dispute the extent of the national competence in relevance. In brief, under the new circumstances, national prohibitions may be proved inefficient.

Apart from that issue, surrogacy, when accepted, involves serious questions regarding the protection of human rights, particularly of the surrogate mother. These refer to a potentially excessive restriction to the surrogate's autonomy due to intentional parents' concerns for the progress of pregnancy and the embryo's or fetus's health conditions. A reasonable standard for judging the nature and relevance of restrictions is the medical assessment of the risks that may emerge for the embryo/fetus, given the conditions of a concrete pregnancy. If restrictions are found excessive for the surrogate (like prohibitions on working, moving, etc..), the relevant contract should not be enforceable (or approved by the court, if that procedure is applicable).

A crucial matter, here, refers to the surrogate's freedom to terminate the pregnancy when the contract is in force. Abortion cannot be disputed legally as a necessary means for protecting the life and health of the surrogate. The problem concerns, mostly, abortion as a legitimate free choice during the time framework that specific national legislation may set out. In that case, contract obligations certainly restrict the freedom of the surrogate on that matter. On the other hand, even if she remains liable for the contract's violation, which entails subsequent obligations on compensation, the surrogate cannot be deprived entirely of that freedom, as she cannot be "forced" to continue an unwanted pregnancy on the basis of her human dignity.

7.3.3 The Embryo Selection

The right to reproduce is also related to "what" children we desire to have. In the context of abortion, we have already seen that some national laws accept that dimension by permitting pregnancy termination for reasons of negative eugenics. With the IVF methodology and the progress in Genetics, eugenic options have been

[38] See in relevance, art. 8 of the UN Convention on the Rights of the Child (1989). Nevertheless, the Court insists, first, on strict implementation of the margin of appreciation view (see ECtHR, D., and Others v. Belgium (2014)) and, second, on the existence of real family life (see ECtHR, Paradiso, and Campanelli v. Italy (2017)). The child's rights do not affect official registers (ECtHR, Advisory opinion concerning the recognition in the domestic law of a legal parent-child relationship between a child born through a gestational surrogacy arrangement abroad and the intended mother, requested by the French Court of Cassation (2019), C. and E. v. France (2019)).

[39] Against the Court's intervention in surrogacy matters, see Ní Shúilleabháin (2019).

multiplied significantly,[40] to the extent that, for some scholars, we are now in place to "design" our offspring.[41]

As mentioned, the diagnosis of human genetic profiles is possible long before a child's birth, during the embryonic life. From the moment of conception, we can already foresee the phenotypic characteristics of the future organism, based exclusively on genetic data. That knowledge may guide an embryo's handling *in vitro*, created through IVF, before its transfer into the woman's body. Given that the embryo is not considered a person legally, the gametes' donors may evaluate crucial elements of its genetic profile for deciding whether the reproduction process should be continued or stopped.

More specifically, with the method of preimplantation genetic diagnosis (PGD), we can perform genetic testing on embryos in vitro for detecting possible genetic disorders, that is, chromosomic abnormalities (like trisomy 21—the Down syndrome), or genetic predispositions for the manifestation of serious diseases due to mutations on responsible genes (like thalassemia, cystic fibrosis, neurological monogenetic diseases, or multifactorial diseases, like cancer, etc..).[42] We are in place, also, to identify non-pathological phenotypes, like for instance, the sex of the embryo; yet, given the present state-of-the-art, we cannot identify the genetic basis of more specific characteristics also, like the hair or the eye color, the muscular formation, or even certain cognitive properties.

So far, based on PGD findings, we can only select or reject an embryo in a particular reproductive process but not intervene in its genetic profile seeking to modify genes. However, genetic engineering methods are now in ongoing development, particularly with the recent techniques of genome editing (CRISPR Cas9, etc..). On the other hand, compared to embryo selection *in vivo* through abortion, the selection of *in vitro* embryos is, in any sense, easier since it does not involve interventions in the woman's body.

Given these technical options, the question is to what extent they conform with the European law's provisions.[43]

The Oviedo Convention undoubtedly covers the selection of embryos for avoiding the presence of pathological genetic predispositions in future children,

[40] On the relevance of current genetic technology to eugenic practices, see Buchanan et al. (2001), pp. 104 seq., 333 seq.

[41] In favor of positive eugenics (see *infra*), cf. Savulescu (2001), pp. 413 seq. *Contra*, De Melo Martin (2004), pp. 80 seq., where the issue of social inequalities, that could be the outcome of positive eugenics, is underlined.

[42] This does not mean that future parents consider phenotypic characteristics in the same manner; they may address a specific characteristic of the future child as non-pathological, contrary to the standard social and medical attitude. See the example referred to as "selection" of disabilities (deafness, etc..) in Kaplan (1993), p. 131. That is why the embryo selection (through abortion or IVF methods) is a matter of autonomous choice and not an ethical duty. See in relevance, Purdy (1996), p. 126. Also, Harris (1993), p. 167. A criticism to PGD and embryo selection, based on the rights of disabled persons, see in Asch and Barlevy (2012).

[43] See a comparative study in Lau (2019), pp. 213 et seq, proposing also a method of compliance with fundamental rights.

according to its general provision of article 12 and the Additional Protocol relevant to Genetic Data for Health Purposes. The wording of article 12 refers to *"tests which are predictive of genetic diseases or which serve either to identify the subject as a carrier of a gene responsible for a disease or to detect a genetic predisposition or susceptibility to a disease,"* which clearly does not cover tests for the detection of other non-pathological genetic characteristics.

More specifically, article 14 makes an explicit reference to assisted reproduction methods by prohibiting embryo selection on the basis of sex (with the exclusion of sex-related diseases). This is an additional indication that the Convention disapproves "positive eugenics." Indeed, sex selection for non-health reasons represents a decision on the future child's "designing". If the scope of that prohibition focuses on "designing" prevention as such, there is no reason to assume that the Convention accepts such a desire for other but sex phenotypes, if linked with genetic characteristics.[44]

7.3.3.1 Selection of Embryos In Vitro for Negative Eugenics: Ethical Issues

In terms of ethics, the acceptance of negative eugenics in IVF procedures by the Oviedo Convention may be subject to criticism.

(a) The relevant provisions seem to support a certain socially "privileged" quality of life for any future person, which justifies even the reproductive process's interruption. This could be challenged by arguing that we cannot judge human life's value with standards of "quality." According to this approach, society and law would have no ethical reasons to reject the existence of persons with pathological issues from birth. On the contrary, they have the moral mission to secure for these persons a dissent life by providing the necessary means for their equal participation in social life, that is, their access to equal opportunities when exercising fundamental rights.[45]

Nevertheless, that approach is not entirely justified. The said moral obligation indeed characterizes a democratic society, on condition though that refers to persons finding themselves accidentally in such a need. There is no ethical ground to cover also pathological situations, which could be avoided by a reasonable intervention or, in other words, to justify even an intentional emergence of such situations. This is because social life, by definition, is a product of our conscious intervention in Nature, following our values, not a passive submission to natural phenomena. In that sense, our apathy in front of a natural fact that we consider pathological contravenes two other social values; the best possible allocation of the limited resources for ensuring social welfare (which needs to consider such predictions

---

[44] See, though, the broad formulation of article 3 para 2 b of the EU Charter of F. Rights, which does not distinguish between negative and positive eugenics, and prohibits selection in general. In line with what the Oviedo Convention accepts, a narrow interpretation of this article is more reasonable, as more consistent with other provisions of the Charter (the right to healthcare in particular).

[45] See art. 3a, 4 c, e, 5 para 2, 8 para 1a of the UN Convention on the Rights of Persons with Disabilities (2007). On the problem of negative discriminations against persons living with pathologies that may be detected by PGD, see Nationaler Ethikrat (2003), pp. 119 seq., 126 seq.

and potential interventions), and mostly the rights to reproduce and create a family of those interested in having healthy children. Indeed, these rights would be severely undermined (and the outcome would negatively influence these persons' overall autonomy), if we rejected the exceptional possibility for them to select healthy characteristics for their future children, before starting the process of pregnancy, to the extent that current technology allows.[46]

The legal acceptance of negative eugenics presupposes that the gamete donors (the future parents) are those who decide over the embryo selection, as subjects of the rights to reproduce and create a family. The doctor has a duty to provide appropriate information about genetic testing results for enabling them to reach a reasonable decision. Nonetheless, on this occasion, the doctors' position is ethically sensitive, given that the information for any genetic predisposition related to health may lead to the rejection of embryos by parents seeking desperately a genetically "perfect" future child. Therefore, ethically speaking, it is suitable the medical advice that limits tests' performance only on certain severe genetic diseases, and not on diseases with uncertain manifestation (i.e., dependent on particular environmental factors) or delayed in the lifespan (i.e., the Alzheimer's disease). On the other hand, there is no issue of medical liability in legal terms regarding the content of that information since the law does not distinguish between genetic diseases.

7.3.3.2 Selection of Embryos for the Sake of a Third Person's Health ("Savior Sibling")

A different case of selection is that intending to protect the health of a third person and not of the future child. Thanks to the PGD methodology, parents of an existing child suffering from a serious blood disease, after the IVF procedure, may select a histocompatible with that child embryo for continuing the reproduction process; the ultimate purpose, here, is to obtain primordial blood cells from the umbilical cord, to use them as transplants and eventually cure with safety the patient child. The issue was first raised in several cases in the UK, encountered by the Human Fertilization and Embryology Authority (HFEA) and the courts.[47]

In such cases, we address the selection of embryos for health reasons but not purposing to negative eugenics. Do the provisions mentioned above of the Oviedo Convention cover that kind of embryo selection in IVF?

From an ethical approach, the issue presents two opposing aspects. On the one hand, selection seems to instrumentalize a future person for serving therapeutic means, which violates that person's human dignity. On the other, the intending use of the embryonic biological material seems favorable for the future child's

---

[46] The issue emerged in French, Danish and Portuguese case-law, as well as in the case-law of state courts in the U.S. See, in particular, the case Perruche of the French Cour de Cassation (2000) that acknowledged this right raising strong legal, social and political protests though. As a result, the French legislation banned similar considerations for the future (l. No 2002-303).

[47] The HFEA has accepted that option under strict conditions, on a case-by-case basis. See "HFEA to change its mind on savior siblings?" 19.7.2004, at http://www.bionews.org.uk. In Norway, that option has been legalized after a change of the relevant legislation. See Solbakk (2005), pp. 99 seq.

human dignity, since from the very beginning of his/her existence, this child will be considered the "savior" of a fellow person.

Nevertheless, legally speaking, what is essential regarding the protection of human dignity is to avoid any instrumentalization of the person concerned and not a certain "enhancement" of that dignity. This is because, to the law, human dignity is not subject to "measurement" (that is, to enhancement also). Otherwise, the equal protection of all humans, regardless of natural or social characteristics, would be collapsed; no person has "more" or "less" dignity than others before the law. For this reason, the second of the above ethical arguments is not legally relevant.

Yet, nor the first argument seems convincing. The fact that a child is born for serving an openly admitted aim does not necessarily result in loss or violation of human dignity. Indeed, at least unconsciously, any reproduction serves certain aims, beyond the self-interest of the child concerned, being either the fulfillment of a family plan or individual pleasure or social reputation of parents, etc.. This represents the real motivation for reproducing, which is closely associated with personal lifestyles, plans, and individual desires of the prospective parents; it is not a matter of "luck" or "destiny" beyond self-determination. Evidence of that link between reproduction and personal aspirations or plans is already the typical expression of a simple desire on "what" children we would like to have and "how" we intend to raise them. It seems absurd to illustrate that common desire, as a sort of violation of the child's human dignity, by alleging their "instrumentalization" for serving our "selfish" intentions.

In conclusion, the selection of embryos as "savior siblings" may be considered an acceptable motivation for the exercise of the right to reproduce, not entailing the violation of the child's dignity to be born. Although not referring explicitly to that possibility, the Oviedo Convention does not exclude it, given that article 12 does not specify the health purposes for which genetic testing is allowed.

7.3.3.3 "Family Balancing" and Other Reasons of "Positive Eugenics," de Lege Ferenda

Currently, article 14 of the Oviedo Convention prohibits explicitly selecting the embryos' sex before implantation in order to achieve a sex-balanced family model (when more children of the same sex already exist).

Nevertheless, some ethicists advocate that possibility referring to social or psychological reasons for supporting the quality of family life, and also the encouragement of reproduction overall, for addressing the demographic problem of western societies.[48] Others disagree, stressing that selection of sex would result in extreme implications, particularly in societies that traditionally manifest their preference mostly for male children.[49]

---

[48] See in relevance, Pennings (1996), pp. 2339 seq.

[49] As it happens in India. Cf. Malpani and Modi (2002), pp. 11 seq. On the general debate about sex selection, see Warren (1999), passim.

However, *de lege ferenda*, sex selection for family balancing is an open issue. That "designer babies" problem needs to be analyzed by distinguishing the case where selection does not influence the future child's autonomy and the case where such influence occurs indeed. The selection of sex for family balancing belongs to the first category because the value that mostly serves is not the right to reproduce and create a family (art. 12 ECHR) but rather the right to family life (art. 8 ECHR). "Balancing" the family constitution means a possibility for someone to live in conditions that, according to their view, favor the quality of relationships with the other family members, as it happens, for instance, with the conditions of family housing. In that aspect, a predetermination of the future children sex does not influence their personal autonomy if a problem of balancing really exists and negatively affects the relationships between members of a particular family. Indeed, the future child is selected, not for favoring a preferred sex but due to the existence of children of the opposite sex in that family; if these children belonged to the opposite sex, the preference would be different.

Focusing then on the right to family life, under that approach, a future amendment of the Oviedo Convention on that point would not be unreasonable.

On the contrary, embryo selection on the basis of specific phenotypic characteristics of the future child influences the conditions of personal autonomy directly. This could happen if we were in place to identify with accuracy genetic causes for developing desirable or not characteristics after performing PGD on embryos in vitro.

However, the problem of "designing" future persons already exists, regardless of futuristic visions. Indeed, as we mentioned, even in the field of negative eugenics for securing the health of children to be born, "designing" considerations are also involved. The assumption on whether genetic diseases are "serious" or not is, to some degree, a matter of subjective preference. For some, if we are in place to diagnose genetic predispositions related to future health, then there are no reasons for restricting the option to reject embryos *in vitro* (or *in vivo*, through abortion) in order to have the "perfect" child in terms of health safety, which would be a "designer baby" option also.

Nevertheless, there is no question that health protection has a different aim than that of positive eugenics. By securing the health, we intend to defend the future person's autonomy against natural impediments. On the contrary, when we select phenotypes, we always take the risk to trouble autonomy by imposing predeterminations on the lifestyle of the individual concerned or even a predefined lifestyle, decided by others and not by that concrete individual.[50] In this view, only a very specific selection of characteristics, based on precise knowledge of the genotype involved, being irrelevant to any determination of the future person's lifestyle,

[50] See, in particular, the argumentation presented by Habermas (2003), pp. 81 seq., and the approach defending a right to an "open future" in Buchanan et al. (2001), pp. 170 seq. In contrast, for a justification of eugenics through genetic engineering with no influence on personal lifestyle choices, following the Rawlsian theory of Justice, see Agar (1998), pp. 171 seq.

could be ethically acceptable. So far, we are not able to accomplish these conditions; therefore, the risk for autonomy excludes any discussion about that, particularly considering changes in the existing legal framework.

Legally speaking, the selection of embryos on the ground of negative eugenics corresponds to a special confirmation of the human dignity principle, whereas selection for positive eugenics involves severe risks for that principle. This difference could drive the interpretation of legal provisions and their application insofar as the ongoing development of technological methods provides unforeseen control options in human reproduction.

7.3.4 The Case of Reproductive Cloning

Assisted reproduction has also raised ethical questions regarding the potential application of reproductive cloning in humans. With that method, the nucleus of an oocyte is removed and replaced by the nucleus of a somatic cell; thus, the partial genetic information of 23 chromosomes that contained the oocyte originally is replaced by the full genetic information of 46 chromosomes of the somatic cell. As proved by relevant experiments in mammals and other animal species, if the reproductive process continues after transferring that new biological entity into a female womb, the organism to be created is phenotypically identical to the donor of the used somatic cell. Theoretically, then, we cannot exclude the application of that method to humans also.

Modern Bioethics discusses that possibility extensively, even if, at present, prospects of successful application in humans look too remote for technical reasons. However, the first human clone's supposed appearance could certainly result in radical changes regarding social attitudes and, moreover, in reconsiderations of social relationships, due to the extent of control over human reproduction.

There are arguments for and against that futuristic assumption. The first underline, mostly, the possibility to overcome the complete inability of certain persons to reproduce naturally. The arguments opposed include various theoretical approaches stressing that the clone, being in purpose identical to the "original" donor of the somatic cell, would inevitably experience a violation of human dignity.

The European Biolaw, however, prohibits explicitly reproductive cloning in humans, according to the first Additional Protocol of the Oviedo Convention.[51] Also, the international soft law of the UNESCO Declarations on Bioethics and Human Rights[52] and on the Human Genome consider explicitly reproductive cloning as contrary to human dignity.[53] This is explained in two aspects.

On the one hand, the involvement of volunteer women, for receiving into their body cloned human embryos, with no scientific evidence about potential health implications during a supposed pregnancy, would instrumentalize these persons; this

[51] See, also, art. 3 para 2 d of the EU Charter of F. Rights.

[52] Cf. Teboul (2004), pp. 85 seq.

[53] Contra and Wu (1998), Harris (1997), pp. 143 seq.

is contrary to the human dignity principle, also given that consent is not a sufficient condition for research, according to the Oviedo Convention.

On the other, even under the assumption that the method would be proved successful in humans, the creation of clones would be considered an example of positive eugenics, with the relevant problems that we mentioned previously. Indeed, being a "product" of intentional "designing," a clone faces the risk to be perceived socially as an "instrument" of parents/ "designers"; the clone, thus, would carry a permanent burden to prove his/her uniqueness, which would be unaffordable in terms of personal autonomy. This element is incompatible with the human dignity principle, a subject of which would be a human clone as well.[54]

7.4 Further Studying...

ECtHR: A., B. and C. v. Ireland, 16 December 2010 (Grand Chamber)

"Three women living in Ireland, who became pregnant unintentionally, complained that, because of the impossibility of obtaining a legal abortion in Ireland1, they had to go to the United Kingdom for an abortion and that the procedure was humiliating, stigmatising and risked damaging their health. One of the applicants in particular, in remission from a rare form of cancer and unaware that she was pregnant, underwent checkups contraindicated in pregnancy. She understood that her pregnancy could provoke a relapse and believed that it put her life at risk.

The Court found that Ireland had failed to implement the constitutional right to a legal abortion. There had therefore been a violation of Article 8 (right to respect for private and family life) of the Convention concerning the applicant in remission from cancer (the Court held there had been no violation of Article 8 concerning the other two applicants), because she was unable to establish her right to a legal abortion either through the courts or the medical services available in Ireland. The Court noted in particular the uncertainty surrounding the process of establishing whether a woman's pregnancy posed a risk to her life and that the threat of criminal prosecution had a "significant chilling" effect both on doctors and the women concerned."[55]

Question

Comment on the following paragraphs, under the light of the ECHR, and the Oviedo Convention:

[54] Note the difference with the case of homozygotic twins, which, although genetically identical, have not been "designed" intentionally; therefore, they do not carry that burden in their autonomy.

[55] ECtHR, Press Unit.

253. "... the Court has a number of concerns as to the effectiveness of this consultation procedure as a means of establishing the third applicant's eligibility for a lawful abortion in Ireland.

It is first noted that the ground upon which a woman can seek a lawful abortion in Ireland is expressed in broad terms: Article 40.3.3, as interpreted by the Supreme Court in the X case, provides that an abortion is available in Ireland if it is established as a matter of probability that there is a real and substantial risk to the life, as distinct from the health, of the mother, including a risk of self-harm, which can only be avoided by a termination of the pregnancy (see the X case, cited at paragraphs 39–44 above). While a constitutional provision of this scope is not unusual, no criteria or procedures have been subsequently laid down in Irish law, whether in legislation, case-law or otherwise, by which that risk is to be measured or determined, leading to uncertainty as to its precise application. Indeed, while this constitutional provision (as interpreted by the Supreme Court in the X case) qualified sections 58 and 59 of the earlier 1861 Act (see paragraph 145 above), those sections have never been amended so that, on their face, they remain in force with their absolute prohibition on abortion and the associated serious criminal offences, thereby contributing to the lack of certainty for a woman seeking a lawful abortion in Ireland.

Moreover, whether or not the broad right to a lawful abortion in Ireland for which Article 40.3.3 provides could be clarified by Irish professional medical guidelines as suggested by the Government (see also the High Court judgment in MR v. TR and Others at paragraph 97 above), the guidelines do not in any event provide any relevant precision as to the criteria by which a doctor is to assess that risk. The Court cannot accept the Government's argument that the oral submissions to the Committee on the Constitution, and still less obstetric guidelines on ectopic pregnancies from another State, could constitute relevant clarification of Irish law. In any event, the three conditions noted in those oral submissions as accepted conditions requiring medical intervention to save a woman's life (pre-eclampsia, cancer of the cervix and ectopic pregnancies) were not pertinent to the third applicant's case.

Furthermore, there is no framework whereby any difference of opinion between the woman and her doctor or between different doctors consulted, or whereby an understandable hesitancy on the part of a woman or doctor, could be examined and resolved through a decision which would establish as a matter of law whether a particular case presented a qualifying risk to a woman's life such that a lawful abortion might be performed.

254. Against this background of substantial uncertainty, the Court considers it evident that the criminal provisions of the 1861 Act would constitute a significant chilling factor for both women and doctors in the medical consultation process, regardless of whether or not prosecutions have in fact been pursued under that Act. Both the third applicant and any doctor ran a risk of a serious criminal conviction and imprisonment in the event that a decision taken in medical consultation, that the woman was entitled to an abortion in Ireland given the risk to her life, was later found not to accord with Article 40.3.3 of the Constitution. Doctors also risked professional disciplinary proceedings and serious sanctions. The Government have not indicated whether disciplinary action has ever been taken against a doctor in this regard. The Review Group Report 1996, the Green Paper 1999 and the Fifth Progress Report on Abortion 2000 each expressed concerns about the lack of legal protection for medical personnel. As to the Government's reliance on A and B v. Eastern Health Board, Judge Mary Fahy and C, and the Attorney General (notice party), [1998] 1 IR 464 ("the C case"), doctors consulted by women such as the third applicant were not in the same legal situation as those in the C case who were providing opinions as regards a rape victim who was a suicide risk, a situation falling clearly within the ambit of the X case.

255. Accordingly, and referring also to McCarthy J.'s judgment in the X case (see paragraph 44 above), the Court does not consider that the normal process of medical consultation could be considered an effective means of determining whether an abortion may be lawfully performed in Ireland on the ground of a risk to life.

...

264. The Court considers that the uncertainty generated by the lack of legislative implementation of Article 40.3.3, and more particularly by the lack of effective and accessible procedures to establish a right to an abortion under that provision, has resulted in a striking discordance between the theoretical right to a lawful abortion in Ireland on the ground of a relevant risk to a woman's life and the reality of its practical implementation (see Christine Goodwin, §§ 77–78, and S. H. and Others, § 74, both cited above. See also the Commissioner for Human Rights, paragraph 109 above).

ECtHR: Evans v. United Kingdom, 10 April 2007 (Grand Chamber)

"The applicant, who was suffering from ovarian cancer, underwent in-vitro fertilisation (IVF) with her then partner before having her ovaries removed. Six embryos were created and placed in storage. When the couple's relationship ended, her ex-partner withdrew his consent for the embryos to be used, not wanting to be the genetic parent of the applicant's child. National law consequently required that the eggs be destroyed. The applicant complained that domestic law permitted her former partner effectively to withdraw his consent to the storage and use by her of embryos created jointly by them, preventing her from ever having a child to whom she would be genetically related.

For the reasons given by the Chamber in its judgment of 7 March 2006, namely that the issue of when the right to life began came within the State's margin of appreciation, the Grand Chamber found that the embryos created by the applicant and her former partner did not have a right to life. It therefore held that there had been no violation of Article 2 (right to life) of the Convention. The Grand Chamber further considered that, given the lack of European consensus, the fact that the domestic rules had been clear and brought to the attention of the applicant and that they had struck a fair balance between the competing interests, there had been no violation of Article 8 (right to respect for private and family life) of the Convention. Lastly, the Grand Chamber held that there had been no violation of Article 14 (prohibition of discrimination) taken in conjunction with Article 8 of the Convention."[56]

[56] ECtHR, Press Unit.

Question

Comment on the following paragraphs, under the light of the ECHR, and the Oviedo Convention:

89. While the applicant criticised the national rules on consent for the fact that they could not be disapplied in any circumstances, the Court does not find that the absolute nature of the law is, in itself, necessarily inconsistent with Article 8 (see also the Pretty and Odièvre cases cited in paragraph 60 above). Respect for human dignity and free will, as well as a desire to ensure a fair balance between the parties to IVF treatment, underlay the legislature's decision to enact provisions permitting of no exception to ensure that every person donating gametes for the purpose of IVF treatment would know in advance that no use could be made of his or her genetic material without his or her continuing consent. In addition to the principle at stake, the absolute nature of the rule served to promote legal certainty and to avoid the problems of arbitrariness and inconsistency inherent in weighing, on a case by case basis, what the Court of Appeal described as "entirely incommensurable" interests (see paragraphs 25–26 above). In the Court's view, these general interests pursued by the legislation are legitimate and consistent with Article 8.

90. As regards the balance struck between the conflicting Article 8 rights of the parties to the IVF treatment, the Grand Chamber, in common with every other court which has examined this case, has great sympathy for the applicant, who clearly desires a genetically related child above all else. However, given the above considerations, including the lack of any European consensus on this point (see paragraph 79 above), it does not consider that the applicant's right to respect for the decision to become a parent in the genetic sense should be accorded greater weight than J's right to respect for his decision not to have a genetically-related child with her.

ECtHR: Costa and Pavan v. Italy, 28 August 2012

"This case concerned an Italian couple who are healthy carriers of cystic fibrosis and wanted, with the help of medically-assisted procreation and genetic screening, to avoid transmitting the disease to their offspring.

The Court held that there had been a violation of Article 8 (right to respect for private and family life) of the Convention, finding that the interference with the applicants' right to respect for their private and family life had been disproportionate. It noted in particular the inconsistency in Italian law that denied the couple access to embryo screening but authorised medically-assisted termination of pregnancy if the foetus showed symptoms of the same disease. The Court also stressed the difference between this case, which concerned preimplantation diagnosis (PID) and homologous insemination4, and that of *S.H. and Others v. Austria* (see above), which concerned access to donor insemination. Although the question of access to PID raised delicate issues of a moral and ethical nature, the legislative choices made by Parliament in the matter did not elude the Court's supervision."[57]

[57] ECtHR, Press Unit.

Question

Comment on the following paragraphs, under the light of the ECHR, and the Oviedo Convention:

60. The Court notes at the outset that the applicants' complaint does not concern the question whether, taken alone, the ban on their recourse to PGD is compatible with Article 8 of the Convention. The applicants complain of a lack of proportionality of such a measure given that Italian law does allow them to abort the foetus if it is affected by the disease of which they are carriers.

61. In order to justify this interference, the Government refer to the concern to protect the health of "the child" and the woman, the dignity and freedom of conscience of the medical professions and the interest in precluding a risk of eugenic selection.

62. The Court is not persuaded by those arguments. While stressing that the concept of "child" cannot be put in the same category as that of "embryo", it fails to see how the protection of the interests referred to by the Government can be reconciled with the possibility available to the applicants of having an abortion on medical grounds if the fœtus turns out to be affected by the disease, having regard in particular to the consequences of this both for the fœtus, which is clearly far further developed than an embryo, and for the parents, in particular the woman (see the report of the CDBI of the Council of Europe and the information contained in the Belgian Bill, paragraphs 25 and 34 above).

63. Furthermore, the Government have failed to explain how the risk of eugenic selection and affecting the dignity and freedom of conscience of the medical professions would be averted in the event of an abortion being carried out on medical grounds.

64. The Court cannot but note that the Italian legislation lacks consistency in this area. On the one hand it bans implantation limited to those embryos unaffected by the disease of which the applicants are healthy carriers, while on the other hand it allows the applicants to abort a fœtus affected by the disease (see also the report of the European Commission, paragraph 27 above).

65. The consequences of such legislation for the right to respect for the applicants' private and family life are self-evident. In order to protect their right to have a child unaffected by the disease of which they are healthy carriers, the only possibility available to them is to start a pregnancy by natural means and then terminate it if the prenatal test shows that the fœtus is unhealthy. In the instant case the applicants have already terminated one pregnancy for that reason, in February 2010.

66. In these circumstances the Court should not underestimate either the anxiety experienced by the first applicant, whose only hope of having another child, since she is unable to have recourse to PGD, carries the concomitant risk that the child will be born with the disease or the suffering inherent in the painful decision to undergo, as the case may be, an abortion on medical grounds.

67. The Court also notes that in the case of S.H. (cited above, § 96), the Grand Chamber established that, in cases of heterologous insemination, having regard to medical and scientific developments, the State's margin of appreciation could not be decisively narrowed.

68. While acknowledging that the question of access to PGD raises sensitivie moral and ethical questions, the Court notes that the solutions reached by the legislature are not beyond the scrutiny of the Court (see, mutatis mutandis, S.H., cited above, § 97).

69. In the present case the Court reiterates that, unlike the case of S.H. (cited above), where the Court assessed the compatibility of Austrian law prohibiting heterologous insemination with Article 8 of the Convention, its task in this case, which concerns homologous insemination, is to verify the proportionality of the measure in question in the light of the fact that termination of pregnancy on medical grounds is an option for the applicants (see paragraph 60 above).

70. It is therefore a specific situation which, according to the comparative-law materials in the Court's possession, apart from Italy, concerns only two of the thirty-two States studied, namely, Austria and Switzerland. Moreover, with regard to the latter State, the Court notes that a draft amendment to the current ban on PGD, to provide for regulated access, is now being examined ...

ECtHR: Mennesson and Others v. France and Labassee v. France, 26 June 2014

"These cases concerned the refusal to grant legal recognition in France to parent-child relationships that had been legally established in the United States between children born as a result of surrogacy treatment and the couples who had had the treatment. The applicants complained in particular of the fact that, to the detriment of the children's best interests, they were unable to obtain recognition in France of parent-child relationships that had been legally established abroad.

In both cases the Court held that there had been no violation of Article 8 (right to respect for private and family life) of the Convention concerning the applicants' right to respect for their family life. It further held in both cases that there had been a violation of Article 8 concerning the children's right to respect for their private life. The Court observed that the French authorities, despite being aware that the children had been identified in the United States as the children of Mr. and Mrs. Mennesson and Mr. and Mrs. Labassee, had nevertheless denied them that status under French law. It considered that this contradiction undermined the children's identity within French society. The Court further noted that the case-law completely precluded the establishment of a legal relationship between children born as a result of – lawful – surrogacy treatment abroad and their biological father. This overstepped the wide margin of appreciation left to States in the sphere of decisions relating to surrogacy."[58]

[58] ECtHR, Press Unit.

Question

Comment on the following paragraphs, under the light of the ECHR, and the Oviedo Convention:

96. As the Court has observed, respect for private life requires that everyone should be able to establish details of their identity as individual human beings, which includes the legal parent-child relationship ...; an essential aspect of the identity of individuals is at stake where the legal parent-child relationship is concerned (see paragraph 80 above). As domestic law currently stands, the third and fourth applicants are in a position of legal uncertainty. While it is true that a legal parent-child relationship with the first and second applicants is acknowledged by the French courts in so far as it has been established under Californian law, the refusal to grant any effect to the US judgment and to record the details of the birth certificates accordingly shows that the relationship is not recognised under the French legal system. In other words, although aware that the children have been identified in another country as the children of the first and second applicants, France nonetheless denies them that status under French law. The Court considers that a contradiction of that nature undermines the children's identity within French society.

97. Whilst Article 8 of the Convention does not guarantee a right to acquire a particular nationality, the fact remains that nationality is an element of a person's identity (see *Genovese v. Malta*, no. *53124/09*, § 33, 11 October 2011). As the Court has already pointed out, although their biological father is French the third and fourth applicants face a worrying uncertainty as to the possibility of obtaining recognition of French nationality under Article 18 of the Civil Code ... That uncertainty is liable to have negative repercussions on the definition of their personal identity.

98. The Court also observes that the fact that the third and fourth applicants are not identified under French law as the children of the first and second applicants has consequences for their inheritance rights. It notes that the Government deny this, but observes that the *Conseil d'État* has ruled that in the absence of recognition in France of a legal parent-child relationship established abroad with regard to the intended mother, a child born abroad as the result of a surrogacy agreement cannot inherit under the mother's estate unless the latter has named the child as a legatee, the death duties then being calculated in the same way as for a third party ..., that is, less favourably. The same situation arises in the context of inheritance under the intended father's estate, even if he is the biological father as in this case. This is also a component of their identity in relation to their parentage of which children born as the result of a surrogacy agreement performed abroad are deprived.

99. The Court can accept that France may wish to deter its nationals from going abroad to take advantage of methods of assisted reproduction that are prohibited on its own territory (see paragraph 62 above). Having regard to the foregoing, however, the effects of non-recognition in French law of the legal parent-child relationship between children thus conceived and the intended parents are not limited to the parents alone, who have chosen a particular method of assisted reproduction prohibited by the French authorities. They also affect the children themselves, whose right to respect for their private life—which implies that everyone must be able to establish the substance of his or her identity, including the legal parent-child relationship—is substantially affected. Accordingly, a serious question arises as to the compatibility of that situation with the children's best interests, respect for which must guide any decision in their regard.

Provide Legally Appropriate Solutions

1. In a fertility clinic, a couple is interested in fertilization with a donor asking for the use of oocytes derived from

- The woman's sister or
- A sportswoman with blue eyes preferably

Also, the couple asks the physician in charge to perform PGD on the embryos in vitro in order to select for transfer only

- Female embryos
- With no Down syndrome
- With no genetic predisposition for Alzheimer's disease
- With no genetic predisposition for breast cancer (genes BRCA 1, 2)

Comment on the legal validity of the above wishes under the current European Biolaw.

2. Elen A. agrees with a couple of intending parents to carry an embryo created through IVF with their reproductive material. After birth, Elen A. refuses to deliver the baby, arguing that she developed a strong affection for the baby during pregnancy and wishes to undertake parental care. Under the current European Biolaw

- Is the initial contract between Elen A. and the couple of intending parents legally valid?
- Is the claim of Elen A. legally sound, given that she has no genetic ties with the baby?

References

Agar N (1998) Liberal Eugenics (Public Affairs Quarterly 12: 137–155). In: Kuhse H, Singer P (eds) Bioethics. An anthology. Blackwell, Oxford, 1999, pp 171–181

Annas GJ (2007) The supreme court, and abortion rights. N Eng J Med 356:2201–2207

Asch A, Barlevy D (2012) Disability and Genetics: A Disability Critique of Pre-natal Testing and Pre-implantation Genetic Diagnosis (PGD), in doi:https://doi.org/10.1002/9780470015902.a0005212.pub2

Baylor College of Medicine (2000) Evolution and Revolution: The Past, Present, and Future of Contraception, Contraception Online. in http://www.contraceptiononline.org

Boys SK, Harris EM (2019) IVF and the anti-abortion movement: considerations for advocacy against overturning roe v. Wade. Adv Soc Work 19:518–533

Buchanan A, Brock DW, Daniels N, Wikler D (2001) From chance to choice. Genetics & justice. Cambridge University Press, Cambridge

Comité Consultatif National d'Éthique (CCNE) (1991) Avis sur les réductions embryonnaires et fœtales, No 24

Cowley C (2003) The conjoined twins and the limits of rationality in applied ethics. Bioethics 17: 69–88

De Melo Martin I (2004) On our obligation to select the best children: a reply to Savulescu. Bioethics 18:72–83

Dücker S, Hörnle T (2018) German law on surrogacy and egg donation: the legal logic of restrictions. In: Mitra S, Schicktanz S, Patel T (eds) Cross-cultural comparisons on surrogacy and egg donation. Palgrave Macmillan, London, pp 231–253
Fenwick D (2012) The modern abortion jurisprudence under Article 8 of the European convention on human rights. Med Law Int 12:249–276
Habermas J (2003) The future of human nature. Polity Press, Cambridge
Harris J (1993) Is gene therapy a form of eugenics? (Bioethics 7: 178–187). In: Kuhse H, Singer P (eds) Bioethics. An anthology. Blackwell, Oxford, 1999, pp 165–170
Harris J (1997) "Goodbye Dolly?" The Ethics of Human Cloning (J. of Medical Ethics 23: 353–360). In: Kuhse H, Singer P (eds) Bioethics. An anthology. Blackwell, Oxford 1999, pp 143–152
Hodson N, Townley L, Earp BD (2019) Removing harmful options: the law and ethics of international commercial surrogacy. Med Law Rev 27:597–622
Horsey K (2018) Surrogacy in the UK: further evidence for reform: Second Report of the Surrogacy UK Working Group on Surrogacy Law Reform, London
Hume DA (1739/1984) Treatise on Human Nature, III, I. Penguin, London
Kaplan D (1993) Prenatal screening, and its impact on persons with disabilities (Clinical Obstetrics and Gynecology 36: 605–612). In: Kuhse H, Singer P (eds) Bioethics. An anthology. Blackwell, Oxford 1999, pp 130–136
Kass LR (1973) Implications of Prenatal Diagnosis for the Human Right to Life (Hilton B et al. ed., Ethical Issues in Human Genetics, Plenum Pub. Corp.). In: Mappes TA, Degrazia D (eds) Biomedical ethics, 5th edn. McGraw-Hill, New York, pp 516–520
Lau PL (2019) Comparative legal frameworks for pre-implantation embryonic genetic interventions. Springer, Cham
Malpani A, Modi D (2002) Preimplantation sex selection for family balancing in India. Hum Reprod 17:11–12
Marquis D (1989) Why abortion is immoral (J. of Philosophy 86: 183–202). In: Kuhse H, Singer P (eds) Bioethics. An anthology. Blackwell, Oxford, 1999, pp 46–57
Muller-Hill B (1994) Lessons from a dark and distant past [Clarke A. (ed.), Genetic counseling: practice and principles, Routledge, London, pp 133–141]. In: Kuhse H, Singer P (eds) Bioethics. An anthology. Blackwell, Oxford 1999, pp 182–187
Nationaler Ethikrat (2003) Genetic diagnosis before and during pregnancy. Berlin
Nelson L (2018) Provider conscientious refusal of abortion, obstetrical emergencies, and criminal homicide law. Am J Bioeth 18:43–50
Ní Shúilleabháin M (2019) Surrogacy, system shopping, and article 8 of the European convention on human rights. Int J Law Policy Fam 33:104–122
Obermeyer CM (1995) A cross-cultural perspective on reproductive rights. Hum Rights Q 17:366–381
Pennings G (1996) Ethics of Sex Selection for Family Balancing. Family Balancing as a Morally Acceptable Application of Sex Selection. Human Reprod 11:2339–2342
Piersanti V, Consalvo F, Signore F, Del Rio A, Zaami S (2021) Surrogacy and "procreative tourism". What does the future hold from the ethical and legal perspectives?. Medicina 57:47. https://doi.org/10.3390/medicina57010047
Priaulx N (2017) The social life of abortion law: on personal and political pedagogy. Med Law Rev 25:73–98
Purdy LM (1990) Are pregnant women fetal containers? (Bioethics 4: 273–291). In: Kuhse H, Singer P (eds) Bioethics. An anthology. Blackwell, Oxford, 1999, pp 71–81
Purdy LM (1996) Genetics and reproductive risk: can having children be immoral? (Reproducing persons: issues in feminist bioethics, Cornell U.P., N. York, pp. 39–49). In: Kuhse H, Singer P (eds) Bioethics. An anthology, Blackwell, Oxford 1999, pp 123–129
Ramm T (1989) Die Fortpflanzung – ein Freiheitsrecht? Juristenzeitung 44:861–874
Robertson JA (2001) Precommitment strategies for disposition of frozen embryos. Emory Law J 50: 989–1046

Rotabi KS, Mapp S, Cheney K, Fong R, McRoy R (2017) Regulating commercial global surrogacy: the best interests of the child. J Hum Rights Soc Work 2:64–73

Sandor J. (2018). Transnational surrogacy: an overview of legal and ethical issues. In: Mitra S, Schicktanz S, Patel T (eds) Cross-cultural comparisons on surrogacy and egg donation. Palgrave Macmillan, London, pp 35–55

Savulescu J (2001) Procreative beneficence: why we should select the best children. Bioethics 15: 413–426

Simonstein F (2005) Artificial Reproduction Technologies – All the Way Towards the Artificial Womb? Paper presented at the 19th European Conference on Philosophy of Medicine and Health Care and 22nd EACME Conference "Ethics and Philosophy of Emerging Medical Technologies," Barcelona

Solbakk JH (2005) The bio-politics of pre-implantation genetic diagnosis. A Norwegian Case, in Jonsdottir I., PGD and Embryo Selection. Report from an International Conference on Preimplantation Genetic Diagnosis and Embryo Selection, TermaNord, Copenhagen, pp 99–104

Sonnenberg–Schwan U, Gilling–Smith C, Weigel M (2005) HIV and wish for parenthood. In: Hoffmann C, Rockstroh JK, Kamps BS (eds) HIV Medicine, Steinhauser verl. & Flying Publ., Paris

Stretton D (2004) The deprivation argument against abortion. Bioethics 18:144–180

Teboul G (2004) Un instrument international prohibant le clonage humain reproductif? J Int Bioethique 15:85–93

Trawick E, Sampson A, Goldman K, Campo-Engelstein L, Caplan A, Keefe DL, Quinn GP (2020) Posthumous assisted reproduction policies among a cohort of United States' in vitro fertilization clinics. F&S Rep 1:66–70

Vidalis T (2007) Regulating assisted reproduction in the EU: time for a new directive? J Int Biotechnol Law 4:12–15

Warren MA (1999) Sex selection: individual choice or cultural coercion? In Kuhse H, Singer P (eds) Bioethics. An anthology. Blackwell, Oxford, pp 137–142

Wilkinson S (2003) The exploitation argument against commercial surrogacy. Bioethics 17:169–187

Wu L (1998) Family planning through human cloning: is there a fundamental right? vol 98, Columbia Law Rev, pp 1461–1515

Enhancement 8

8.1 General

Based on continuous research development, the rapid progress in medical technology and pharmacology refer currently to interventions intending not only for the protection of health but also for the "enhancement" of the human organism. Such interventions are not performed for reestablishing a normal state of troubled biological functions but for improving phenotypic characteristics or capabilities at various levels.

Enhancement includes, first, a variety of pharmaceutical treatments accessible by healthy persons also. These treatments are used mostly for increasing bodily skills in order to pursue certain achievements, like the use of doping substances in sports competitions,[1] or the use of drugs for enhancing either specific cognitive skills (concentration, memory, etc..) or psychological reactions (control of stress, sedation, etc..).[2] In a broader sense, we can consider "enhancement" even interventions aiming to change external phenotypic characteristics, like aesthetic operations that may address situations of personal self-confidence, and mostly change of characteristics that may influence others, like sex reassignment surgery and face transplantations.

As much more interventional acts of enhancement could be considered those of intentional genetic modifications performed either for changing the genome of a

[1] The list of enhancing substances is constantly enriched, comprising growth hormone, botox injections, Viagra, and, in sports, various additives with proteins, vitamins, etc.., and also illicit substances and drugs (steroids, etc.., see in detail: https://www.wada-ama.org/en/content/what-is-prohibited/prohibited-at-all-times/diuretics-and-masking-agents). See also Van de Ven (2020), Chap. 1.

[2] See Mehlman (2005), Wolpe (2002), Farah (2005). The most known cases are serotonin compounds (SSRIs—Prozac etc.) of the memory-enhancing Gingko Biloba and Ritalin for addressing concentration issues, particularly in children. See Kramer (1993).

T. Vidalis, *The Emergence of Biolaw*, Springer Textbooks in Law,
https://doi.org/10.1007/978-3-031-02359-0_8

particular person only or possibly for affecting the genetic profile of future persons, that is, the descendants of the person undergoing genetic intervention.

Finally, as an enhancement, we can characterize treatments for aging, intending to prolong the lifespan, namely to postpone the natural, non-pathological decline of the human organism, to the extent possible.

Enhancement is distinguished, thus, from health recovery,[3] to the measure that it does not cope with threats for health-damaging, but affects our normal healthy skills and characteristics. That distinction creates a field of reflection in modern Bioethics, and moreover, it has opened a discussion over the beginning of a course towards the future evolution of our species (trans-humanism, post-humanism), historically the first to be based entirely on our technology. But if health protection represents an undoubted ethical and legal value, what is the value of enhancement? That question is crucial, particularly if we take into account the significance of other potentially conflicting values.[4]

In that problem, we presuppose that a clear distinction between "health" and "enhancement" is possible; the concept of health as such, regardless of its narrower or broader content (following what we mentioned in Chap. 3 on "pathology"), cannot cover the concept of enhancement, that is, "something more" than a healthy condition of the organism. With that remark, we examine, in the following, particular examples of enhancement, from the aspect of biological autonomy and its balance with other values in the European legal system.

8.2 Increasing Physical and Cognitive Capabilities

The use of pharmaceuticals for increasing the physical and cognitive capabilities of a particular person is of interest from two aspects; that of protection of this person's individual health and the aspect concerning the protection of other persons' rights and legitimate interests in contexts of competitive relationships (sports, tests in education, etc..).

Regarding the first, no specific dilemmas arise, given that both health and enhancement fall within the area of biological autonomy of the person concerned; therefore, it is up to the person's free decisions on how to harmonize these potentially conflicting choices. An argument against enhancing interventions that may be dangerous for individual health would oppose health as an object of individual freedom. Even if the use of pharmaceuticals may influence deeply mental characteristics, and subsequently the personality as a whole (like in the case of anti-depression treatment, where phenomena of schizoid personality occurred, with

[3] See Buchanan et al. (2001), pp. 110 seq., esp. 149 seq. Also, McGee (2020), Savulescu (2019).

[4] A philosophical defense of enhancement, see, for example, in Savulescu (2019). For a brief presentation of arguments in favor and against enhancement, under the view of human dignity, see Bostrom (2005), pp. 202 seq. The most known objection to the post-humanism approach has been expressed by Fukuyama (2002), which insists on a firm conceptualization of human dignity denying any form of enhancement that could entail a relativistic interpretation of that principle.

the Prozac experience, etc..),[5] this cannot be a legitimate reason for denying free choices insofar as other persons are not affected.

However, since the use of enhancing treatments also represents a medical intervention, the rule of informed consent needs to be followed by the physicians involved, as in any medical act. This means that a medical doctor may be found liable for not providing appropriate information to the interested person before deciding on such an intervention. Medical liability regarding information is more intense in the field of enhancement due to the non-therapeutic purpose of a relevant act. Indeed, the attending physician needs to ensure the genuine nature of the person's will for avoiding misunderstandings or even frivolous impressions about the expected results and potential implications. The difference with conventional medical acts is that, here, no health damage needs to be addressed; on the contrary, insufficient or misleading information about the use of pharmaceuticals may entail the transformation of a healthy person into a patient.[6]

In sports, intensive pressure on athletes' autonomy occurs due to the context of competition that also involves economic, etc. complex interests, particularly in international games. Third persons, trainers, doctors, parents, etc.., may exercise that pressure by prioritizing success in athletic performance over the protection of the athlete's health and biological autonomy in general.[7] From the aspect we examine, this may jeopardize the accuracy of the information provided and mislead the athlete when deciding. The medical liability of physicians involved is also, here, unquestionable.

On the other hand, in competitive contexts, a more serious problem is the influence that the use of pharmaceuticals for the enhancement of individual specific skills may have on other persons' legitimate interests. Here, the question is whether an athletic or educational competition always presuppose limitations on biological autonomy for participant athletes or students. Because, one might argue that, on the contrary, personal choices must be unlimited since every person can control freely his/her nutrition, physical exercise, etc.., and might have access even to additional lawful means of enhancement (with the use of vitamins, nutrition additives, etc..) as well. If, for instance, we take into account that with intense exercise in altitude conditions, the athlete's stamina and relevant skills are increased drastically, it looks reasonable to equate that enhancing practice with the use of pharmaceuticals.

We can answer that question by noting that certain limitations, even on biological autonomy, are necessary, if suitable to the competition's scope, namely assessing more persons' performance regarding their skills. The accuracy of that assessment requires a comparison between these persons in terms of parity; this is feasible only

[5] Affecting possibly the "creativity" of a person which may challenge the free (and original) will as such. That problem raises questions even in the field of intellectual property rights. See Dunagan et al. (2020), referring to brain enhancement interventions.

[6] This is a difficult appraisal since it is related to social and ethical approaches regarding, in particular, the concepts of personality and selfhood. See in relevance, Wolpe, *supra*, pp. 393 seq., and Farah, *supra*, pp. 37 seq.

[7] Cf. European Group on Ethics (EGE) (1999) 1.6, 2.10.

when we presuppose limitations on biological autonomy, that is, on methods for improving physical or cognitive skills, in order to ensure equal treatment for everybody.[8]

A further question is related to the nature of limitations and, more specifically, to the standards according to which we can identify enhancing pharmaceuticals as acceptable or not, in reference to particular competition contexts. A reasonable standard may be the scientific evidence about potential damages to the user's health, suggesting the prohibition of certain substances. Other objective standards cannot be proposed with certainty, however. Given that an enhancing intervention does not necessarily involve risks for the individual health,[9] we can equate that intervention with any legitimate improvement of skills through nutrition, hard and long exercise, or intensive study. Thus, the assessment of parity in competition between participants reaches a limit since there is no substantial difference among the available means of enhancement.

Under that view, additional limitations on pharmaceutical use can be adopted only on the basis of standard agreements between participants in concrete contexts, following the specific objective of competition. A characteristic example is that of the use of substances for therapeutic reasons when these have also enhancing results for the individual skills of the person concerned.[10] In terms of biological autonomy, that particular use of substances cannot be excluded if supporting individual health protection. Yet, if that person is an athlete, and the terms of competition exclude substances use, his/her participation presupposes compliance with that restriction for ensuring the legitimate interests of the other athletes, even if this may be opposed to our athlete's health.

In sports, although a European "anti-doping" legislation does not exist,[11] international associations (IOC, FIFA, FIBA, etc..) have adopted binding non-legal rules that determine prohibited pharmaceuticals and substances, as well as control procedures, based on the protection both of the athletes' health and the conditions of competition.[12]

---

[8]The "fairness" question plays a crucial role in common disapproval of cognitive enhancement. See a relevant empirical study in Faber et al. (2016).

[9]EGE, *supra*, 1.3.

[10]EGE, *supra*, 1.4.

[11]Perhaps, this gap could be covered by the EU legislation on work relationships if professional competition in sports would be considered within that context. See EGE, *supra*, 2.9, 2.10.

[12]See. EGE, *supra*, 1.5.

8.3 Altering External Characteristics

Contrary to aesthetic interventions through plastic surgery, which meant as manifestations of the person's biological autonomy not affecting the rights or legitimate interests of others, two cases of external characteristics' alteration, these of face transplantation and sex-changing, are of interest under this latter view.

8.3.1 Face Transplantation

Face transplantation from deceased donors to receivers experiencing extensive face damage, usually due to serious accidents or diseases, may be performed for therapeutic or aesthetic reasons. Generally, this practice should be addressed in legal terms, like conventional transplantations. Regarding the attending doctor's role, though, medical liability is more intense, particularly concerning the patient's information, due to potential implications that may occur in such a complicated operation and the high possibility of failure as to the aesthetic result.

Aesthetic face transplantations, in particular, intending to enable the healthy transplant recipient to live a normal social life, raise different issues affecting other people, from the fact that it results in extensive alteration of the recipient's external characteristics.[13]

First, a question referring to the transplant recipient is whether his/her biological autonomy may justify even a complete change of the external "identity," which is vital for a person's social relationships. An argument against would insist on legitimate interests of others or even the state (for securing certainty of law, where personal identity matters, or even for public security reasons in air transports, etc..) that require permanency of the original face characteristics, as much as possible, throughout any person's lifespan. To that problem, priority should be given to the person's biological autonomy. Indeed, any interests of third persons or the state cannot be evaluated as more vital, compared to the reparation of a basic element of personality, the lack of which may disrupt the entire lifestyle of the individual concerned. Besides, existing malformations may already have resulted in violating the person's identification radically, as appearing in public documents (passports, IDs, etc..), therefore, no certainty exists to contacts with the state or social relationships of that person whatsoever.

A second issue refers to the relatives of the deceased face donor. Is removing the face transplant in compliance with the due respect to the deceased person's body, according to the Oviedo Convention, or does it mean a complete change of that body beyond accepted limits? Furthermore, are the emotions of the deceased person's relatives considered when that person's image is to be "transferred" to another person?

[13] Cf. Comite Consultatif National d' Éthique (CCNE) (2004), IV 3, and Huxtable, Woodley (2005).

Here, we need to recall that, what is at stake, is not a deceased person's "right" to their body, but only the family members' rights to their own personality, that include their emotions over the deceased person. In the case where the latter, while living, had not expressed a wish to become a donor, the relatives decide on the potential donation of transplants following their emotions and attitude over what they consider "respect" for the dead body. "Respect" reflects, precisely, a dimension of the right to their own personality, not an "objective" stance determined by others or the society as a whole; otherwise, it would be absolutely unclear a standard for "respect" and, consequently, whether the family members may decide over transplant donation. In the case we examine, the free decision of relatives to donate the "face" of their beloved dead functions as a guarantee that they have accepted the removal and transplantation of the face transplant consciously, in conformity with their right to personality.[14]

8.3.2 Interventions of Sex Change

Reassigning sex interventions may also be seen as transforming the external characteristics of a person for enhancement reasons. Indeed, the state of deviation between the external characteristics of sex and the way that the person concerned experiences them (known as a difference between "sex" and "gender"), cannot be considered a "disease" since it does not involve any harm to a healthy organism. Sex reassignment, as a free choice, aims mostly at the restitution of a psychosocial balance for ensuring a steady development of personality, frequently threatened by negative social discrimination that the person concerned experiences.

Again, here, we need to cope with a problem of legal certainty over the limits of that person's biological autonomy versus the legitimate interests of others or the public interest itself; sex change means a radical transformation of the person's identity as expressed publicly and influences social interactions.

In that case, also, we need to recognize priority to personal autonomy. This is because the person interested experiences a state of inconsistency between their sex and gender that results in unaffordable restrictions on many potential manifestations of individual freedom, including sexual life, friendship relations, the rights to marriage and family, and even professional relationships to the measure that sex and gender affect their determination and development. Moreover, human dignity itself is jeopardized when the person is obliged to reveal their diversity in non-tolerant social contexts. Therefore, a complete transformation of the person's initial identity, through sex-change operations, as a legitimate expression of biological autonomy, is legally justified, even if this may have implications for third persons or the state, in terms of legal certainty; as the ECtHR has accepted, that conclusion may also lead to an obligation of the state for updating public

[14]However, a restoration of the deceased person's face even after the transplant removal is also possible. See CCNE, *supra*, IV 4a.

inscriptions in relevant registries, concerning the personal status, after a sex change intervention.[15]

8.4 Genome "Enhancement"

Enhancement has been associated with possible interventions in the human genome also. Already applied in other species, the development of genetic engineering methods promises, for the future, modifications of the human genetic profiles in order not only to "fortify" them against serious genetic diseases but also to achieve desirable phenotypic characteristics, physical or even cognitive ones. This prospect looks nowadays feasible, due to the rapid enrichment of our knowledge on the nature and interactions of genes and to recent methods of "genome editing" (CRISPR Cas9, etc..), permitting very accurate and relatively more straightforward modifications of genomes, Compared to conventional methods of gene therapy.[16]

Questions on whether gene interventions for enhancement purposes, producing irreversible effects, may reduce the personal autonomy of persons undergoing them, or of future persons, if performed in human gametes or embryos, are crucial in modern Bioethics, and they occur in practical terms, following the rapid progress of genetic technology. For its part, the European Biolaw addresses them in the light of the Oviedo Convention.

8.4.1 Uncertainty and Accepted Interventions: The Oviedo Convention

Following article 13, the Oviedo Convention does not accept genetic enhancement in humans. As we already mentioned in Chap. 3 (3.8.4), the wording of this article does not cover a purpose of gene therapy other than the protection of health in its narrow meaning, that is, what could be defined as a diagnostic, preventive, or therapeutic act; interventions for "designing" physical or cognitive characteristics are, thus, excluded. The rationale for that limitation is related to the current condition of uncertainty regarding the functions of genes; this condition may negatively impact a person undergoing such an intervention, including a possible reduction of his/her overall autonomy.

Nevertheless, the condition of uncertainty cannot be considered as a reason for the "absolute" prohibition of genetic enhancement, in the sense that it is dependent directly on the progress of our knowledge about human genetics. To the extent that,

[15] See ECtHR, Goodwin v. the U.K. (2002), a judgment that changed the previous negative stance of the Strasbourg Court, as expressed in Rees v. U.K. (1986), Cossey v. U.K. (1990), Sheffield, Horsham v. U.K. (1997).

[16] See a strong legal objection against interventions in the human genome, characterizing these even as crimes against humanity, in Annas (2000), esp. pp. 776 seq.

progressively, we can gain certainty about the outcome of particular genetic interventions, the rationale of that prohibition is ethically weakened. The current "genome editing" technologies seem very promising to that end due to their accuracy and easiness of performance. If in the future, these expectations are confirmed, we will face a problem with that explicit wording of the Oviedo Convention; at that time its amendment should be considered, since performing safe genetic interventions for enhancement purposes could not be reasonably excluded from the spectrum of adult individuals' biological autonomy.

8.4.2 Interventions Affecting Future Generations

Regarding interventions affecting future persons, we address again a potential degradation of their autonomy. However, we already determine the limits of our descendants' autonomy when, for instance, we decide on the nature of education that we provide to them or even the general conditions of their living (housing, etc..). What could be the difference if we had the option to move one step beyond by designing their DNA also, based on an assumed solid and accurate knowledge about the influence of genes on specific skills or physical or even cognitive characteristics?

Again, a futuristic question is whether we need to reconsider article 13 of the Oviedo Convention in case the current uncertainty about the functions of genes would be overcome. As we have seen, regarding therapeutic interventions or interventions intending to "fortify" the organism against the potential manifestation of genetic diseases, there is no specific reason to insist on that prohibition if we can expect beneficial genetic consequences to descendants with certainty.[17]

In contrast, regarding genetic enhancement interventions not related to health reasons, a supposed certainty about the biological outcome (the influence on future persons' physical or cognitive characteristics) does not guarantee their personal autonomy status. Here, we need to recall the objections against positive eugenics, as a matter of principle; the fact that the future person would experience their state as a product of "designing", with no option of change, would mean a permanent burden for the practical exercise of their personal autonomy. We cannot compare this situation to social conditions of care that parents "design" and decide according to their preferences and beliefs for raising their children. Indeed, any social determinations may become subject to critical control by the person experiencing them, either during or after a socialization process; in contrast to biological determinations that are irreversible (at least, so far), the outcome of any socialization

[17] Against the current moratoria for heritable genome editing applications, see in particular, Knoppers and Kleiderman (2019). See also, in relevance, Harris (2015), pp. 30–34. Regarding the need for adjustments in regulation, given the difficulty to distinguish between interventions in somatic and in germ line when applying genome editing, see Sándor (2019). However, see Beriain et al. (2019), arguing that article 13 does not prohibit genome editing in embryos for basic research purposes, if implantation and the process of reproduction are excluded. Against a detailed legal regulation in these matters, see for example, Anomaly et al. (2020).

may be changed freely by the individual concerned, at any time. As noted previously concerning the issue of embryo selection, the only genetic intervention that could be accepted in the context of future persons' enhancement is that of gene modifications performed on the ground of accurate knowledge about their phenotypic outcome, which would not entail a predetermined lifestyle for these persons; given the current state-of-the-art, this is an unfeasible condition at present.[18]

8.5 Aging Issues: Enhancement and Natural Decline

The use of technology and the access to medicinal treatments for "delaying" the natural decline of a human organism when getting old may also be seen as a matter of enhancement. The aim, here, is the conservation of an active and healthy life for as long as possible.

Although the biological process of natural aging remains largely unknown in its details, we have identified certain elements of that process at the molecular level, leading to correlations with specific genes. Certainly, at present, we are not in place to manage aging by using technological applications; there are still indications that this would be possible in the future. On the other hand, the relative expansion of the average lifespan in modern societies, thanks to significant improvements in living conditions and healthcare services, has overturned the populations' age balance in favor of older people. This fact raises complex socioeconomic implications relevant to the social management of aging, from the insufficient coverage of insurance and healthcare needs, to the social marginalization of the elderly (which involves experiences of undermined dignity) or to increasing neurological and psychiatric pathologies. These issues are investigated legally, as objectives related to the equal treatment of persons and the implementation of social rights.[19]

What is interesting for Biolaw specifically, is the question of whether, given the societal issues mentioned above, we need to recognize limits to the biological autonomy of persons wishing to utilize the available technological means for prolonging their lifespan, particularly when this may influence the autonomy of others (family members, mostly). Indeed, the artificial prolongation of a person's lifespan could be criticized as ethically unacceptable for two reasons. First, because this makes the problem of age-balancing in a certain society worst, and second, because that person, if already retired from any productive activity, would burden his/her close family members, restraining thus their own autonomy and lifestyle.

[18]Those opposed to the modification of germline, besides the danger of eugenics for future generations, note an additional issue related to the distant future of humanity; by restricting the genetic diversity of our species, we may make humans less able to adapt to environmental changes, therefore more vulnerable to diseases, epidemics, etc.., in the long run. See on that, Baylis and Robert (2004), pp. 7 seq. (The authors disagree with that thought, pointing out that genetic enhancement methods are "inevitable").

[19]See in general, Comite Consultatif National d' Ethique (CCNE) (1998).

Nonetheless, although these are undeniable problems, it would be unreasonable to observe limitations on the biological autonomy of the person concerned. This is because, due to the human dignity principle, we cannot impose a "socially accepted" limit on the individual lifespan, meaning an end-point to the life of a person, for the sake of other persons' interests.[20] To ban the availability of technological applications for prolonging the individual lifespan would violate human dignity, leading to a sort of the person's instrumentalization since that person would be deprived of an option to postpone his/her death.

From that argumentation, we can draw a more general conclusion; the constant expansion of biological autonomy, with the multiplication of options permitting control of certain functions of our organism, should be considered an essential feature of human value, notably regarding the conservation of individual life. Not only pathologies occurring during the human lifespan, but also the natural decline of the organism with aging should be treated as legitimate objectives of personal control when we dispose of the knowledge about mechanisms and available methods of that control. From this particular point of view, no human biological decline should be considered "natural", and "fatal."

To summarize, in terms of Biolaw, interventions intending to enhance the human organism in order to postpone the natural decline of aging belong to the area of biological autonomy and should not be subject to limitations. As medical acts, they are also governed by the rule of informed consent, expressed in the European law mostly by the relevant provisions of the Oviedo Convention and the EU Charter of Fundamental Rights.

8.6 Further Studying...

ECtHR: C. Goodwin v. U.K., 11 July 2002

"The applicant complained about the lack of legal recognition of her post-operative sex and about the legal status of transsexuals in the United Kingdom. She complained, in particular, about her treatment in relation to employment, social security and pensions and her inability to marry.

Although the applicant had undergone gender re-assignment surgery provided by the national health service and lived in society as a female, she remained for legal purposes a male. This had effects on her life where sex was of legal relevance, such as in the area of pensions, retirement age etc.. A serious interference with private life also arose from the conflict between social reality and law which placed the transsexuals in an anomalous position

(continued)

[20] See, in relevance, the "rights of the elderly," recognized by art. 25 of the EU Charter of Rights, which refers explicitly to "dignity," "independence," and "participation" of aged persons.

in which they could experience feelings of vulnerability, humiliation and anxiety.

The Court considered that the lack of legal recognition of the change of gender of a postoperative transsexual lay at the heart of the applicant's complaints under Article 14 of the Convention. These issues had been examined under Article 8 and resulted in the finding of a violation of that provision."[21]

Question

Comment on the following paragraphs under the light of the ECHR, and the GDPR:

90. (...), the very essence of the Convention is respect for human dignity and human freedom. Under Article 8 of the Convention in particular, where the notion of personal autonomy is an important principle underlying the interpretation of its guarantees, protection is given to the personal sphere of each individual, including the right to establish details of their identity as individual human beings (see, inter alia, Pretty v. the United Kingdom, no. 2346/02, judgment of 29 April 2002, § 62, and Mikulić v. Croatia, no. 53176/99, judgment of 7 February 2002, § 53, both to be published in ECtHR 2002-...). In the twenty first century the right of transsexuals to personal development and to physical and moral security in the full sense enjoyed by others in society cannot be regarded as a matter of controversy requiring the lapse of time to cast clearer light on the issues involved. In short, the unsatisfactory situation in which post-operative transsexuals live in an intermediate zone as not quite one gender or the other is no longer sustainable. Domestic recognition of this evaluation may be found in the report of the Interdepartmental Working Group and the Court of Appeal's judgment of Bellinger v. Bellinger (see paragraphs 50, 52–53).

91. The Court does not underestimate the difficulties posed or the important repercussions which any major change in the system will inevitably have, not only in the field of birth registration, but also in the areas of access to records, family law, affiliation, inheritance, criminal justice, employment, social security and insurance. However, as is made clear by the report of the Interdepartmental Working Group, these problems are far from insuperable, to the extent that the Working Group felt able to propose as one of the options full legal recognition of the new gender, subject to certain criteria and procedures. As Lord Justice Thorpe observed in the Bellinger case, any "spectral difficulties", particularly in the field of family law, are both manageable and acceptable if confined to the case of fully achieved and post-operative transsexuals. Nor is the Court convinced by arguments that allowing the applicant to fall under the rules applicable to women, which would also change the date of eligibility for her state pension, would cause any injustice to others in the national insurance and state pension systems as alleged by the Government. No concrete or substantial hardship or detriment to the public interest has indeed been demonstrated as likely to flow from any change to the status of transsexuals and, as regards other possible consequences, the Court considers that society may reasonably be expected to tolerate a certain inconvenience to

[21] ECtHR, Press release issued by the Registrar.

enable individuals to live in dignity and worth in accordance with the sexual identity chosen by them at great personal cost.

92. In the previous cases from the United Kingdom, this Court has since 1986 emphasised the importance of keeping the need for appropriate legal measures under review having regard to scientific and societal developments (see references at paragraph 73). Most recently in the Sheffield and Horsham case in 1998, it observed that the respondent State had not yet taken any steps to do so despite an increase in the social acceptance of the phenomenon of transsexualism and a growing recognition of the problems with which transsexuals are confronted (cited above, paragraph 60). Even though it found no violation in that case, the need to keep this area under review was expressly re-iterated. Since then, a report has been issued in April 2000 by the Interdepartmental Working Group which set out a survey of the current position of transsexuals in inter alia criminal law, family and employment matters and identified various options for reform. Nothing has effectively been done to further these proposals and in July 2001 the Court of Appeal noted that there were no plans to do so (see paragraphs 52–53). It may be observed that the only legislative reform of note, applying certain non-discrimination provisions to transsexuals, flowed from a decision of the European Court of Justice of 30 April 1996 which held that discrimination based on a change of gender was equivalent to discrimination on grounds of sex (see paragraphs 43–45 above).

93. Having regard to the above considerations, the Court finds that the respondent Government can no longer claim that the matter falls within their margin of appreciation, save as regards the appropriate means of achieving recognition of the right protected under the Convention. Since there are no significant factors of public interest to weigh against the interest of this individual applicant in obtaining legal recognition of her gender re-assignment, it reaches the conclusion that the fair balance that is inherent in the Convention now tilts decisively in favour of the applicant. There has, accordingly, been a failure to respect her right to private life in breach of Article 8 of the Convention

Provide Legally Appropriate Solutions

1. Robert B., a big pharmaceutical company's CEO, decides to undergo a serious operation on his brain for attaching an implant that enhances memory and concentration at rates double the normal ones. Colleagues of Robert B. disagree, stressing that such medical intervention is probably illegal since a) it results in an "abnormal" health condition, and b) a radical change in the CEO's cognitive abilities may entail unforeseen risks for the company.
   Robert B. argues that no legal problem occurs, on condition that appropriate information by the attending physician has been provided for potential side effects and implications related to other normal functions of his organism.
   Comment on these arguments in the light of the current European Biolaw.
2. Referring to medical doctors' scientific freedom, the plastic surgeon Sally X refuses to perform a complicated operation of external characteristics changes. The surgeon considers absolutely unreasonable the persistence of her client James F to look like a Hollywood celebrity, despite the obvious risks of that intervention already explained to James F. James F argues that, legally, Sally X needs to comply with any of her client's free will referring to the informed consent rule,

and for supporting this, he presents a psychiatric assessment relevant to his will ability.

Comment on the legal validity of these arguments.

References

Annas GJ (2000) The man on the moon, immortality, and other millennial myths: the prospects and perils of human genetic engineering. Emory Law J 49:753–782

Anomaly J, Gyngell C, Savulescu J (2020) Great minds think different: preserving cognitive diversity in an age of gene editing. Bioethics 34:81–89

Baylis F, Robert JS (2004) The inevitability of genetic enhancement technologies. Bioethics 18:1–26

Beriain IDM, Armaza EA, Sánchez AD (2019) Human germline editing is not prohibited by the Oviedo convention: an argument. Med Law Int 19:226–232

Bostrom N (2005) In defense of Posthuman dignity. Bioethics 19:202–214

Buchanan A, Brock DW, Daniels N, Wikler D (2001) From chance to choice. Genetics & justice. Cambridge University Press, Cambridge

Comité Consultatif National d' Éthique (CCNE) (1998) Rapport sur le vieillissement, Opinion No 59

Comité Consultatif National d' Éthique (CCNE) (2004) L' allotransplantation de tissu composite (ATC) au niveau de la façe (Greffe totale ou partielle d' un visage), Opinion No 82

Dunagan J, Grove J, Halbert D (2020) The Neuropolitics of brain science and its implications for human enhancement and intellectual property law. Philosophies 5. https://doi.org/10.3390/philosophies5040033

European Group on Ethics (EGE) (1999) Ethical Aspects Arising from Doping in Sport (Opinion No 14)

Faber NS, Savulescu J, Douglas T (2016) Why is cognitive enhancement deemed unacceptable? The role of fairness, deservingness and hollow achievements. Front Psychol 7. https://doi.org/10.3389/fpsyg.2016.00232

Farah MJ (2005) Neuroethics: the practical and the philosophical. Trends Cogn Sci 9:34–40

Fukuyama F (2002) Our posthuman future: consequences of the biotechnology revolution. Farrar, Strauss, and Giroux, New York

Harris J (2015) Germline manipulation and our future worlds. Am J Bioeth 15:30–34

Huxtable R, Woodley J (2005) Gaining face or losing face? Framing the debate on face transplants. Bioethics 19:505–522

Knoppers BM, Kleiderman E (2019) Heritable genome editing: who speaks for "future" children? CRISPR J 2:285–292. https://doi.org/10.1089/crispr.2019.0019

Kramer PO (1993) Listening to Prozac. Penguin Books, New York

McGee A (2020) Using the therapy and enhancement distinction in law and policy. Bioethics 34: 70–80

Mehlman M (2005) Cognition-enhancing Drugs: An Ethical and Policy Analysis, Paper presented at the 19th European Conference on Philosophy of Medicine and Health Care and 22nd EACME Conference "Ethics and Philosophy of Emerging Medical Technologies," Barcelona

Sándor J (2019). Editing human reproduction? Legal and ethical aspects of genome editing. In: Sills E, Palermo G (eds) Human embryos and preimplantation genetic technologies. Academic Press, Elsevier, pp 185–197

Savulescu J (2019) Human enhancement. In: Edmonds D (ed) Ethics and the contemporary world. Routledge, London, pp 319–334

Van de Ven K, Mulrooney KJ, McVeigh J (eds) (2020) Human enhancement drugs. Routledge, London

Wolpe PR (2002) Treatment, enhancement, and the ethics of Neurotherapeutics. Brain Cogn 50: 387–395

Self-Ownership 9

9.1 General

In the previous analysis, we presented the main elements of the current European Biolaw regarding the natural person, the "complete" human being. We defined biological autonomy, namely the legal ability to control all data and functions of our organism, as the crucial characteristic at the level of modern law that distinguishes humans from other living beings. We indicated its consequences, mainly when biological autonomy is tested under certain conditions.

Nevertheless, personal "control" does not mean just an exercise of fundamental rights to reveal, maintain, or develop the organism's natural abilities. "Control" also means the option to manage products, parts, or data of our biology when these have been isolated but still maintain a certain utility.

From a legal approach, all collected products and removed parts of the human organism are considered mere "objects" since they are not characterized by the unique coexistence of biological and cognitive elements appearing in "complete" persons. Similarly, legal "objects" are also considered biological data of humans collected and processed for various purposes. All the above means that we can handle, use, or exploit products, parts, and data removed from a human organism, as they remain useful even after the death of the person concerned. For that reason, "ownership" of these objects is legally a valid concept.[1]

The person from which products, parts, or data have been obtained is recognized as their owner. This is because their existence is originally dependent on his/her biological existence.[2] That person also could be considered as the "total" of these elements. This is the difference from other objects of ownership, the nature of which

[1] See, in detail, Arnoux (2003), pp. 149 seq. This conclusion also fits the wording of art. 17 of the EU Charter of F. Rights (right to Possessions).

[2] *Contra*, see the Moore case in the U.S., where ownership over human biological material was recognized only to the university that performed research and not to the donor of that material (Moore v. Regents of the University of California, 51 Cal. 3d 120, 1990).

T. Vidalis, *The Emergence of Biolaw*, Springer Textbooks in Law,
https://doi.org/10.1007/978-3-031-02359-0_9

is never so determinative to the owner's personhood. For that reason, parts in particular, and also some products of the human body, are considered as something more than mere objects, defined as "components of personhood" as well, which has some consequences regarding the extent of the owner's discretion, as we shall see later.

Even so, the alienation of certain products, parts, or data from the human body is lawful to the extent that it does not mean the degradation of the person concerned in terms of human dignity. Hence, the expression "ownership of ourselves" is legally valid, as a conventional determination of the relationship between a person and their biological components.[3] In other words, in terms of Biolaw, we are owners of the total biological material and data that form our natural existence. This logical confirmation is necessary for determining specific legal correlations from the moment of these components' removal.

Still, the main problem that occurs is whether ownership of removed products,[4] parts or data of the human organism makes legally possible any use or exploitation of them, or, on the contrary, their close association with the owner's personhood imposes strict limitations in relevance. We will examine that problem, in its most interesting dimension, that of the commercialization of human components, separately for each category of them.

9.2 Gametes as Products

In terms of commercialization, gametes (oocytes and sperm) are the most interesting products of the human organism. Unlike some national laws, at the level of the European law, no specific provisions exist in relevance.[5] Here, provisions of general application are those of article 22 of the Oviedo Convention and 3 para 2 c of the EU Charter of Fundamental Rights, prohibiting commercialization of the human body or parts of it. Thus, the question is whether gametes could be considered "parts" of the human body.

In principle, products of the body are not "parts" of it, insofar as they represent renewable biological material, created permanently by the human organism. This means that any loss of products does not result in potential damage to individual health. The rationale of the articles mentioned above focuses precisely on this point: The removal of a human body's "part" may cause damage to individual health;

[3]In that sense, "ownership" covers a broad spectrum of human natural properties, corresponding to the term "personality" as used by continental legal systems. On the philosophical grounds of self-ownership, see in general, Locke (1689/1965), p. 329.

[4]A reference to the most known example: Milk, as a product of the human body, was in the past a beneficial commodity for many women not able to feed their children by themselves. As "products," we mean here, in general, any material deriving from the human organism that is not necessary to maintain its structure and functions.

[5]Beyond Europe, a comparison between the (permissive) U.S. and the (prohibiting) Canadian law, see in Ikemoto (2017).

therefore, any commodification should be prohibited since it creates conditions of humans' exploitation violating directly the human dignity principle. By contrast, removing renewable material (like saliva, urine, milk, hormones, etc.) and tissue (blood, bone marrow, sperm) has no damaging effects on the human organism, therefore, the rationale of the articles we mentioned is irrelevant here.

A question still stands regarding oocytes since they are not renewable material in that sense. Oocytes are created and stored in the woman's body in a limited amount, which is continuously diminished throughout the age of fertility. Still, this constant loss is a natural process not affecting the woman's organism's normal condition unless a violent intentional removal of oocytes occurs; this is the case of intensive hormone stimulation for obtaining multiple oocytes to be available in IVF procedures. Insofar as this medical practice may conclude to severe damages to the woman's organism (including the emergence of cancer), the removal of oocytes may be equated to the removal of "parts" of the human body, and thus falls within the rationale of the relevant articles of the Oviedo Convention and the EU Charter, prohibiting their commodification.

9.3 Parts of the Human Organism: Cells, Tissues, Organs

A person is also the owner of cells, tissues, and organs after their removal from his/her organism. Transferring that ownership to others presupposes the person's previous explicit consent, based on information regarding the purpose of that biological material's use. The current European legislation, recognizing the close connection of biological material with personhood, provides specific terms for that use.[6] As a general rule, article 22 of the Oviedo Convention states that

> When in the course of an intervention any part of a human body is removed, it may be stored and used for a purpose other than that for which it was removed, only if this is done in conformity with appropriate information and consent procedures

In the case of tissues and organs, the provisions pertinent to transplantation from living donors are also applicable. As mentioned, the protection of the donor is ensured (a) with the requirement on using tissues and organs from living donors only, if no transplants from deceased persons or an alternative treatment exist, (b) with a restriction of putative donors, and (c) with the provision requiring formal consent, which can be withdrawn at any time. All these conditions need to be accomplished to transfer the biological material's ownership to the recipient.

Transfusion of blood, which is also human tissue, is legitimate if the donor's formal informed consent exists as a sole necessary condition, with no need for the other two conditions to be met. This is reasonable, given that blood is a kind of

[6] In the U.S. case-law, see in relevance Glantz et al. (2008).

renewable tissue (like sperm, also); therefore, its connection with the donor‘s personality is considered weak.[7]

A special category of tissue is that of the placenta and umbilical cord after birth. A useless material in the past nowadays provides primordial hemopoietic cells useful for transplantations. Following the above analysis, the ownership of that material belongs to the newborn; this means that parents exercise the cells' management as legal representatives. Therefore, although no specific provision in legislation exists, an apparent condition for that material's transfer and use is the parents' informed consent.

There is also the case of collection and use of human biological material for research purposes, where, again, no specific legislation at the European level exists. This case usually refers to further experimental processing of human tissue, initially collected for clinical reasons (diagnostic, mostly), not excluding the possibility of free donation of that material for research purposes.[8]

Here, any use for research purposes is allowed, if informed consent by the donor exists, specifically for each particular research purpose ("specific consent"), or at least pertaining to a concrete field of research, for covering any purpose falling within that field. This conclusion derives from the general provision of article 22 of the Oviedo Convention mentioned above if combined with article 9 para 2 (a) of the GDPR (interpreted under the light of recital 33); the latter is related to the protection of the donor's medical and genetic data, which is the real object of that kind of research. On the contrary, general consent is not sufficient to fulfill this condition since it would remain unknown to the donor any potential use of the collected material and relevant sensitive data.

It is worth noting a particular point that affects the ownership issue. The transfer of ownership to the investigator or the research sponsor does not mean that, in future research, they are free from the obligation to obtain consent from the former owner of the biological material. The latter's sensitive data are expressed in that material and can be obtained and used at any time, regardless of who the owner is, since their association with the donor's personhood cannot be canceled. Therefore, the only solution for allowing full control of the material by the new owner is its anonymization in conditions of safety, prior to the transfer, in order to completely exclude access to the donor's personal data.

---

[7]The renewable nature of the tissue is crucial. It is not identical to the artificial replacement of parts of the body, which presupposes supply of biological material from other persons on the basis of specific agreement. The natural renewal of tissue is not dependent on any contract condition whatsoever, which guarantees the protection of human dignity. In favor of commercialization of renewable components of the body, see Cherry (2000).

[8]Regarding research use of biological material from deceased persons, and its commercialization based upon the family members' consent, see Champney et al. (2019).

9.4 Biological Data from the Human Organism

Besides various sorts of biological material, all data deriving from a human organism's composition and functions may also be considered a subject of management, even of economic exploitation under certain terms. We have already examined the issue regarding the management of genetic data. Other sorts of biological data may be valuable also in employment and insurance contexts or in the field of biomedical research, as we have seen in the example of biobanking.

Biological data is considered sensitive personal data, subjected basically to the exclusive control of the individual concerned, according to the GDPR legislation. Nevertheless, that legislation does not refer to the potential exploitation of personal data by its subject, as it focuses only on the "defensive" protection of data from unauthorized interventions of other persons. Yet, there is no doubt that, as happens with other personal data categories, biological data may also be valuable for economic activities. Thus, they are not only sensitive information requiring adequate protection but also elements of ownership for their subject.

Indeed, this is an example of a non-materialized object's ownership, similar to the example of intellectual property, with the difference that, here, we do not have a "creator" or "inventor" but a mere "owner" naturally possessing this object. That difference is crucial, though, because it indicates that biological data, along with biological material deriving from a certain person, constitute parts of that person's very personhood, a fact that profoundly influences their potential management as elements of ownership.

More specifically, as we have already mentioned regarding the management of tissues, any research on biological data presupposes, in principle, the previous informed consent of the person concerned, that is, of their owner. This means that a supposed transmission of the data ownership to the investigator or the research sponsor may be observed only on the condition of data anonymity. The new owner should have no access to the identity of the old one; otherwise, all requirements regarding consent and other means of data protection (confidential processing, etc..) would remain applicable since the biological connection of these data with the person of their origin would not be affected by any ownership changes.

In any case, commercial exploitation of biological data is not excluded, at least on the GDPR's ground.[9] Anyone has the right to exploit his/her biological data, on the condition that this would not affect the due protection of them as sensitive data. Commercial exploitation (performed, for example, in the form of a contract between the person concerned and a biobank) should not be considered contrary to the human dignity principle. The latter would happen only if the data protection condition is not fulfilled, namely when data still remain identifiable after the transaction.

---

[9] According to art. 4 of the UNESCO's Human Genome Declaration: *"The human genome in its natural state shall not give rise to financial gains."* It is clear that processing of genetic data for research or clinical purposes which may involve commercial applications (diagnostic tests, gene therapy, etc..), cannot be considered as commercialization of genome "in its natural state"; this makes sense only as a prohibition referred to the commodification of the human body as a whole.

9.5 Further Studying...

California Supreme Court: Moore v. Regents of University of California, 793 P.2d 479 (Cal. 1990)
The Court held that J. Moore had no property rights over biological material extracted from his body for treatment purposes that has been used for commercial purposes also, after medical research performed without the plaintiff's informed consent.

Question

Comment on the following paragraphs, under the light of the ECHR, and the Oviedo Convention (assuming that a European court is competent for this case):

Neither the Court of Appeal's opinion, the parties' briefs, nor our research discloses a case holding that a person retains a sufficient interest in excised cells to support a cause of action for conversion. We do not find this surprising, since the laws governing such things as human tissues, transplantable organs, blood, fetuses, pituitary glands, corneal tissue, and dead bodies' deal with human biological materials as objects sui generis, regulating their disposition to achieve policy goals rather than abandoning them to the general law of personal property. It is these specialized statutes, not the law of conversion, to which courts ordinarily should and do look for guidance on the disposition of human biological materials.

Lacking direct authority for importing the law of conversion into this context, Moore relies, as did the Court of Appeal, primarily on decisions addressing privacy rights. One line of cases involves unwanted publicity. ... These opinions hold that every person has a proprietary interest in his own likeness and that unauthorized, business use of a likeness is redressible as a tort. But in neither opinion did the authoring court expressly base its holding on property law. ... Each court stated, following Prosser, that it was "pointless" to debate the proper characterization of the proprietary interest in a likeness. ... For purposes of determining whether the tort of conversion lies, however, the characterization of the right in question is far from pointless. Only property can be converted.

Not only are the wrongful-publicity cases irrelevant to the issue of conversion, but the analogy to them seriously misconceives the nature of the genetic materials and research involved in this case. Moore, adopting the analogy originally advanced by the Court of Appeal, argues that "[i]f the courts have found a sufficient proprietary interest in one's persona, how could one not have a right in one's own genetic material, something far more profoundly the essence of one's human uniqueness than a name or a face?" However, as the defendants' patent makes clear and the complaint, too, if read with an understanding of the scientific terms which it has borrowed from the patent the goal and result of defendants' efforts has been to manufacture lymphokines. Lymphokines, unlike a name or a face, have the same molecular structure in every human being and the same, important functions in every human being's immune system. Moreover, the particular genetic material which is responsible for the natural production of lymphokines, and which defendants use to manufacture lymphokines in the laboratory, is also the same in every person; it is no more unique to Moore than the number of vertebrae in the spine or the chemical formula of hemoglobin.

Another privacy case offered by analogy to support Moore's claim establishes only that patients have a right to refuse medical treatment. . . . In this context the court in Bouvia wrote that "'[e]very human being of adult years and sound mind has a right to determine what shall be done with his own body. . . .'" . . . Relying on this language to support the proposition that a patient has a continuing right to control the use of excised cells, the Court of Appeal in this case concluded that "[a] patient must have the ultimate power to control what becomes of his or her tissues. To hold otherwise would open the door to a massive invasion of human privacy and dignity in the name of medical progress." Yet one may earnestly wish to protect privacy and dignity without accepting the extremely problematic conclusion that interference with those interests amounts to a conversion of personal property. Nor is it necessary to force the round pegs of "privacy" and "dignity" into the square hole of "property" in order to protect the patient, since the fiduciary-duty and informed-consent theories protect these interests directly by requiring full disclosure.

. . .

Finally, the subject matter of the Regents' patent the patented cell line and the products derived from it cannot be Moore's property. This is because the patented cell line is both factually and legally distinct from the cells taken from Moore's body. Federal law permits the patenting of organisms that represent the product of "human ingenuity," but not naturally occurring organisms. . . . Human cell lines are patentable because "[l]ong-term adaptation and growth of human tissues and cells in culture is difficult often considered an art . . .," and the probability of success is low. . . . It is this inventive effort that patent law rewards, not the discovery of naturally occurring raw materials. Thus, Moore's allegations that he owns the cell line and the products derived from it are inconsistent with the patent, which constitutes an authoritative determination that the cell line is the product of invention. Since such allegations are nothing more than arguments or conclusions of law, they of course do not bind us. [the Court's references have been removed]

Provide Legally Appropriate Solutions

In order to ensure the supply of biological material for therapeutic purposes, a national law allows

- Selling human sperm
- Selling bone marrow
- Prioritization of organ donors and members of their families in the public list of transplant receivers

Comment on these stipulations under the light of the European Biolaw.

References

Arnoux I (2003) Les droits de l'être humain sur son corps. Presses Universitaires de Bordeaux, Bordeaux

Champney TH, Hildebrandt S, Gareth Jones D, Winkelmann A (2019) Bodies R us: ethical views on the commercialization of the dead in medical education and research. Anat Sci Educ 12:317–325. https://doi.org/10.1002/ase.1809

Cherry MJ (2000) Bodies and minds in the philosophy of medicine: organ sales and the lived body. In: Tristram Engelhardt H Jr (ed) The philosophy of medicine. Springer, Dodrecht, pp 57–78

Glantz L, Roche P, Annas GJ (2008) Rules for donations to tissue banks — what next? N Engl J Med 358:298–303

Ikemoto IC (2017) Assisted reproductive technology use among Neighbours: commercialization concerns in Canada and the United States, in the global context. In: Lemmens T, Martin AF, Milne C, Lee IB (eds) Regulating creation: the law, ethics, and policy of assisted human reproduction. University of Toronto Press, Toronto, pp 253–273

Locke J (1689/1965) Two treatises of Government. A Mentor Book, London

Part II

From Biodiversity To Intelligent Machines

10 Introduction: Biodiversity as a Legal Value

Beyond the status of persons, pertinent to our own species, the value of life[1] remains vague for modern law. In legal terms, the common characteristic of the enormous variety of living beings, from the simplest prokaryotic microorganisms to the non-human primates (the big apes), is their property as legal "objects." Contrary to human persons, microorganisms, plants, and animals have no legal capacity, that is, no rights and obligations. Their value is relative, legally acknowledged only to the extent that it may serve the human value, which is considered the only legal value *per se.*

Still, living "objects" are not treated by the law as similar to the non-living things of our world. From numerous legal provisions (particularly of international instruments), we can deduce a certain prioritization of the value acknowledged in various forms of life. Perhaps the clearest example is that of the Rio de Janeiro Convention on Biodiversity, adopted by the UN in 1992. According to Article 1:

> The objectives of this Convention, to be pursued in accordance with its relevant provisions, are the conservation of biological diversity, the sustainable use of its components and the fair and equitable sharing of the benefits arising out of the utilization of genetic resources . . .

Furthermore, Article 2 states in relevance:

> For the purposes of this Convention:
>
> 'Biological diversity' means the variability among living organisms from all sources including, inter alia, terrestrial, marine and other aquatic ecosystems and the ecological complexes of which they are part; this includes diversity within species, between species and of ecosystems.

[1] The definition of life presents difficulties for life sciences. See Tsokolov (2009), pp. 401 seq., Jagers op Akkerhuis (2010), pp. 245 seq. In general, however, a condition of life presupposes at least a single cellular structure. In that case, we can refer to "organisms," a generic concept that includes microbes, plants, and animals.

T. Vidalis, *The Emergence of Biolaw*, Springer Textbooks in Law,
https://doi.org/10.1007/978-3-031-02359-0_10

The Rio Convention remains the most important binding international legal instrument that ensures a status of general legal protection for non-human species. Its provisions support a conclusion according to which living objects are legally significant since they constitute the concept of "biodiversity," which is recognized as a special legal value.[2] Several other binding instruments of the international law and the EU legislation[3] describe the specific status of protection for concrete species also. This common status of protection pertains to the total of non-human species in their immense variety, not to particular species. In that sense, the special value of "biodiversity" requires the conservation of all existing or future living organisms.

"Conservation of biodiversity" does not mean, of course, that we are able to prevent the natural process of individual species' extinction or the appearance of new ones. That process corresponds to the natural norms of Evolution, beyond any human control. We know, thus, that biodiversity changes continuously from the dawn of life under these norms, and that process includes the human species, as well.[4] Still, ethical reasons oblige contemporary democratic societies to care and take adequate measures for conserving the current conditions of biodiversity, to the extent that these are dependent upon human activities also.[5]

These ethical reasons constitute precisely the basis for accepting biodiversity not only as a "fact," an objective condition of Nature, but also as a distinct value that the modern law adopts. We will examine the two most important of these reasons, as expressed in modern theory, and in environmental policies.

10.1 Serving Humans

10.1.1 The Existing Generations

The dominant justification of the non-human species' value focuses on their utility for humans. According to that approach, we should address the worth of all life forms as dependent on the only value *per se*, the human value.[6]

---

[2]In soft law, see art. 17 of the UNESCO's Bioethics Declaration.

[3]Cf. Jans and Veder (2008), pp. 451 seq. Legal provisions related to liability on environmental harms are a significant part of that legislation and relevant to the conservation of biodiversity (ibid., pp. 340 seq.). On the determination of crimes against the natural environment ("green criminology"), cf. White (2008). On the international law of biodiversity, cf. Louka (2006), pp. 288 seq.

[4]Regarding significant changes in biodiversity that occurred at the geological time scale, see Raup (1988), pp. 51 seq. Nature, certainly, has no "respect" for values or rights related to biodiversity; see Schrader – Frechette (1998), p. 21. For the current state of biodiversity, cf. Mora et al. (2011).

[5]On the economic aspects of biodiversity as a common good, cf. Perrings (2006), pp. 137 seq., Balmford et al. (2002), pp. 950 seq. On protective policies, that also include the establishment of gene banks and seed banks for the conservation of rare species, cf. Louka (2002), pp. 33 seq., 93 seq. At the EU level, see Ingrassia et al. (2005), pp. 40 seq.

[6]On moral duties for the protection of Nature, see Jonas (1984). In general, regarding approaches on Society/Nature relationships, see Benton (1994), pp. 28 seq.

However, this justification does not exclude differentiation in value between species. For instance, the priority of animals (of vertebrate animals, particularly) is generally accepted, based mainly on emotions we develop for them, on their morphological resemblances with our species, and even on some distant kinship we believe that biologically exists. However, all animals share an inferior ethical status, remaining at the service of human interests.

In that scheme, the human person enjoys a top position in a specific hierarchy of values, no matter if a biological connection with other species (particularly with animals) has been scientifically confirmed already from Darwin's research. This hierarchical position of human persons is exclusive; no other species could be eligible for that, and no person could lose it for any reason.[7]

A consequence of this approach is that, since we accept the relative value of any non-human species, we consider ethically valid the instrumentalization of all other species for serving the interests of humans.[8] Therefore, the value of non-human species (and of their individual members) is determined in accordance with functions useful for us in a variety of circumstances, and it is never considered a value *per se*. Historically, the law always accepted the hierarchical supremacy of human value; the legal recognition of the non-human species' relative value is only a recent step of modern law, practically meaning measures for their legal protection.

A more cohesive approach suggests that biodiversity conservation serves human life by supporting vital functions of existing persons—the present generations.

Life sciences have shown that the quality of the atmosphere, the soil, and subsoil, the marine environment, as well as the equilibrium of the food chain, are directly dependent upon the existence and behavior of non-human species. Hence, the human organism's essential functions would be impossible without the involvement of other living beings. Not only our nutrition comprises many animal and plant species, but our respiration is also related to the exchange of oxygen, and carbon dioxide in plant species, our digestion and the general condition of health are dependent on the existence of beneficial microorganisms in our body, to mention just a few characteristic examples.[9]

This close relationship of our species with other living organisms indicates that their value is not an accidental but a necessary component of the human value, suggesting that a serious harm to biodiversity may violate the latter, as well. In that sense, the value of non-human species, although relative, has an inviolable core in terms of ethics and law.

---

[7] Although genomic research blurs that distinction guiding to legal challenges. On a relevant discussion, see Knoppers and Greely (2019), pp. 1455–1457.

[8] On the deterministic view of human dominance over Nature, see Benton, supra, pp. 31 seq., Palsson (1996), pp. 67 seq.

[9] See in relevance, O'Hara and Shanahan (2006), pp. 688 seq., Petrova et al. (2015). Also, Van Der Werf et al. (2009), pp. 737–738.

10.1.2 Responsibility Towards Future Generations

In principle, non-living species serve the present human generations of existing persons. Yet, there are also opinions according to which the use of natural resources (including other species) meets a moral limit that we can describe as our responsibility towards future human generations.[10] Consequently, the protection of biodiversity is not founded on the needs and interests of individual persons, living in concrete time and site; it should be understood in reference to its eternal usefulness for the human species. The concept of "sustainability" reveals, in fact, that moral limit, which obliges us to moderate our needs for ensuring long-lasting support for our close or distant descendants' needs.

In that rationale, again, the value of biodiversity has as ultimate moral ground the humans' support, even if not associated with concrete, existing persons, but to the humankind as a whole. However, that approach accepts a strong status of the non-human species' legal value as well; this value justifies restrictions on present generations' fundamental rights, particularly regarding economic freedom and property. Therefore, responsibility towards future generations attracts the attention of modern law, even if it poses certain questions requiring answers.[11]

First, there is a question on terminology. Some make use of the term "rights of future generations"[12] expanding, thus, the concept of the "subject of right" to cover non-existing persons as well. We are familiar with a similar enlargement of that concept, in the case of embryo/fetus "rights," where a "complete" human person does not exist yet (even if there is a potential for that). The concept of "rights" for future generations faces a similar problem.[13]

One can argue that the use of the term "right" is valid indeed, since the individual persons of future generations exist already and live among us; these are our children and grandchildren. But children and grandchildren certainly enjoy rights as existing, real persons, not as future representatives of the human species. The question focuses on distant generations, the members of which are not even conceived. For them, is it valid to use that terminology?

The problem, here, is not that we need to presuppose a real ability for exercising rights, to recognize someone as a subject of them (therefore, to presuppose the actual existence of that person).[14] We know that many existing persons are not able to exercise several fundamental rights (due to their age, mental condition, etc.), but still, to the law, they remain "subjects" of these. The real problem with that enlargement of the concept of "right" is that we risk losing its connection with a certain

---

[10] See art. 16 of the UNESCO's Bioethics Declaration.

[11] Cf. Belaidi (2008), pp. 139, 142 seq. For the constitutional recognition, see Gosseries (2008), p. 448. A sceptical approach, due to the limited legal elaboration of the "inter-generational equity", see in Dellaux (2021).

[12] Cf. Chong (2006), pp. 103 seq.

[13] See Gosseries (2008), pp. 450 seq.

[14] See on that issue, criticism in Partridge (1990).

individual. Indeed, to the law, a necessary component of that concept is precisely that concrete connection, not the reference to an uncertain entity (like the future generations, or the human species, or the humankind as a whole); and this is so, because we need to ensure the practical worth of rights, their suitability for pursuing individual aims. A right is nothing but a legal tool for practically serving its subject's interests by producing substantial effects for third persons; otherwise, it is anything but a meaningless rhetoric scheme.

Under this view, the term "responsibility" of the present generations towards the future human beings is more accurate because it is associated with existing persons; it determines, also, a standard for legitimate restrictions on their rights in order to ensure the sustainability of resources for the sake of those who will come after us in the planet.

The other question in relevance refers just to this last conclusion: Is it correct to assume that such a general reason for restricting fundamental rights of real persons, on the basis of future generations' interests, really exists? The answer is positive if we approach biodiversity as a necessary condition for the existence of human life as a natural phenomenon. Legally this is confirmed indirectly through the concept of "sustainable development," which requires our permanent care to conserve Nature's components. Indeed, "sustainability" would be meaningless if the legislator did not imply specific care for future generations.

In that sense, the conservation of biodiversity stands as a general limit for any activity of present persons, even for activities corresponding to the exercise of fundamental rights, such as economic freedom or property rights. In other words, its significance for the life of humans cannot be balanced with other values or interests deriving from these rights, but it always prevails over them. This reference to the concept of "sustainable development," explicitly recognized by the European law,[15] constitutes the basis for understanding our responsibility towards future generations not only in ethical but in legal terms also.

Still, we need an objective standard to determine the extent of that legal limitation on fundamental rights. To what extent is it legitimate to "sacrifice" the needs, the desires, the initiatives, or even the pleasures of present generations, for the sake of the future ones? The answer is difficult since a general definition of that standard seems impossible, given the variety of conditions that may emerge in a certain society. Nevertheless, we can assume at least two elements that need to coexist for justifying restrictions.

First, the legal concept of "biodiversity" should be well defined and analyzed. The Rio de Janeiro Convention provides a clear definition, but there are still open questions. Do we violate biodiversity if we destroy individual plants or animals, or if we extinct a whole population of them in a restricted area? As we will see in the following chapters, the current law provides useful guidance in relevance.

Second, the concept of "sustainability" involves necessarily a comparison of needs between the present and future generations. Future generations may have

[15] See art. 37 of the EU Charter of F. Rights.

needs that we can assume with certainty and other needs, unforeseen at present. To restrict the individual rights of present persons just for avoiding hypothetical threats relevant to unforeseen needs seems outside any valid legal rationale. On the contrary, if we rely on objective data predicting with accuracy future conditions (for example, a trend for increasing population concluding to insufficient means of nutrition, or a gradual shortage of natural resources due to the climate change[16]), a specific limitation of economic rights for the sake of future generations is legally valid.

Apparently, a legal safeguard for any accepted limitation of rights complying with the above prerequisites is the specific documentation of state measures in relevance at any level of regulation. However, the important issue here is that our responsibility towards future generations has a legal dimension, suggesting that any unlimited wastefulness of natural resources, including the life of non-human species, in the name of fundamental rights, remains out of the question, given the existing positive law in Europe.

10.2 The Inner Value of Species

Opposite to the above "speciesist" approach is the "deep ecology" standpoint that defends the inner value of all species.[17] Its roots can be found back in the nineteenth century's philosophical contributions and activist movements, but it has been developed particularly since the 1960s.[18]

According to its basic position, the value of all species is a value *per se*[19] and not a mere reflection of human value. Regardless of their classification, any animal or plant represents a distinct and complete moral entity, equal to the human moral entity. We must, therefore, consider non-human species, in legal terms, not as "objects" but as "subjects" of law enjoying real "rights" as those recognized for human individuals.[20]

[16] Cf. Heller and Zavaleta (2009), Araujo et al. (2011). The issue is relevant particularly to global warming; see studies on that in Emil Morhardt (ed.) (2014). Also: Liberatore (1994), pp. 190 seq. Adam (1994), pp. 101 seq., and Turner et al. (2009), pp. 278–279. On the global issue of water management, cf. Primentel et al. (1998), pp. 104 seq., esp. 119 seq.

[17] See Schrader – Frechette, supra, pp. 16 seq., Benton, supra, pp. 38 seq. Against a dualistic approach Human/Nature, Palsson, supra, pp. 72 seq.

[18] See the classic argumentation by Stone (1974).

[19] A value per se for the environment in general (not limited to living species) is also associated with conventional notions of common interest (like that of "ordre public"), not necessarily defended by deep ecology. See Belaidi, supra, pp. 463 seq.

[20] An approach associated with utilitarianism. See the classic Bentham (1789/2005), pp. 282–283. Nevertheless, H. Jonas also uses concepts like "dignity," "interests," and "rights" of Nature (supra, pp. 136 seq.). Martha Nussbaum (2007) adopts an opposed to utilitarianism approach, insisting on social behavior and cognitive abilities of animals, on which a value per se is grounded. Beyond disagreements between defendants of utilitarianism, animal rights, and human-oriented trends, see the argumentation of Degrazia (1996). For a criticism of ownership over animals, see Waintraub

Moreover, for some supporters of "deep ecology," that approach is even independent of life as a natural quality and must be extended to non-living natural beings also (mountains, rivers, etc..).[21] A similar legal approach defends the attribution of rights to non-living entities or, in other words, their recognition as "legal persons.".[22] In some countries, there is already legislation adopting this approach even at the constitutional level. Ecuador and Bolivia recognize, thus, the rights of Nature ("Pacha Mama") or "Mother Earth" in their constitutions, and a famous law in N. Zealand attributed the status of a legal "person" to a river (Wanganui river), which is considered sacred by the local Maori populations.[23]

An argument for justifying the recognition of non-living entities as legal persons could be that, traditionally, the legal systems accept that, not only human "natural" persons, but also fictional entities (like private companies or associations and municipalities or even the state itself, as well as, international organizations and NGOs) have legal personality, and they are recognized as holders of rights for their own interests. Nevertheless, this argument disregards the fact that these legal "persons" exist only for serving the interests and respective rights of natural persons, the humans. Although selfish interests of the latter may conflict with the collective interests represented by entities as legal persons, we cannot imagine a status of them independently of natural persons; fictional entities with legal personality are nothing but mere means for pursuing collective goals in a certain society. Therefore, we cannot equate the strong status of a natural person with that of a fictional legal person that, in reality, does not have self-interests as such; indeed, the "interests" and "rights" of all existing fictional entities are destined to serve natural persons as mere instruments, with no value *per se*.

To associate the concept of "rights" with interests *per se* of a certain individual seems, thus, presupposing a biological basis that enables that individual, at least, to experience their necessity, through its sentient or cognitive abilities.[24] Otherwise, the absence of that biological minimum instrumentalizes by default a non-living legal person and its rights for the sake of living holders of rights.

Nevertheless, this remark does not mean that, to attribute legal personality and recognize as holders of rights natural non-living objects or entities (or even the Nature as a whole), similarly to what we already adopt for fictional entities, would not strengthen the environmental protection in terms of law. There is no doubt that

(2009), and Francione (2008), pp. 25 seq., where any use (including humane ones) of animals as objects is rejected.

[21] The most known approach to Earth as an integrated living organism is the "Gaia" hypothesis by J. Lovelock (1979/2000).

[22] See in detail, Shelton (2015). From a legal point of view, see also the L. H. Tribe's characteristic suggestion on the enlargement of the ethical concern, in order to encompass even non-living objects [Tribe (1974)]. Contra, cf. Sagoff (1974), p. 205. The idea of a new Amendment to the U.-S. Constitution for recognizing "wildlife rights" was expressed by Favre (1978), pp. 279 seq.

[23] See Shelton, supra, para 24, 25, 31. More examples see in Chapron et al. (2019).

[24] Shelton, supra, para 23.

this would represent an advanced legal status for natural objects or entities, which moreover is based upon a known and functional standard of all legal systems.

We can summarize the political program of "deep ecology"as an appeal for the species' or Nature's "liberation." In some works, that claim is considered the next step of a series of historical liberations that occurred within the human species itself (in particular, of women and people of color), with the difference that it will be the first referring to other species or natural beings.[25] We address, then, an essential reaction against the distinction between human/non-human species and the alleged absolute supremacy of the human value (speciesism),[26] also denying the distinction between "subject"/"object" in ethics and law.

A strong element of this attitude is that it corresponds perfectly to the Evolution Theory by translating the findings of the latter into ethical and legal assumptions. Indeed, we cannot justify the distinct position and, moreover, the moral supremacy of the human species by the evidence we have from the species' Evolution. Relevant data indicate that all life forms share a common kinship from a distant "ancestor" (the pre-DNA macromolecules produced by coincidental chemical reactions in the archaic era of young Earth), and they started a course of gradual differentiation throughout millions of centuries, following their adaptation to continuously changed environmental conditions. Findings of modern Genetics confirm that conclusion nowadays, as they reveal extensive similarities between all species' genomes, including the human genome, leading to almost identical genetic profiles for species belonging to similar biological taxonomy (for example, among mammals, even more among primates, where humans and great apes belong).[27]

The fact that observations in Nature demonstrate that the existing genotypic or phenotypic differences between species are only quantitative ones, with no biological "gaps", leads us to make similar considerations at the level both of ethics and law, rejecting absolute distinctions in terms of species' value, moreover an image of a sole specie's hierarchical supremacy. Under this view, the human value is not associated with the status of a "subject" meant to be served by the other living "objects" of inferior value. In other words, if we assume that absolute value distinctions among species do exist, and one of them should be treated as "supreme," we accept an ethical maxim not supported by the reality of life as a natural phenomenon; this inconsistency weakens, eventually, the argumentation of speciesism.[28]

---

[25] Supporters' activism presents either moderate or radical forms, using peaceful or violent means. See, in general, Yearley (1994), pp. 150 seq.

[26] Cf. the classic criticism by Singer (1975/2009)], Ch. 5, 6.

[27] Cf. Chen and Li (2001), pp. 444 seq.

[28] According to Social Anthropology findings, some communities do not distinguish between animals and humans in terms of values and ethics (the case of Chewong tribe in Malaysia). Cf. Howell (1996), pp. 127 seq. An argumentation in favor of "justice" towards animals, inspired by the human-oriented respect of "diversity" and "discrimination" prohibitions, is presented by Bryant (2007), pp. 207 seq.

On the other hand, if we abolish the distinction between subject/objects that supports the moral priority of humans, we accept an unprecedented subversion of a normative order which functions regularly already from the era of the Agricultural Revolution at the dawn of the Neolithic.[29]

Indeed, since humans acquired specific knowledge for controlling the life of other species in order to serve basic needs (establishing permanent agricultural production and residence), they considered themselves a dominant species. In the beginning, unconsciously, intentionally at a later time (following the development of religion), and then wholly rationally, with the progress of philosophical reasoning and the attempts to justify an authentic moral authority,[30] the human species lives in societies with this particular normative determination. What could replace that as a reliable alternative? Could be, perhaps, a new plan of coexistence with species considered morally equal,[31] knowing though, that humans will need to "step back" consciously from the exploitation of other living beings, even if they did so throughout millennia? A realistic prospect for that remains distant, and practically with no representation in contemporary legal systems; the modern law is still based upon the first moral reason defending the non-human species' protection, which presupposes logically the human dominance under certain normative conditions over all the other living beings.

10.3 The Legal Approach

The adoption of the human-oriented position by modern legal systems has two important consequences that determine the relevant regulation.

10.3.1 A Selective Protection of Species

The first of these indicates that, by separating humans from the other living beings, the law accepts an approach based on a discriminative image of biodiversity, an image that refuses to understand biological similarities and kinships between species, believing rather in the existence of ontological "gaps" in their classification.

Not only humans but other species as well are separated from others in that view. However, non-human species share a common status of equal treatment, according to their general classification. In that sense, as we will see later on, microorganisms are treated legally under a common status, as well as the vast majority of plant species, with the exemption of rare plants. Among animal species, all non-vertebrate animals, representing the largest category, share a common status also, with very few exemptions (rare species, cephalopods, etc.). In contrast, we can observe

[29] Cf. Barker (2009), esp. pp. 42 seq.

[30] With the Kantian approach as the most characteristic, here. Cf. Kant (1790/1999), 459 seq.

[31] Singer, supra, Ch. 1.

differentiations among vertebrate animals, following various criteria (the size, the rareness, or the utility for humans). A special position is attributed to non-human primates (the big apes), the status of which is distinct from that of the other mammals or vertebrates in general.

There are two interesting elements relevant to that discriminative regulation.

First, distinctions express a hierarchical protection of species considered unequal between them, in terms of value. Within that classification, we find microorganisms at the lowest level and vertebrate animals at the superior one.

Second, this selective perception of species weakens in a sense the legal value of biodiversity as a whole, which in principle is not consistent with differentiation between them. Indeed, the fact that plants or non-vertebrate animals do not enjoy the same degree of legal protection as vertebrates (due to the lack of specific legislation providing concrete guarantees), makes a large part of biodiversity more vulnerable to risks of intentional harm and even species' extinction. Since the very concept of biodiversity is structured on the basis of species' interdependence (as expressed through the food chain etc.), this unequal treatment could undermine its legal significance. To that problem, current legislation has provided some solutions, as we will see later.

10.3.2 A Relative Protection of Non-human Species

The second consequence of the human-oriented approach is relativity regarding the protection of the other species, Compared to humans. This characteristic is expressed in two directions.

First, differentiations in the non-human species' protection are justified, usually in reference to humans and their interests. The relatively strong status of protection for mammals is associated with the fulfillment they offer concerning various human needs (feeding, working, entertaining, etc..) and with a sense of "closeness" with us that yields emotions of love, admiration, confidence, sorrow, pleasure, etc. In contrast, our indifferent attitude regarding species like the plants, the non-vertebrates, and even the vast majority of vertebrate animals, is due to our impression that they do not support our interests or needs; therefore, the law has no reason to recognize a similar to mammals' legal value. In brief, the discriminative legal perception of non-human species is directly linked to the superiority of human value, requiring differences in their treatment, according to the nature of human interests and needs to be served.

On the other hand, the relativity of the species' protection emerges in situations of conflict with human goods (life, health, property, etc..), when the value of non-human beings is taken into account, but not as equal to the value of these goods. Thus, the law accepts even the extinction of animals, no matter if they enjoy a comparatively strong status of protection, if human health, or property are at stake, with no implications for the humans involved. Moreover, the law accepts violations of the animals' corporal integrity or restrictions to their moving in similar conditions (especially in the case of wild animals in captivity).

Relativity in the non-human species' protection inevitably creates a general normative condition permitting considerable discretion for humans when handling other living beings. This condition entails, though, normative uncertainty about the real significance of protective policies and the efficiency of relevant legal provisions, as the protection of other living beings is always dependent on specific human interests, activities, and needs. In the following chapters, we attempt to approach the law in Europe, under the light of fundamental principles that guarantee the value of biodiversity and its components, to respond to that problem of normative uncertainty.

Chapter 11 focuses on legal rules intending for the conservation of the existing biodiversity or, in other words, for the protection of the main categories of species, namely microorganisms, plants, and animals.

In Chap. 12, we examine the correlation of biodiversity with the species' reproduction, which is affected either by priorities set in economic development or by a persistent lack of knowledge regarding potential implications of biotechnological products.

The last Chap. 13 is dedicated to life's commercialization; some aspects of that problem regarding the human body's components have been discussed already in Part I.

10.4 Further Studying...

ECtHR: PETA Deutschland v. Germany, 8 November 2012

"The case concerned a civil injunction which prevented the animal rights organisation PETA from publishing a poster campaign featuring photos of concentration camp inmates along with pictures of animals kept in mass stocks. The Court held in particular that a reference to the Holocaust had to be seen in the specific context of the German past. In that light, the Court accepted that the German courts had given relevant and sufficient reasons for granting the civil injunction".[32]

Question

Comment on the following paragraphs under the light of the ECHR, the CBD, and the EU Charter of Fundamental Rights:

[32] ECtHR, Press release.

46. The Court recalls that the domestic authorities have a variety of tools at their disposal allowing them to strike a fair balance between the various interests at stake. In assessing the proportionality of the measure at issue they have, beyond the complete prohibition or authorisation of the expression of an opinion, in particular the option of setting specific limits to the authorisation or to the prohibition. Moreover, freedom of expression is applicable to not only "information" or "ideas" that are favourably received or regarded as inoffensive or as a matter of indifference, but also to those that offend, shock or disturb. As set forth in Article 10, this freedom is subject to exceptions, which must, however, be construed strictly, and the need for any restrictions must be established convincingly (see Axel Springer AG v. Germany [GC], no. 39954/08, § 78, 7 February 2012, with further references). Furthermore, there is little scope under Article 10 § 2 of the Convention for restrictions on political speech or on debate on matters of public interest (see, among other authorities, Ceylan v. Turkey [GC], no. 23556/94, § 33, ECT HR 1999-IV; Verein gegen Tierfabriken Schweiz (VgT) v. Switzerland (no. 2) [GC], no. 32772/02, § 92, ECT HR 2009 and Mouvement raëlien v. Switzerland [GC], no. 16354/06, § 61, 13 July 2012).

47. Turning to the circumstances of the instant case, the Court observes, at the outset, that the applicant association's intended poster campaign concerned battery animal-farming. Accordingly, as it related to animal and environmental protection, it was undeniably in the public interest (Compare Verein gegen Tierfabriken Schweiz, ibid.). It follows that only weighty reasons can justify the interference with the applicant's right to freedom of expression in this context. The Court further observes that the domestic courts adjudicating the applicant's case carefully examined whether the issue of the requested civil injunction would violate the applicant association's right to freedom of expression. In doing so, the domestic courts applied the standards developed by the Court as set out above. They expressly accepted that the impugned representations constituted expressions of opinion and where thus protected under the right to freedom of expression. They further acknowledged that this right protected expressions even if they were formulated in a polemic or offensive way (Compare paragraph 10, above) and that the impugned posters related to questions of public interest, as they were obviously intended to criticise the conditions under which animals were kept (Compare paragraph 11, above).

48. The Court further observes that the domestic courts considered that the intended poster campaign did not pursue the aim to debase the depicted concentration camp inmates, as the pictures merely implied that the suffering inflicted upon the depicted humans and animals was equal. The domestic courts considered, however, that the applicant association confronted the plaintiffs with their suffering and their fate of persecution in the interest of animal protection. It was this "instrumentalisation" of the plaintiffs' suffering that violated their personality rights in their capacity as Jews living in Germany and as survivors of the Holocaust. This violation was aggravated by the fact that the depicted Holocaust victims were shown in a most vulnerable state. Having regard to the seriousness of this violation, the courts considered that the applicant association's interests in publishing the impugned pictures had to cede. While expressing its doubts as to whether the intended campaign violated the human dignity of either the depicted persons or the plaintiffs, the Federal Constitutional Court endorsed the lower courts' assessment that the campaign banalised the fate of the Holocaust victims and that the violation of the plaintiffs' personal honour was particularly serious.

49. The Court considers that the facts of this case cannot be detached from the historical and social context in which the expression of opinion takes place (Compare Hoffer and Annen, cited above, § 48 and Rekvényi v. Hungary [GC], no. 25390/94, §§ 46 et seq., ECT HR 1999-III). It observes that a reference to the Holocaust must also be seen in the specific context of the German past (see Hoffer and Annen, ibid.) and respects the Government's

stance that they deem themselves under a special obligation towards the Jews living in Germany (Compare paragraph 36, above). In the light of this, the Court considers that the domestic courts gave relevant and sufficient reasons for granting the civil injunction against the publication of the posters. This is not called into question by the fact that courts in other jurisdictions might address similar issues in a different way (also Compare Müller v. Switzerland, 24 May 1988, § 36, Series A no. 133).

50. The Court further recalls that the nature and severity of any sanction imposed are also factors to be taken into account when assessing the proportionality of the interference (see, among other authorities, Ceylan, cited above, § 37 and Annen II v. Germany (dec.), nos. 2373/07 and 2396/07, 30 March 2010). Turning to the circumstances of the instant case, the Court notes that the proceedings at issue did not concern any criminal sanctions, but a civil injunction preventing the applicant association from publishing seven specified posters. The Court finally observes that the applicant has not established that it did not have other means at their disposal of drawing public attention to the issue of animal protection.

51. Having regard to the foregoing considerations and, in particular, to the careful examination of the case by the domestic courts, the Court accepts that the civil injunctions issued against the applicant association were a proportionate means to protect the plaintiffs' personality rights.

There has accordingly been no violation of Article 10 of the Convention.

References

Adam B (1994) Running out of time: global crisis and human engagement. In: Redclift M, Benton T (eds) Social theory and the global environment. Routledge, London

Araujo M.B., Alagador D., Cabeza M., Nogues–Bravo D., Thuiller W. (2011), Climate change threatens European conservation areas, Ecol Lett 14: 484–492

Balmford A, Bruner A, Cooper P, Costanza R, Farber S, Green RE, Jenkins M, Jefferiss P, Jessamy V, Madden J, Munro K, Myers N, Naeem S, Paavola J, Rayment M, Rosendo S, Roughgarden J, Trumper K, Turner RK (2002) Economic reasons for conserving wild nature. Science 297:950–953

Barker G (2009) The agricultural revolution in prehistory. Why did foragers become farmers? Oxford University Press, Oxford

Belaidi N (2008) La lutte contre les atteintes globales a l' environnement: vers un ordre public écologique. Bruylant, Bruxelles

Bentham J (1789/2005) An introduction to the principles of morals and legislation. Oxford University Press, Oxford

Benton T (1994) Biology and social theory in the environmental debate. In: Redclift M, Benton T (eds) Social theory and the global environment. Routledge, London

Bryant TT (2007) Similarity of difference as a basis for justice: must animals be like humans to be legally protected from humans? Law Contemp Probl 70:207–254

Chapron G, Epstein Y, López-Bao JV (2019) A rights revolution for nature. Science 363:1392–1393

Chen F-C, Li W-H (2001) Genomic divergences between humans and other hominoids and the effective population size of the common ancestor of humans and chimpanzees. Am J Hum Genet 68:444–456. https://www.ncbi.nlm.nih.gov/pmc/articles/PMC1235277

Chong C (2006) Restoring the rights of future generations. Int J Green Econ 1:103–120

Degrazia D (1996) Taking animals seriously: mental life and moral status. Cambridge University Press, Cambridge

Dellaux J (2021) Future generations. In: Türkelli GE, De Moerloose S (eds) De Feyter K. Edward Elgar Publishing, Encyclopedia of law and development, pp 91–94
Emil Morhardt J (ed) (2014) Consequences of global climate change. Claremont Acadmic Press, Claermont California
Favre DF (1978) Wildlife rights: the ever-widening circle. Environ Law 9:241–281
Francione GL (2008) Animals as persons. Essays on the abolition of animal exploitation. Columbia University Press, New York
Gosseries A (2008) On future generations' future rights. J Polit Philos 16:446–474
Heller NE, Zavaleta ES (2009) Biodiversity management in the face of climate change: a review of 22 years of recommendations. Biol Conserv 142:14–32
Howell S (1996) Nature in culture or culture in nature? Chewong ideas of "humans" and other species. In: Descola Ph, Palsson G (eds) Nature, and society. Anthropological perspectives. Routledge, London
Ingrassia A, Manzella D, Martyniuk E (2005) The legal framework for the management of animal genetic resources, FAO, Rome, in MON-079894.pdf
Jagers op Akkerhuis G. A. J. M. (2010) Towards a hierarchical definition of life, the organism and death. Found Sci 15:245–262
Jans JH, Veder HHB (2008) European environmental law, 3rd edn. Europa Law Publ, Groningen
Jonas H (1984) The imperative of responsibility. In: Search of an ethics for the technological age. University of Chicago Press, Chicago
Kant I (1790/1999), Duties Towards Animals (from Lectures on Ethics). In Kuhse H, Singer P (eds) Bioethics. An anthology. Blackwell, Oxford, p 459
Knoppers BM, Greely HT (2019) Biotechnologies nibbling at the legal "human". Science 366: 1455–1457
Liberatore A (1994) Facing global warming. The interactions between science and policy-making in the European community. In: Redclift M, Benton T (eds) Social theory and the global environment. Routledge, London
Louka E (2002) Biodiversity & human rights. The international rules for the protection of biodiversity. Transnational Publishers, New York
Louka E (2006) International environmental law. Fairness, effectiveness and world order. Cambridge University Press, Cambridge
Lovelock J (1979/2000) Gaia: A new look at life on earth, 3rd edn. Oxford University Press, Oxford
Mora C, Tittensor DP, Adl S, Simpson AGB, Worm B (2011) How many species are there on Earth and in the ocean? PLoS Biology 9, http://journals.plos.org/plosbiology/article?id=10.1371/journal.pbio.1001127
Nussbaum M (2007) Facing animal complexity. Harvard Review of Philosophy Lecture. https://bordeure.files.wordpress.com/2008/09/facing_animals-nussbaum.pdf
O'Hara AM, Shanahan F (2006) The gut flora as a forgotten organ. EMBO Rep 7:688–693
Palsson G (1996) Human – environmental relations. Orientalism, paternalism and communalism. In: Descola P, Palsson G (eds) Nature, and society. Anthropological perspectives. Routledge, London
Partridge E (1990) On the rights of future generations, Upstream/downstream: issues in environmental ethics 40:56–58, and http://gadfly.igc.org/papers/orfg.htm
Perrings C (2006) Biodiversity as a local public good. In: Basili M, Franzini M, Vercelli A (eds) Environment, inequality, and collective action. Routledge, New York, pp 131–147
Petrova MI, Lievens E, Malik S, Imholz N, Lebeer S (2015) Lactobacillus species as biomarkers and agents that can promote various aspects of vaginal health. Front Physiol 6. https://doi.org/10.3389/fphys.2015.00081
Primentel D, Houser J, Preiss E, White O, Fang H, Mesnick L, Barksy T, Tariche S, Schreck J, Alpert S (1998) Water resources: agriculture, the environment and ethics. In: Lemons J, Westra L, Goodland R (eds) Sustainability and integrity: concepts and approaches. Springer, Dordrecht, pp 104–134

Raup DM (1988) Diversity crises in the geological past. In: Wilson EO (ed) Biodiversity. National Academy Press, Washington DC, pp 51–57

Sagoff M (1974) On preserving the natural environment. Yale Law J 84:205–267

Schrader – Frechette K. (1998) Sustainability and environmental ethics. In: Lemons J, Westra L, Goodland R (eds) Sustainability and integrity: concepts and approaches. Springer, Dordrecht, pp 16–30

Shelton D (2015) Nature as a legal person. VertigO 22. https://doi.org/10.4000/vertigo.16188

Singer P (1975/2009) Animal liberation. Harper Collins Publ., New York

Stone CD (1974), Should trees have standing? Toward legal rights for natural objects. W. Kaufmann, Los Altos, California

Tribe LH (1974) Ways not to think about plastic trees: new foundations for environmental law. Yale Law J 83:1315–1348

Tsokolov SA (2009) Why is the definition of life so elusive? Epistemological considerations. Astrobiology 9:401–412

Turner WR, Oppenheimer M, Wilcove DS (2009) A force to fight global warming. Nature 462: 278–279

Van Der Werf GR, Morton DC, DeFries RS, Olivier JGJ, Kasibhatla PS, Jackson RB, Collatz GJ, Randerson JT (2009) CO2 emissions from forest loss. Nat Geosci 2:737–738

Waintraub E (2009) An analysis of the legal classification of animals: toward a step-wise deconstruction of the property status of animals. J Philos Sci Law 9:1–60

White R (2008) Crimes against nature: environmental criminology and ecological justice. Willan Publ, Devon

Yearley S (1994) Social movements, and environmental change. In: Redclift M, Benton T (eds) Social theory and the global environment. Routledge, London

The Conservation of Species 11

11.1 Microorganisms

Microorganisms represent the vast majority of non-human species. Most of them are unicellular, invisible to the naked eye, and are characterized by astonishing diversity. We can classify them from the simplest prokaryotic, with no cell nucleus (i.e., bacteria), to the eukaryotic microorganisms (i.e., fungi, algae, protozoa) having a cellular structure with a distinct nucleus, like in superior species. A separate category is that of viruses, not considered living organisms, but reproduced themselves as parasites by exploiting other organisms' genetic material.

Microorganisms contribute substantially to many natural processes affecting the individual human organism and even our social interactions. Precisely as such, they attract the interest of law. We can identify some major areas of relevance.

(a) The area of health is mostly structured around protection from pathogen microbes, causing diseases. Protection is organized with the management of environmental factors to avoid the spreading of such microbes, involving regulations, and restrictions on human activities. Similarly, the management of the environment is necessary for protecting plant and animal health, where the presence of pathogen microbes is also a serious concern.
(b) The field of nutrition is directly associated with the adequate exploitation of microorganisms' properties, either for food production (i.e., through fermentation) or for avoiding food alteration that may harm the consumers' health.
(c) The area of environmental protection, especially for coping with pollution and ecological harm, focuses nowadays on the use of microorganisms functioning as biocatalysts for cleaning.[1] Besides, technologies of "active microorganisms"

[1] Cf. Cappitelli and Sorlini (2008), pp. 564 seq, and studies presented in Diaz (2008).

T. Vidalis, *The Emergence of Biolaw*, Springer Textbooks in Law,
https://doi.org/10.1007/978-3-031-02359-0_11

mixtures are now used as biological fertilizers, instead of chemical compounds, and attract the interest of scientists and farmers.[2]

(d) Finally, in the field of energy, we have the exploitation of microorganisms, as a renewable source, based on their biochemical properties. For that reason, we produce biofuels in environmental conditions suitable for the development of such beneficial species.[3]

Negative or positive, this significant involvement of microorganisms in vital human interests implies the necessity of legal regulation. However, the presence of microorganisms is, to the law, almost exclusively, associated with potential risks that may occur to the natural environment and human health.

Given that, the essential instruments of the European Biolaw regulating the microorganisms' management are those of the EU legislation.

First, Directive 2000/54/EC governs the protection of workers from exposure to biological agents at the workplace. This Directive has codified previous legislation of the 1990s and determines the basic directions for addressing infections from harmful microorganisms. Its basic elements are the classification of these directions in four categories, according to the seriousness of potential risks for human health, and the provision of employers' obligations for taking preventive measures in relevance.[4]

More specifically, that legislation contains several basic requirements for safety at the workplace, which are:

- Establishing adequate risk assessment
- Deciding on control measures
- Ensuring information relevant to health and safety
- Providing adequate training to employees
- Establishing a system of registries to enable control
- Implementing measures of good working practice

It is worth noting that the Directive's provisions do not refer only to protection from a random presence of pathogen microbes but also to safety measures in cases of intentional presence of these (for example, an intentional infection of laboratory animals for experimental purposes).[5] According to that legislation, management of pathogen microorganisms means not only avoidance or extinction, but also conservation of microbes, under certain conditions of control, in the framework of a research project.

Here, a question emerges regarding the extent of that discretion to the intentional use of pathogen microorganisms. The freedom of Research is not sufficient to justify

[2]Cf. Yamada and Xu (2001), pp. 255 seq.

[3]Cf. Christie (2007), pp. 294 seq., Rittmann et al. (2008), pp. 604 seq.

[4]See art. 2.

[5]See art. 16 para 1.

absolute discretion but needs to be balanced with the social right to Health of the EU Charter of Fundamental Rights (art. 35), and the European Social Charter (art. 11). That necessity requires measures for avoiding contaminations, as already provided by the Directive (especially in art. 6).

The Directive does not require cessation of relevant activity in cases of risk uncertainty. In general, its provisions refer to detectable risks for designing appropriate measures related to the employees' safety, as described in its annexes.

Transferring pathogen microorganisms under conditions of safety is also of legislative interest. Regarding the use of such microorganisms for research purposes, the regulation of their transfer refers mostly to scientists working in microbiological laboratories.

In the EU legislation, this matter is governed by Regulation 3381/1994, concerning the export of "dual-use" products from the EU, meaning products that may be used for both peaceful and non-peaceful purposes. According to the rationale of this Regulation, any such export is subject to a system of preventive control by national authorities, presupposing a relevant notification on the specific product by the exporter and a subsequent export license.[6] Procedures of cooperation between national controlling authorities, as well as information and data registration procedures, are also included in that Regulation.[7]

Regarding microorganisms, the same law focuses on the danger from the potential development of biological weapons, a matter of specific interest for international law, as well. Indeed, the Geneva Protocol (1925), generated from the experience of World War I, and especially the UN Biological Weapons Convention (BWC, 1972), are here the basic international instruments with binding force.[8] They include an absolute prohibition of activities aiming "indirectly or directly" at developing biological and toxic weapons. However, this provision cannot prevent transferring of "dual-use" products since these are not destined exclusively for non-peaceful purposes.

In the food sector, Regulation 882/2004 organizes protection from pathogen microorganisms, requiring controls to food products for detecting the potential presence of microbes.[9] In addition, Directive 98/83 refers to quality controls on drinkable water. Controls are in place to identify concentrations of bacteria, e-coli, etc., beyond a certain level of safety.[10] EU legislation also exists regarding remediation, especially of drinkable and marine water, requiring decontamination after detecting pathogen microorganisms. Directive 98/83 is important, here, as relevant to drinkable water.

---

[6] See art. 6 seq.

[7] See art. 13 seq.

[8] Cf. Dando (2013), pp. 185 seq., Fidler and Gostin (2008), pp. 187 seq. On the anthrax incidents in the U.S., after the 9/11's terrorist attacks, see Cole (2009).

[9] Cf. European Commission (2006).

[10] Safety is also an issue that we need to address in relation to the presence of beneficial microorganisms (probiotics, etc..). Cf. Rijkers et al. (2011), pp. 1291 seq.

Microbiological control is also provided for the quality of animal feed, according to Regulation 183/2005.[11] This legislation includes measures for the good operation of companies acting in the field, regarding their facilities, equipment, quality controls, etc.., in order to ensure feed safety and traceability of their ingredients.

Considering the above provisions, a general conclusion we can draw is that, in the case of microorganisms, the law does not express particular interest in protecting concrete species. This is so, even if microorganisms cover the main part of the existing biodiversity and support the conditions and functions of living of all other organisms, including humans. In that sense, we can say that the life of microorganisms is placed at the lowest level of legal evaluation; it is protected only in a few cases and never as such, but in relation to direct benefits that we humans expect from its conservation.

11.2 Plants

The law illustrates the status of plants as superior Compared to the status of microorganisms since, in plants, we distinguish species deserving special protection. Although that protection is not generated by a value *per se*, as related (directly or indirectly) to certain benefits for humans, here we have for the first time a clear expression of biodiversity as a legal value.

Indeed, the legal significance of certain plant species expresses the rationale of their conservation, as inalienable components of the existing biodiversity; a human intervention concluding to the limitation or even the extinction of these species inevitably violates biodiversity as a legal value and is subjected to specific legal responsibility backed with concrete sanctions. In that sense, any such human intervention is of legal interest and cannot be equated with the natural extinction of plant species throughout the Evolutionary process, a phenomenon with no legal relevance.

In particular, legal responsibility is reflecting the fact that, in principle, humans can control the conservation of species, acting as guardians of their protection. Therefore, human activities that may affect a certain species must be guided by the necessity to protect it as a component of the whole biodiversity. As we have seen, a similar normative approach is not applied to microorganisms since, there, the conservation of biodiversity remains legally indifferent.[12] On the other hand, the same legal interest regarding biodiversity exists in animals, as we will see in the relevant paragraph.

For both plants and animals, we accept an additional aspect of protection; we care for the protection of ecological systems, that is, of larger or smaller concentrations of

[11] See in relevance, also, Regulation 1520/2007.

[12] In a futuristic scenario, assuming that we would be in place to develop an artificial alternative for replacing the functions of beneficial for humans microorganisms, a potential extinction of these would not be legally significant for the protection of biodiversity.

individuals from certain species, along with non-living natural components, forming a "microenvironment" that deserves special legal acknowledgment.[13]

Usually, the protection of such ecosystems (like forests, for instance) is directly associated with biodiversity (like in article 2 of the CBD). As a natural condition, biodiversity is promoted in ecosystems suitable for the emergence and development of life. Many rare species of plants are flourishing almost exclusively in these "microenvironments." However, this is a matter of specific natural conditions, and we need to avoid confusion with the value of biodiversity as such. This is clear in the example of forests, which are explicitly protected with specific legal provisions as entire ecosystems, although isolated trees belonging to the same forest species and existing in urban areas do not share the same protection. What matters, then, is the concentration of individuals from certain plant species in an entity defined as a "forest" and not isolated individuals outside of that entity, or random concentrations not defined as "forests". This special entity is essential, as yielding concrete benefits for human life (release of oxygen, absorption of carbon dioxide, exploitation of forest products, etc..), in which the conservation of biodiversity is indeed observed but not exclusively. It is, then, more accurate to address the protection of ecosystems as one of the necessary conditions for the conservation of biodiversity and not to equate it with the latter.

We also need to distinguish the value of biodiversity from the protection of cultivations of plant species serving a variety of our needs, from food and clothing to the production of energy.[14] Farming is also a human activity that contributes to species conservation but does not focus on that purpose. On the contrary, we know that significant disasters in ecosystems occurred in industrial agriculture involving intensive exploitation of the soil, extensive use of chemical compounds, and oriented management of natural sources of water, also influence negatively the conditions of local biodiversity of plant and animal species. For Biolaw then, all laws regulating agriculture are of interest only to the extent that their provisions may affect the conservation of biodiversity directly.

In conclusion, there are two main categories of legal rules concerning plant species that belong to the object of Biolaw due to their pertinence to the legal value of biodiversity: (a) rules for the conservation of particular plant species or plant individuals, (b) rules for the conservation of ecosystems suitable for plant flourishing. The former rules aim to protect the biodiversity value directly, while the latter serve that value indirectly, by ensuring special environmental conditions for concrete flora species.

---

[13] For the concept of ecosystems, where plants are considered as energy and food providers and animals as consumers, see Moran et al. (1986), p. 30.

[14] Or, even cleaning of the environment. See Meagher (2000), pp. 153 seq.

11.2.1 Conservation of Concrete Plant Species or Plant Individuals

Legal rules belonging to the first category are:

(i) The UN Convention on Biological Diversity (CBD -Rio de Janeiro, 1992), already mentioned
(ii) The international Convention on International Trade in Endangered Species of Wild Fauna and Flora (CITES – Washington, 1973)
(iii) The international *Convention* on the Conservation of European Wildlife and Natural Habitats (Bern, 1979)
(iv) Regulation 338/1997, incorporating the CITES instrument into the EU legal order
(v) Directives 92/43/EC and 97/62/EC on the conservation of natural habitats and of wild fauna and flora
(vi) Finally, the International Plant Protection Convention (FAO 1951, 1997), an instrument relevant to specific measures of plant protection from pathogen organisms, especially in the field of international trading transports

An integrated approach to these texts leads us to the following remarks:

I. The legislation adopts the already presented content of biodiversity provided by the CBD in relation to future generations' interests. Biodiversity is also considered genetic diversity (diversity of genes), where genetically modified organisms become an issue. Special attention is paid to "rare," "wild", and "endangered" species.
II. The law protects concrete species of flora, corresponding to the scientific classification of "species," "subspecies," and "populations" (existing in a specific location).[15] Plant species of special interest for the EU law are the "endangered," the "vulnerable" (potentially endangered in the future), the "rare" (potentially vulnerable in the future), and the "endemic" (dependent on the protection of their habitat). The responsibility of the EU competent organs focuses particularly on endangered species (art 1, Directive 92/43)
III. The third level of protection refers to plant individuals. The concept of "specimen" is crucial in this respect, meaning any individual plant dead or alive, as well as any part or derivative of it that may be plainly determined. (Directive 92/43, art. 1 m). The CITES instrument and Regulation 338/1997 make specific reference to the protection of specimens meant for trading purposes, in the sense that, if protection is lacking at the level of plant individuals, the entire species could be considered endangered. Annexes to these instruments indicate concrete species of flora in relevance. According to Regulation 338/1997, in terms of protection, is sufficient for the plant to be derived from one "parent" belonging to the annexed list of species; therefore, hybrid plants are also included (article 2 t)

[15] See art. 2 s of Reg. 338/1997.

It is worth mentioning that the relevant annexes correspond to specific classified protection of plant species. The essential standard for attributing the highest degree of protection is the emerging danger from trading certain rare species (article 3 para 1, Regul.) A lower degree of protection is recognized for indigenous species threatened by the introduction and cultivation of non-indigenous live specimens (article 3 para 2 d, Regul.)

Control in imports or exports of protected plant species by national authorities includes the suitability of space for the conservation of specimens, measures for avoiding contamination from diseases, the evaluation of imported species' impact on indigenous plants, etc. (article 4).[16] Licensing from the state of export (granted according to its national legislation) is a crucial presupposition in relevance, as a measure of preventive control. The Regulation provides relatively soft measures for transfers of specimens among scientists, or scientific institutes, in order to encourage research (article 7 para 4). A lower degree of protection is also provided for artificially reproduced specimens since practically no extinction risk occurs. (article 7 para 1 b).

Two main principles characterize both the Regulation and the CITES instrument: (a) a principle recognizing the possibility of more strict protection by national legislation,[17] (b) a requirement for an evidence-based register of species in relevant annexes for justifying protection.

Finally, the Regulation mentions a specific obligation for mutual information between competent national organs and the European Commission (article 15).

The most significant conclusion we can reach from the above presentation, is that the law establishes classified protection, first at the level of the entire biodiversity, second, at the level of concrete species, and third, at the elementary level of specimens (individual plants or parts of them). On that basis, we can indicate the standards of protection:

(a) The central value is that of biodiversity as a whole; the values of species and specimens are considered only as secondary ones. Contrary to the legal consideration of humans, where the value of each and every individual/person is the crucial one, the legal consideration of plants focuses on individual plants (attributing relative protection), only if a danger for the concrete species' extinction occurs, entailing a direct violation of biodiversity. This means that the extinction of a plant specimen remains legally indifferent insofar as no danger exists for the species.
(b) Apart from the case in which destruction of specimens or risks for species are directly harmful to the entire biodiversity, the values of species and specimens are considered relative and need to be balanced with other legal values potentially involved. For example, the protection of public health or the freedom of research and education are values that usually prevail over the conservation of

[16] Cf. Mooney (1999), pp. 665 seq., Levine et al. (2003), pp. 775 seq.

[17] See recit. 3 of the Regulation's preamble.

plant specimens and may justify even extensive destruction of plant concentrations.

11.2.2 Conservation of Ecosystems

Furthermore, the current law provides indirect protection of plants, relevant to the conservation of their natural environment (a location where they live and reproduce), namely the natural ecosystems.

Some of the legal instruments already mentioned include such provisions. In addition, essential instruments in relevance are the UN Convention on Wetlands (Ramsar, 1971), referring to a specific type of ecosystems, and Directive 92/43/EC on the *conservation of natural habitats and of wild fauna and flora* (Directive "Natura 2000").[18] Moreover, following certain classifications, national laws in Europe provide more comprehensive protective measures for the conservation of land, marine, and mixed ecosystems (national parks, protected landscapes, etc..).

The conservation of natural ecosystems is associated with the protection of plants, either because a danger to plant biodiversity occurs (extinction of species) or because there are benefits for human communities (aesthetic benefits, in particular). However, the management of these areas must take into account other values and legitimate interests (economic growth, housing, agriculture, etc..) that may be prioritized in concrete conditions.[19] The general principle, again, attributes the highest protection to the conservation of plant biodiversity as a whole; if biodiversity remains safe, interventions in particular plant species are allowed (excepting "important" species that enjoy enhanced protection), and if a particular species is not threatened, no protection of individual plants of it is required.

The Directive establishes a network of protected ecosystems across Europe and describes national obligations for their conservation, including a system of information for ensuring the overall protection of these areas. Specific types of ecosystems are mentioned in its annexes, as concrete measures need to be taken, pertinent to each type.

Still, protection is only relative. This is confirmed by the provision of Article 4 para 2, permitting a flexible implementation of the Directive's measures, in agreement with the European Commission, if protected ecosystems represent more than 5% of the national territory. Moreover, exceptions for reasons of "overriding" public interest (including economic or social interests) are accepted, on the condition that the member state will take compensatory measures to ensure the overall

---

[18]On protected ecosystems at a global scale, cf. IUCN & UNEP–WCMC (2011). On crucial ecosystems that sustain global biodiversity ("hotspots"), cf. Mittermeier et al. (2011), pp. 3 seq. See, also, the determination of protected lands' percentages, as a threshold for the conservation of global biodiversity ("Aichi Biodiversity Targets"), in Larsen et al. (2015), pp. 74 seq.

[19]However, no "acquis communautaire" is accepted in favor of the environment. See ECJ, C-57/89 [Commission v. Germany, 1991, I-883), which required stricter measures for protecting biodiversity, than those established by Directive 92/43 afterward.

protection of the network (article 6 para 4). Yet, such measures cannot prevent damages to local biodiversity and the potential extinction of plant species in a particular site.

Forests also represent natural ecosystems of special interest, although their protection is ensured mostly by national legislation. At the EU level, relevant policies are developed as well, aiming at sustainable forest management, and promoting the multifunctional role of forests regarding the conservation of biodiversity, the mitigation of climate change, and the exploitation of natural forest products.

The general regulatory framework, here, is mostly that of the "Natura 2000" Directive (92/43/EC), making detailed reference to forests as *"natural habitat types of Community interest whose conservation requires the designation of special areas of conservation"* (Annex I, 9), and of Regulation 1293/2013 ("LIFE"), which establishes a "Programme for the Environment and Climate Action," also related to the "Natura 2000" objectives.

In the past, a special Regulation was enacted (2152/2003) regarding a monitoring system of "forests and environmental interactions in the Community". Its explicit purposes referred to *"monitoring of air pollution and air pollution effects and of other agents and factors that have an impact on forests, such as biotic and abiotic factors and factors of anthropogenic origin, monitoring of forest fires and their causes and effects, forest fire-prevention"; "monitoring of soils, carbon sequestration, climate change effects and biodiversity, as well as protective functions of forests"; "monitoring activities in the assessment of the condition of forests and the further development of monitoring activity"* (art. 1). This Regulation provided definitions about protected territories (art. 2), which include "forests" and "other wooded lands" (following the characteristics of trees (height), and the percentage of the covered area). This program remained in force for a limited period of 3 years, which could be prolonged after the Commission's assessment (art. 18, 19). It is now repealed by Regulation 1293/2013 mentioned above.

11.3 Animals

11.3.1 The Main Distinction

Being closer to the status of "persons", animals represent a category with normative variations in the law's ontology. However, we can notice the main distinction characterizing their legal status. On the one hand, there are animal species hardly different from plants in their natural existence. These are classified biologically in a grey zone between the plant and the animal realm. Besides, we can observe the enormous category of invertebrate animals (insects, worms, arthropods, etc..), representing the vast majority of animal species.

All species belonging to the above categories are addressed under the same legal view, which is identical to the legal approach of plants, prioritizing the general value of biodiversity. Only species under threat of extinction are explicitly protected by the existing legislation, precisely because a danger exists for the entire biodiversity.

Legal provisions that describe the status of animal protection are those already mentioned regarding plants, particularly provisions related to rare species and vulnerable ecosystems.

On the other hand, the legal approach to vertebrate animal species, a category that includes all species of birds, fishes, reptiles, amphibians, and mammals, is completely different. The current European law provides a status of special protection for them, acknowledging thus a separate category of superior animal species, in terms of their comparative legal value, representing the minority of the existing fauna.

Apart from the general legislation pertaining to all animal species, there are specific legal instruments of protection relevant to various classifications of vertebrate animals, usually following their living conditions. Therefore, there are instruments concerning wild animals, pet animals, farm animals, and laboratory animals. Primates other than humans, that is, the great apes, are ethically considered a special category of vertebrates, deserving a higher degree of protection, very similar to the status of humans.[20] This is also expressed by the current European legislation, particularly regarding the use of primates in experimental procedures.[21]

The sharp distinction that the law adopts between invertebrate and vertebrate animals reflects a fundamental ethical preference for living beings phenotypically closer to the human species. The advanced legal protection of vertebrate animals can be explained under this view, if focusing on their developed neural system. This makes them capable of experiencing sentiments of pain, discomfort, pleasure, etc.., when responding to environmental stimuli and exercising essential cognitive functions, like memory. It is true that there are also invertebrate animals with advanced neurophysiologic properties (like the cephalopods[22]), but these represent only limited exceptions.

It is worth mentioning that this distinction does not correspond to a specific Evolutionary process driving from "incomplete" to "complete" living beings. A vast number of vertebrate animal species, including numerous superior mammals, had been extinct for many million years, long before the appearance of non-vertebrates living in our days. Indeed, the term "Evolution" refers only to the process of life forms' adaptation to various environmental conditions, and it is not related to an "advanced" nature of a particular living organism or to any phenotypic "proximity" to the human species; a dinosaur, a mammoth, and moreover the "Homo erectus" (a standing ape, considered our direct ancestor), although closer biologically to our species than invertebrates of our time, belong to the past of Evolution. Starting from

[20]On the advanced abilities of non-human primates, see Tomasello et al. (2003), pp. 153 seq. Restrictions on using these animals in research procedures have been adopted in the U.K., the Netherlands, Sweden, Germany, Austria, N. Zealand, etc. Cf. Taylor (2001), pp. 35 seq. Following that trend, in the U.S., see the report by the Institute of Medicine of the National Academies (2011), which has been adopted by the NIH. On relevant reservations, though, see European Commission/ Scientific Steering Committee (2002).

[21]Cf. Seelig (2007), pp. 21 seq. From an activist view: Laboratory Primate Advocacy Group (2005).

[22]Cf. Vitti (2013), pp. 393 seq.

the first pre-DNA macro-molecules, Nature yields continuously simple or more complex living organisms, following the sole criterion of their conformity with changing environmental conditions. If we evaluate some species as more important than others by attributing higher legal protection, this has nothing to do with their "success" as natural "models", but only reflects our attitude of natural "proximity" to them, according to a purely human-based range of values.

11.3.2 Protection of Invertebrate Animals

Normative differences between the vast category of invertebrates and that of plant species are not clear. There are several criteria for classifying a living organism either to flora or to fauna, such as the ability of movement, the process of nutrition (photosynthesis in plants), the cellular structure, the sensory properties, the reproductive properties (sexual, mostly, in animals), etc.

However, for the law, problems in classification are not crucial since invertebrate animals are legally treated only as parts of the whole animal biodiversity, the protection of which is prioritized, similarly to what happens with plants and the flora biodiversity. Therefore, from the legal aspect, individual invertebrates are indifferent if the species they belong to remains safe, and interventions at the species' level are lawful if they do not jeopardize the animal biodiversity. Like in plants, there are some exceptions to this rule for species considered as "important."

On that basis, for invertebrate animals, the already mentioned legislation is also relevant, namely:

(i) The Rio de Janeiro Convention on Biological Diversity (CBD)
(ii) The CITES instrument and the relevant Regulation 338/1997 of the EU
(iii) The international *Convention* on the Conservation of European Wildlife and Natural Habitats
(iv) Directives 92/43, 97/62, relevant also to the protection of local fauna.[23]

General provisions included in these legal instruments concerning the protection of biodiversity also apply to invertebrate animals, similar to the example of plant species. In addition, there are some specific provisions pertaining to animals, namely:

[23] Cf. ECJ, C-103/00 [Commission v. Greece, 2002, I-1147 (violation of Caretta-Caretta turtles' habitat], ECJ C-378/01 [Commission v. Italy, 2003, I-2857 (free zone of bird protection)]. Also, ECJ C-67/99 [Commission v. Ireland, 2001, I-5757, on the common protection of Natura areas)].

(i) Certain clauses regarding avoidance of "injury," "damage to health," "cruel treatment"[24] and "captivity"[25] pertain to individual animals. Of course, these clauses refer to vertebrates, but they are also relevant to some invertebrate species having an elementary neurophysiologic structure. In that sense, they reflect a basic legal interest even for individual invertebrates.
(ii) Provisions regarding the transport of individual animals in cases of traveling zoos, circuses, menageries, plant exhibitions, or other traveling exhibitions also pertain mostly to vertebrates, but they may apply to invertebrate species, as well.[26]
(iii) More robust protection for individuals or populations of both vertebrate and invertebrate wild animals is legally provided, prohibiting capture, killing, and disturbance "during the period of breeding, rearing, hibernation, and migration".[27]

11.3.3 Protection of Vertebrate Animals

In vertebrate animals, we can notice a legal interest not only in the value of biodiversity as a whole, or that of certain species but in individual animals, as well. The known model of protection pertaining to non-human living beings (indifference for individuals, if the concrete species is not threatened—accepted interventions in a particular species if biodiversity remains safe) changes in the case of vertebrates; individual vertebrate animals are legally significant also.

We encounter, thus, a situation very similar to the legal status of the human species. If the human value, as a basis of all legal rights and a structural element of any modern legal system, pays key attention to individual humans, it sounds logical to imagine a similar value for vertebrates. Since their individuals also attract the interest of law, perhaps we should presume the existence of normative consequences similar to human "interests," even if we remain reluctant to use this terminology for animals.

Research in current legislation for vertebrate animals supports this hypothesis, yet the image is much more complicated than the human interests/rights correlation. Indeed, although we can identify common properties or needs existing in all vertebrate species (from fishes and reptiles to the closest to humans species of primates), i. e. the need for corporal integrity with no suffering and pain, the need for housing and nest, the sexual reproduction, and the care for their offspring, the law does not recognize the same level of protection for all individual vertebrates,

[24] See CITES art. III para 2 (c), para 4 (b), art. IV para 2 (c), para 4 (b), para 6 (b) etc.

[25] See Reg. 338/97, art. 4 para 6 c. Artificial breeding under captivity conditions is encouraged as a measure of species conservation; therefore, some deviations from strict protection are also observed. See art. 7 para 1 ibid.

[26] CITES art. VII para 7.

[27] Directive 92/43, art. 12 para 1 b.

regarding these needs, as it does for human individuals; since no animal is recognized as a subject of rights, the extent of protection is always determined *in concreto*, as dependent on the prioritization of relevant human interests.[28]

Nevertheless, the European law accepts the concept of "animal welfare" as an expression of the protective framework pertaining especially to vertebrate animals, where individuals matter. Animal welfare is considered a general standard for limitations on the human handling of animals, acknowledging special needs of them, and mostly

- The protection of life and corporal integrity[29]
- The adequate feeding
- The adequate housing

Additional needs are also acknowledged in relation to vertebrates' concrete species, as we will see later. At present, we can highlight certain aspects of animal welfare explicitly mentioned in the European law. These are included in two relevant legal instruments, the Council of Europe's Convention of 1968 and (at the EU level) Regulation 1/2005/EC, concerning the protection of animals during transport.[30] Following their provisions, special attention is paid to the animal welfare, particularly regarding veterinary controls of animals to be transported,[31] the prohibition of transports before and after delivery, the necessary resting of animals during transports, the availability of appropriate space and cages, the necessary cleanness of spaces, the regular feeding and hydration of animals, and even their euthanasia under certain conditions (for avoiding suffering from injuries, diseases, etc..). All these measures are characteristic provisions on welfare, generally required, either for big or for small vertebrates, regardless of their planned use in transport destinations. Nevertheless, animal welfare is not the prevailing standard regarding potential purposes served by the use of animals. Human needs may impose limitations by exposing animals to conditions of risk for their life or health. The use of animals for human pleasure, involving economic profit, like public exhibitions, circus performances, animal races, etc.., is a very characteristic example. Since animals are legally considered as simple objects of handling, there is a large margin for the legislator to justify their use in reference to utilitarian purposes. Given that, in principle, animal welfare remains essential as a legal standard, at least, for evaluating methods pertinent to concrete uses of animals. However, national legislations often move one step beyond by prohibiting specific uses as unacceptable under the view of animal welfare.

---

[28] See Kant, *supra*.

[29] However, there are arguments in favor of the animal life's value, even if animals are destined for specific uses presupposing their killing. See Orlans et al. (1998), p. 233.

[30] Cf. European Parliament (2001), with comments on the previous legislative regime in the EU (Directives 91/628, 95/29, etc..).

[31] See on that issue, Vapnek and Chapman (2010).

11.3.4 Specific Categories of Vertebrate Animals

11.3.4.1 Wild Animals

Wild vertebrates enjoy the highest legal protection of their vital needs. These animals live and move freely in natural environments, a fact that the law recognizes and protects clearly, but not in an absolute sense.[32]

First, protection of life is declared explicitly in several legal instruments,[33] prohibiting, for instance, wild animals' hunting. Nevertheless, exceptions of lawful hunting also exist, on certain conditions about the period, means of hunting, etc.., indicating the relative nature of protection.[34]

Second, the law recognizes the protection of wild animals' corporal integrity, especially when their transport is allowed, and injuries must be avoided. Other legal provisions refer to prohibitions of maltreatment, injuries, damage to health, or cruel treatment.[35]

Third, free movement is also a basic need of interest for wild animals.[36] Still, that need is restricted for different reasons,[37] such as public health, security (like free movement of wild birds that may impede aviation), agricultural and forest economy, professional fishing, scientific research, and even hunting involving animal chasing. A stronger restriction is certainly the lawful captivity of wild animals in zoos or exhibitions. Nevertheless, the protection of free movement is associated with strict limitations and even prohibition of wild animals' trading.[38]

Fourth, free nest is also protected with special provisions ensuring the natural habitats of wild animals. The Natura legislation and the Ramsar Convention represent, here, the most characteristic legal frameworks. Care for the nest, including feeding, overwintering, etc.., may also mean the creation of new habitats[39] (like in

[32] On wild animals as part of the whole Nature (against an approach defending their individualization in protection), see Swart (2005), pp. 251 seq.

[33] See, art. 6 para 1 of Directive 79/409 on wild birds, Dir. 92/43, art. 12 para 1 a.

[34] See. Dir. 79/409, άρθ. 7, 8. Also, in relevance, ECJ C-435/92, [Association pour la protection des animaux sauvages and others v. Prefet de Maine-et-Loire et Prefet de la Loire-Atlantique, I-67 (determining an objective disturbance of wild animals, due to hunting)]. On the problem of wild animals' consumption that causes zoonotic diseases (like in the example of COVID-19), raising problems of food safety and public health, see Ruishi et al. (2021).

[35] See CITES art. III para 2 (c), para 4 (b), art. IV para 2 (c), para 4 (b), art. V para 2 (b) para 4 (b), art. VII para 7 (c), art. VIII para 3, Regul. 338/97, art. 4 para 1 f, art. 5 para 2 c. A comparison with pet animals, see in Spark (2014), pp. 473 seq.

[36] Cf., Chambers (2015), pp. 7 seq.

[37] See, Dir. 79/409, άρθ. 9. Also, ECJ C-3/96 [Commission v. the Netherlands, 1998, I-3031 (appraisal of economic costs for protecting wild birds, following the Directive 79/409). On "annoying" animals, see Prieur (1996), pp. 340–341.

[38] CITES (passim), Regul. 338/97 (passim), Dir. 92/43, art. 12 para 2. Regarding imports of alien species, see Prieur, *supra,* pp. 282–283. On illicit trading of wild animals, see Nellemann et al. (2014), pp. 23 seq. and table in p. 36. A criticism on the effectiveness of the CITES provisions in relevance, see in Lubelska-Sazanów (2018).

[39] Dir. 79/409, art. 3.

the case of migratory birds' stationing). The natural environment of habitats is also protected indirectly, with provisions prohibiting the import of non-indigenous species,[40] possibly provoking threats to the natural survival of wild animal populations.[41]

Fifth, the protection of wild animal calmness is also a vital need that current legislation recognizes, with provisions for avoiding disturbance of living, particularly during sensitive periods of reproduction, overwintering, etc.. Intentional annoyance (by filming, photographing, etc.. animals) is also included here.[42]

Sixth, the law pays special attention to the reproduction of wild animals. Conservation of nests, the prohibition of egg collection, avoidance of disturbance, as well as prohibitions of hunting during the period of reproduction and raising animal offspring are objectives of explicit protection.[43] Furthermore, the law allows the artificial reproduction of wild animals living in exceptional conditions of captivity.[44]

Finally, regarding these exceptional cases of legitimate detention of wild animals for entertainment purposes, the law recognizes certain guarantees ensuring the animal welfare, namely preventive control measures, like requirements for licensing of zoos, responsible personnel, etc.., backed with relevant sanctions in case of violation. That legislation also includes measures for animal health protection, like a requirement for necessary veterinary care.[45]

11.3.4.2 Pet Animals

The Council of Europe's Convention for the Protection of Pet Animals (Strasbourg 1987) provides a definition, according to which: *"By pet animal is meant any animal kept or intended to be kept by man in particular in his household for private enjoyment and companionship"*. This definition underlines the two characteristic elements for that category of animals: (a) their housing within the limits of concrete private space, and (b) their use for satisfying personal feelings of the owner.

Living in permanent captivity conditions, pet animals have specific needs that the owner should cover. The Convention recognizes, thus, the need for providing "suitable and sufficient food and water," for protecting health,[46] for suitable housing, for training on adequate conditions and avoiding suffering and pain, for protecting reproduction in healthy conditions, and mostly for ensuring the corporal integrity of pet animals. In this regard, surgical operations for aesthetic reasons are explicitly

---

[40] See Dir. 79/409, art. 11. See also an analysis of biodiversity conservation against dangers from the import of alien species (including policies on local biodiversity's restoration) in Shine et al. (2000).

[41] Regarding the influence on biodiversity of restrictions to wild animals' habitats, see Reed (2004), pp. 181 seq., Fischer and Lindenmayer (2007), and Krauss et al. (2010), pp. 597 seq.

[42] See Prieur, *supra*, pp. 281–282.

[43] See Dir. 79/409. 7 παρ. 4. A very interesting analysis about influences occurring on wild species' demographic data, due to practices of selective hunting by humans, see in Milner et al. (2007), pp. 36 seq.

[44] See Dir. 92/43, art. 14 para 2 and Regul. 338/97, art. 7.

[45] See Reg. 338/97, art. 9 para 3.

[46] Cf. Puvača and Chantal (2020).

prohibited. Euthanasia is considered the ultimate means in order to release the animal from a painful survival, or even to protect public health if performed according to a specific methodology ensuring avoidance of pain and suffering (previous anesthesia, etc..) (art. 13).

The rationale of law focuses on appropriate control measures, namely living of pet animals under certain restrictions and continuous supervising of their behavior. Specific provisions of the Convention refer to the management of stray animals, under a permanent concern for avoiding suffering and relevant health implications. National legislations are encouraged to set up measures for the identification and registration of stray animals in order to reduce unplanned breeding and to involve citizens in their management (by reporting animals, they may find, to competent authorities). This approach indicates that the condition of stray animals' "freedom" should not be considered as ensuring their welfare, as long as the animal remains uncontrolled; nevertheless, "control" does not necessarily mean conditions of captivity, if stray animals are identifiable and registered.

11.3.4.3 Farm Animals

The Council of Europe's Conventions on Protection of Animals Kept for Farming Purposes (Strasbourg, 1976) and on Protection of Animals for Slaughter (Strasbourg, 1979) govern the status of farm animals.

The first instrument refers to "*animals bred or kept for the production of food, wool, skin or fur or for other farming purposes,*" particularly in facilities using intensive stock-farming systems, involving technical installations and automatic methods (art. 1). Here, the law intends to conciliate the natural needs of animals with the necessary limitations pertaining to intensive farming. Under this framework, special attention is paid to avoiding "unnecessary suffering or injury" (art. 4, 6, 7). The conditions of housing, feeding, free movement, etc.. should conform with the concrete species and its "physiological and ethological needs" (art. 5), providing also continuous veterinary care and control for the good operation of farming installations (art. 7).[47]

The Convention's on slaughtering stipulations are similar, referring to "domestic solipeds, ruminants, pigs, rabbits and poultry" for meat production. Delivery of animals to slaughterhouses, housing, and the conditions of slaughter are regulated in detail. Regarding delivery, the law pays attention to preventing injuries and maltreatment (i.e., prohibitions of striking on or pressure applied to any sensitive part of the body). Also, "animals' tails shall not be crushed, twisted or broken, and their eyes shall not be grasped. Blows and kicks shall not be inflicted" (art. 5).[48]

---

[47] Similar provisions regarding the animal welfare (pertaining not only to farm animals) are included in the two binding instruments on animal transport, that is, the Council of Europe's *Convention* for the Protection *of Animals* during International *Transport* (revised text, 2003), and the EU Regulation 1/2005. These instruments set up mostly controls, licensing, certifications, etc.., covering fitness, means of transport, and other procedures of animal handling.

[48] Regarding the consumers' interest in animal treatment conditions that may influence the marketing behavior, see Leslie and Sunstein (2007), pp. 117 seq.

If necessary, housing must be appropriate (well-ventilated, adequately lit, protected from climate conditions, with sufficient space for movement, etc..—art. 7). The slaughtering conditions must ensure avoidance of suffering, agitation, injuries (art. 6, 13), and pain (stunning methods required, art. 16).

Nevertheless, the ritual slaughter of bovines or slaughters performed for religious purposes are allowed.[49] In these specific cases, and in cases of emergency slaughtering, instantaneous animal death, or animal killing for health control, provisions ensuring animal welfare are not applicable (art. 13, 17).

At the EU level, Regulation 1099/2009 and Directive 98/58 on farm animals are based largely on the above provisions, also establishing relevant mechanisms of control measures.[50]

11.3.4.4 Laboratory Animals

The treatment of vertebrate animals used in experimental procedures (laboratory animals) is specifically governed by detailed European legislation. It is a field characterized until recently by non-legal regulation included in research ethics, which has inspired the current legislation's basic principles.

The EU Directive 2010/63 refers to the use of vertebrates for research purposes, and it is based upon the "3Rs principle". According to that principle, experimentation on animals is allowed (a) if no alternative method of experimentation exists *("replacement")*, (b) only to the necessary number of animals *("reduction"),* and (c) if measures for avoiding unreasonable suffering and pain are in place *("refinement").*[51]

Regarding the first requirement, alternative methods of experimentation may include, in particular, research on biological material in vitro (i.e., animal or human cell lines[52]), or the use of simulation models in a digital environment.[53] Certainly, these methods do not always ensure results of comparable validity with in vivo experimentation; still, the rationale, here, is that this should be demonstrated by the researchers and approved by competent ethics committees reviewing the protocol in relevance. References to similar experiments already published or even specific documentation indicating that alternative methodology does not exist, are sufficient evidence in this respect.

The "reduction" requirement, referring to the strictly needed number of animals for the concrete experiment, should be fulfilled equally. The fact that, by using more than necessary animals, we may ensure better statistical validity of the research results does not mean that this requirement becomes useless. Its scope implies merely the anticipation of scientifically valid results by using a certain number of

[49] On animal sacrifice in the context of ritual habits, see Orlans et al., *supra,* pp. 307 seq.

[50] On additional issues requiring specific regulation, see Buller et al. (2018), Nalon and Stevenson (2019).

[51] See the classic study by Russell and Burch (1959).

[52] See, in particular, novel "organ-on-a-chip" applications. Cf. Zhang et al. (2018).

[53] Cf. Badyal et al. (2009), pp. 191 seq.

animals, not excluding the possibility of statistically trivial errors. Otherwise, we would accept practically unlimited experimentation on animals, in the name of the best possible accuracy, ignoring that the law does not prioritize research over animal welfare but suggests a balance between these two values.[54]

The "refinement" element of the 3Rs principle is analyzed in distinct dimensions.

First, it encompasses the appropriate animal treatment prior to experimentation, which is related to the provision of suitable conditions for housing, feeding, etc.., namely conditions of animal welfare.[55]

Second, refinement refers to the animal treatment throughout the experimentation process, in which the selection of methods must ensure that unreasonable suffering or pain will be avoided. Again, here, the law implies a balance between animal welfare and the research value since a painful intervention may be necessary for obtaining significant scientific evidence if no alternative exists.[56]

The third dimension is related to the animal treatment after the end of the experiment. If the animal needs to be euthanized, a suitable method (painless to the extent possible) should be performed, and with no possibility to be felt by the animal. Relevant methods have been developed, and they are included in specific protocols.[57] If the animal is not to be killed, its reuse in experimentation should be considered, but that possibility is only exceptional according to the law.

A special category of laboratory animals is that of transgenic and genetically modified vertebrates. These are created after genetic interventions focusing on concrete areas of the natural species' DNA in order to develop "model" phenotypes for performing particular experiments. In several cases, genetic interventions intend to facilitate the manifestation of diseases in the animals involved, for enabling the study of their characteristics.[58] This entails, by default, the occurrence of suffering and pain throughout the animal life cycle. Nevertheless, the legislation does not exclude the creation of such "animal – models," considering as prevailing the value of research over animal welfare.

Two remarks are necessary here: (a) The rule requiring to avoid unnecessary pain and suffering still stands for the sake of these animals' welfare. That rule clearly excludes interventions in their physiology that may yield pain or suffering beyond the level associated with their genetic particularities. (b) Patenting of suffering "animal-models" is under the strict requirement of evidence about a real benefit expected in biomedical research. This is an explicit exception to the general rule permitting patenting on artificial forms of life, as we will see in the relevant Chap. 13.

---

[54] Regarding that point, see the criticism on the U.S. legislation by Vanderau (2006), pp. 721 seq.

[55] Cf. e,g., Seelig, *supra.*

[56] See Francione (2007), p. 45.

[57] Again, research on animals may influence consumers. In some cases, companies of cosmetics report that their products do not involve such research. Cf. Winders (2006), pp. 454 seq.

[58] This is the case of the famous Harvard "oncomouse" (a GM mouse, the body of which develops cancer tumors rapidly). See Orlans et al., *supra*, p. 89 seq.

As noted, these provisions refer to all vertebrate laboratory animals and very few invertebrate species (cephalopods, in particular). The use of invertebrate animals (insects, mainly) in biological research is, with no doubt, of great scientific importance, yet the EU Directive does not cover it.

This does not mean that the use of invertebrates is legally indifferent or uncontrolled. Indeed, at least some species are protected by the law relevant to rare or endangered species, already mentioned. Any use in research procedures of these animals is allowed on the condition that the existence of the concrete species is not jeopardized, according to the general rule of biodiversity conservation. This means, practically, that at least the two requirements of the 3Rs principle should be met, namely the "replacement" and the "reduction" requirement. Therefore, alternative methods of experimentation should be preferred instead of the use of rare or endangered invertebrates, and only the necessary number of them (strictly justified by the concrete scopes of experimentation) is legally accepted.[59]

Following these substantial provisions, the Directive establishes a system of preventive control regarding research on laboratory animals, which involves specific administrative organs. This system is characterized (a) by a procedure of licensing breeders, providers and researchers, for evaluating professional qualities pertinent to the treatment of laboratory animals, and (b) by a requirement of research protocols' approval not only regarding their scientific aspect (i.e., their reliability in terms of the expected results) but also regarding their ethical and legal aspect, that is, the expected compliance with the 3 Rs principle and relevant requirements of the Directive.

11.4 Further Studying...

ECJ: Case C-504/14, European Commission v Hellenic Republic, 10 November 2016

(Failure of a Member State to fulfil obligations — Environment — Nature conservation — Directive 92/43/EEC — Article 6(2) and (3) and Article 12(1)(b) and (d) — Wild fauna and flora — Conservation of natural habitats — Sea turtle Caretta caretta — Protection of sea turtles in the Gulf of Kyparissia — 'Dunes of Kyparissia' Site of Community importance — Protection of species)

[59] The "refinement" requirement, related to the avoidance of suffering and pain during experimentation, could also apply if the neural system of some invertebrate species is developed to the degree that it may react to external stimuli. Nevertheless, this is only an ethical consideration; no legal basis exists for invertebrates incorporating the "refinement" requirement, and acknowledging the corporal integrity of these animals, as the Directive does for vertebrates.

Question

Comment on the following paragraphs under the light of the CBD also:

53 "First, the opening up of new roads to the sea facilitates vehicle access to the beach and is therefore a factor in increasing motor traffic. In fact, according to Opinion No 32/2015 of the Symvoulio tis Epikrateias (Council of State), which the Commission has put forward, the unregulated parking of cars at the ends and on the verges of the access roads causes damage to the dune habitats of the species listed for the Kyparissia area. As a result of that easier access and the increased number of motor vehicles, there is an escalation in noise and light, which disturbs the Caretta caretta sea turtle during egg-laying and the hatching of young turtles. The Hellenic Republic itself has, moreover, found that the opening up and construction of those roads was unlawful and has had damaging effects, as is apparent from the file before the Court, and more particularly from the documents in Annexes 17h and 17i to the application initiating these proceedings.

54 Although the Hellenic Republic neither authorised nor carried out the opening up of those roads in the Kyparissia area, it does not challenge the Commission's complaint that the roads are operational despite the fact that criminal proceedings have been brought and various court cases are pending.

55 By confining itself (i) to bringing criminal proceedings against the executives of the company that built the roads in question and imposing administrative penalties on that company and (ii) to claiming, before the national courts, that the roads concerned are illegal and must be removed, the Hellenic Republic has failed to fulfil the specific obligation imposed on it by Article 6(2) of Directive 92/43 (see, by analogy, judgment of 9 November 1999, Commission v Italy, C-365/97, EU:C:1999:544, paragraph 109).

56 As the Commission submits, the Hellenic Republic should have acted to ensure that those thoroughfares did not remain operational and that use of them did not significantly disturb the Caretta caretta sea turtle or impair the dune habitats located in the Kyparissia area.

57 By failing to take provisional measures for the protection of that area in order to restrict use of the thoroughfares at issue until the conclusion of the above-mentioned court proceedings concerning the legality and possible closure of those thoroughfares, when, as the Advocate General has noted at point 77 of her Opinion, there is nothing to suggest that such measures would not be possible either for practical reasons or for reasons relating to EU law, the Hellenic Republic has failed to fulfil its obligations under Article 6(2) of Directive 92/43.

58 Secondly, as regards the road under construction alongside the railway line which would connect the beach at Kalo Nero with Elaia, it should be noted that it follows the existing railway line and that the Hellenic Republic has not been challenged by the Commission as to the fact that, with the exception of two places, that road is located at a distance of over 200 metres from the coast and is not linked to thoroughfares giving access to the beach. In those circumstances, the Commission has not demonstrated that there is a sufficient likelihood that the road under construction alongside the railway line which would connect the beach at Kalo Nero with Elaia will have a significant effect on the dune habitats and the Caretta caretta sea turtle.

59 Thirdly, as regards the asphalting of certain coastal roads, the paving of sandy tracks facilitates their use, especially by motor vehicles; that is likely to have a significant effect on the dune habitats located in the Kyparissia area and to increase appreciably disturbance of the Caretta caretta sea turtle. The Hellenic Republic, by tolerating the asphalting of roads within the site, has thus infringed Article 6(2) of Directive 92/43.

60 Consequently, the Court finds that, by having tolerated the opening up of five new roads to Agiannaki beach and the asphalting of certain existing roads and thoroughfares, the Hellenic Republic has not prevented the development of access routes to that beach, which is located in the Kyparissia area, and has therefore failed to fulfil its obligations under Article 6(2) of Directive 92/43. This complaint must, however, be rejected in so far as it concerns the construction of a road link between the beach at Kalo Nero and Elaia alongside the existing railway line."

ECJ: Case C-182/10 Marie-Noëlle Solvay and Others v Région wallonne (Reference for a Preliminary Ruling from the Cour Constitutionnelle (Belgium), 16 February 2012

(Assessment of the effects of projects on the environment — Concept of legislative act — Force and effect of the guidance in the Aarhus Convention Implementation Guide — Consent for a project given without an appropriate assessment of its effects on the environment — Access to justice in environmental matters — Extent of the right to a review procedure — Habitats Directive — Plan or project affecting the integrity of the site — Imperative reason of overriding public interest)

Question

Comment on the following paragraphs under the light of the CBD also:

75 An interest capable of justifying, within the meaning of Article 6(4) of the Habitats Directive, the implementation of a plan or project must be both 'public' and 'overriding', which means that it must be of such an importance that it can be weighed up against that directive's objective of the conservation of natural habitats and wild fauna and flora.

76 Works intended for the location or expansion of an undertaking satisfy those conditions only in exceptional circumstances.

77 It cannot be ruled out that that is the case where a project, although of a private character, in fact by its very nature and by its economic and social context presents an overriding public interest and it has been shown that there are no alternative solutions.

78 In the light of those criteria, the mere construction of infrastructure designed to accommodate a management centre cannot constitute an imperative reason of overriding public interest within the meaning of Article 6(4) of the Habitats Directive.

79 The answer to Question 6 is therefore that Article 6(4) of the Habitats Directive must be interpreted as meaning that the creation of infrastructure intended to accommodate a management centre cannot be regarded as an imperative reason of overriding public interest, such reasons including those of a social or economic nature, within the meaning of that provision, capable of justifying the implementation of a plan or project that will adversely affect the integrity of the site concerned.

ECJ: Case C-497/17. Œuvre d'assistance aux bêtes d'abattoirs (OABA) v Ministre de l'Agriculture et de l'Alimentation, Bionoor SARL, Ecocert France SAS, Institut national de l'origine et de la qualité (INAO), REQUEST for a Preliminary Ruling Under Article 267 TFEU from the Cour administrative d'appel de Versailles (Administrative Court of Appeal, Versailles, France, 26 February 2019

(Reference for a preliminary ruling — Article 13 TFEU — Animal welfare — Regulation (EC) No 1099/2009 — Protection of animals at the time of killing — Particular methods of slaughter prescribed by religious rites — Regulation (EC) No 834/2007 — Articles 3 and 14(1)(b)(viii) — Compatibility with organic production — Regulation (EC) No 889/2008 — First paragraph of Article 57 — Organic production logo of the European Union)

Question

Comment on the following paragraphs under the light of the Council of Europe's relevant Convention, and the EU Charter of Fundamental Rights also:

45 In that regard, Regulation No 1099/2009 contributes to 'improving the protection of animals at the time of slaughter' and encourages 'stunning methods [that] can lead to death while avoiding pain and minimising distress or suffering for the animals', as is set out in recitals 4 and 24 respectively.

46 In addition, under Article 3 of Regulation No 1099/2009, 'animals shall be spared any avoidable pain, distress or suffering during their killing'. That general requirement applicable to the killing of animals is given particular concrete expression in Article 4(1) of that regulation, which provides, first, that 'animals shall only be killed after stunning' and, second, that 'the loss of consciousness and sensibility shall be maintained until the death of the animal'.

47 Thus, Article 4(1) of Regulation No 1099/2009, read in conjunction with recital 20 of that regulation, lays down the principle that an animal should be stunned prior to its death and goes so far as to establish this as an obligation. As the Advocate General stated, in essence, in point 43 of his Opinion, scientific studies have shown that pre-stunning is the technique that compromises animal welfare the least at the time of killing.

48 While it is true that Article 4(4) of Regulation No 1099/2009, read in the light of recital 18 thereof, permits the practice of ritual slaughter as part of which an animal may be killed

without first being stunned, that form of slaughter, which is authorised only by way of derogation in the European Union and solely in order to ensure observance of the freedom of religion (see, to that effect, judgment of 29 May 2018, Liga van Moskeeën en Islamitische Organisaties Provincie Antwerpen and Others, C-426/16, EU:C:2018:335, paragraphs 55 to 57), is insufficient to remove all of the animal's pain, distress and suffering as effectively as slaughter with pre-stunning, which, in accordance with Article 2(f) of that regulation, read in the light of recital 20 thereof, is necessary to cause the animal to lose consciousness and sensibility in order significantly to reduce its suffering.

49 In that regard, it should be noted that, while recital 43 of Regulation No 1099/2009 states that slaughter without pre-stunning requires an accurate cut of the throat with a sharp knife to 'minimise' the animal's suffering, the use of that technique does not allow the animal's suffering to be kept to 'a minimum' within the meaning of Article 14(1)(b)(viii) of Regulation No 834/2007.

50 Therefore, contrary to what is claimed by both the French Government and the defendants in the main proceedings in their written observations, the particular methods of slaughter prescribed by religious rites that are carried out without pre-stunning and that are permitted by Article 4(4) of Regulation No 1099/2009 are not tantamount, in terms of ensuring a high level of animal welfare at the time of killing, to slaughter with pre-stunning which is, in principle, required by Article 4(1) of that regulation.

51 It should also be noted that recital 3 of Regulation No 834/2007 lays down the objective of 'maintaining and justifying consumer confidence in products labelled as organic'. In that regard, it is important to ensure that consumers are reassured that products bearing the Organic logo of the EU have actually been obtained in observance of the highest standards, in particular in the area of animal welfare.

52 Having regard to the findings above, the answer to the question referred is that Regulation No 834/2007, in particular Article 3 and Article 14(1)(b)(viii) thereof, read in the light of Article 13 TFEU, must be interpreted as not authorising the placing of the Organic logo of the EU on products derived from animals which have been slaughtered in accordance with religious rites without first being stunned, where such slaughter is conducted in accordance with the requirements laid down by Regulation No 1099/2009, in particular Article 4(4) thereof.

Provide Legally Appropriate Solutions

1. A famous movie star buys a baby tiger from a hunter of wild animals, wishing to keep it as a pet on a private jungle-like farm. The authorities of the country of the baby tiger's origin have licensed the export of the animal. Yet, the authorities of the owner's country prohibit that import on the ground of the wild animals' international legal protection, despite the legal claim of the owner grounded on the protection of pet animals.

Comment on this legal confrontation.

2. A research team submits a proposal for funding that involves testing a new pharmaceutical substance in mice to treat several cancer types. The proposal mentions only the number of animals to be used with no other information. According to the current European legislation

a. Should other information be provided as well?
b. Is there any change regarding the legal approach if the research would be performed on
   - A rare species of spiders?
   - Gorillas?
   - Genetically modified or transgenic rabbits?

References

Badyal DK, Modgill V, Kaur J (2009) Computer simulation models are implementable as replacements for animal experiments. Alternativ Laborat Anim 37:191–195

Buller H, Blokhuis H, Jensen P, Keeling L (2018) Towards farm animal welfare and sustainability. Animals 8:81. https://doi.org/10.3390/ani8060081

Cappitelli F, Sorlini C (2008) Microorganisms attack synthetic polymers in items representing our cultural heritage. Appl Environ Microbiol 74:564–569

Chambers B (2015) A watershed for the conservation of migratory species. Environ Policy Law 45:7

Christie Y (2007) Biodiesel from microalgae. Biotechnol Adv 25:294–306

Cole LA (2009) The anthrax letters: a bioterrorism expert investigates the attack that shaked America. Skyhorse Publ, New York

Dando M (2013) Awareness of and education about the biological and Toxin Weapons convention (BTWC): why this is needed by all life scientists and how it might be achieved. In: Coggon J, Gola S (eds) Global health and international community. Ethical, political and regulatory challenges, Bloomsbury, London

Diaz E (ed) (2008) Microbial biodegradation: genomics and molecular biology. Caister Academic Press, Portland

European Commission (2006) Guidance Document on official controls, under Regulation (EC) No 882/2004, concerning microbiological sampling and testing of foodstuffs, Brussels, in http://www.efet.gr/images/efet_res/docs/legislation/food-hygiene/sampling_testing.pdf

European Commission/Scientific Steering Committee (2002) The need for non-human primates in biomedical research in http://europa.eu.int/comm./food/fs/sc/ssc/out253_en.pdf

European Parliament (2001) Resolution on the Commission report on the experience acquired by the member states since the implementation of Council Directive 95/29/EC amending Directive 91/628/EEC concerning the protection of animals during transport, Oceana Publ., N. York, in http://www.europarl.europa.eu/sides/getDoc.do?pubRef=-//EP//TEXT+REPORT+A5-2001-0347+0+DOC+XML+V0//EN

Fidler DP, Gostin LO (2008) Biosecurity in the global age. Biological weapons, public health and the rule of law. Stanford University Press Stanford Cal

Fischer J, Lindenmayer DB (2007) Landscape modification and habitat fragmentation: a synthesis. Glob Ecol Biogeogr 16:265–280

Francione GL (2007) Reflections on animals, property, and the law and rain without thunder. Law Contemp Probl 70:9–57

Institute of Medicine of the National Academies (2011) Chimpanzees in Biomedical and Behavioral Research: Assessing the Necessity. Washington DC

IUCN & UNEP–WCMC (2011) The World Database on Protected Areas (WDPA) UNEP–WCMC, Cambridge, UK

Kant I, Lectures on Ethics (1790/1963), Harper & Row, New York

Krauss J, Bommarco R, Guardiola M, Heikkinen RK, Helm A, Kuussaari M, Lindborg M, Öckinger E, Pärtel M, Pino J, Pöyry J, Raatikainen KM, Sang A, Stefanescu C, Teder T,

Zobel M, Steffan Dewenter I (2010) Habitat fragmentation causes immediate and time-delayed biodiversity loss at different trophic levels. Ecol Lett 13:597–605

Laboratory Primate Advocacy Group (2005) The Failure of Federal Animal Welfare Regulations to Promote the Well-Being of Nonhuman Primates Used in Biomedical Research, http://www.lpag.org/bureaucracy/failure.html

Larsen FW, Turner WR, Mittermeier RA (2015) Will protection of 17% of land by 2020 be enough to safeguard biodiversity and critical ecosystem services? Oryx 49:74–79

Leslie J, Sunstein CR (2007) Animal rights without controversy. Law Contemp Probl 70:117–138

Levine JM, Vila M, D'Antonio CM, Dukes JS, Grigulis K, Lavorel S (2003) Mechanisms underlying the impacts of exotic plant invasions. Proc Royal Soc London B Biol Sci 270: 775–781

Lubelska-Sazanów M (2018) The wild differences in law when trading in wild animals: a US and EU perspective. Am J Trade Policy 5:39–48

Meagher RB (2000) Phytoremediation of toxic elemental and organic pollutants. Curr Opin Plant Biol 3:153–162

Milner JM, Nilsen EB, Andreassen HP (2007) Demographic side effects of selective hunting in ungulates and carnivores. Conserv Biol 21:36–47

Mittermeier RA, Turner WR, Larsen FW, Brooks TM, Gascon C (2011) Global biodiversity conservation: the critical role of hotspots. In: Zachos FE, Habel JC (eds) Biodiversity hotspots. Springer, Berlin, pp 3–22

Mooney HA (1999) Species without frontiers. Nature 397:665–666

Moran JM, Morgan MD, Wiersma JH (1986) Introduction to environmental science, 2nd edn. W.H. Freeman and Co, New York

Nalon E, Stevenson P (2019) Addressing lameness in farmed animals: An urgent need to achieve compliance with EU animal welfare law. Animals 9:576. https://doi.org/10.3390/ani9080576

Nellemann C, Henriksen R, Raxter P, Ash N, Mrema E, (eds) (2014) The environmental crime crisis – threats to sustainable development from illegal exploitation and trade in wildlife and forest resources. A UNEP rapid assessment, UNEP and GRID-Arendal, Nairobi, in MON-088666.pdf

Orlans FB, Beauchamp TL, Dresser R, Morton DB, Gluck JP (1998) The human use of animals. Case studies in ethical choice. Oxford Uniersity Press, New York

Prieur M (1996) Droit de l' Environnement, 3e edn. Dalloz, Paris

Puvača N, Chantal B (2020) Welfare and legal aspects of making decisions on medical treatments of pet animals. Law-Theory Pract 37:55–64. https://doi.org/10.5937/ptp2004055P

Reed DH (2004) Extinction risk in fragmented habitats. Anim Conserv 7: 181-191

Rijkers GT, de Vos WM, Brummer RJ, Morelli L, Corthier G, Marteau P (2011) Health benefits and health claims of probiotics: bridging science and marketing. Br J Nutr 106:1291–1296

Rittmann BE, Krajmalnik – Brown R., Halden R. U. (2008) Pre-genomic, genomic and post-genomic study of microbial communities involved in bioenergy. Nat Rev/Microbiol 6:604–612

Ruishi S, Qian L, Aziz N (2021) Impact of COVID-19 on Residents' Willingness to Consume Wild Animals: Empirical Insights From China. https://doi.org/10.21203/rs.3.rs-208100/v1

Russell WMS, Burch RL (1959) The principles of humane experimental technique. Methuen, London

Seelig D (2007) A tail of two monkeys: social housing for nonhuman primates in the research laboratory setting. J Appl Anim Welfare Sci 10:21–30

Shine C, Williams N, Gundling L (2000) A guide to designing legal and institutional frameworks on alien invasive species. IUCN, Gland Switz., MON-066015(SBSTTAdoc).pdf; MON-066015en.pdf; MON-066015fr.pdf; MON-066015sp.pdf

Spark G (2014) Protecting wild animals from unnecessary suffering. J Environ Law 26:473–494

Swart J (2005) Care for the wild: an integrative view on wild and domesticated animals. Environ Values 14:251–263

Taylor R (2001) A step at a time: New Zealand's progress towards hominid rights. Animal Law Rev 7:35–43

Tomasello M, Call J, Hare B (2003) Chimpanzees understand psychological states, the question is which ones and to what extent. Trends Cogn Sci 7:153–156
Vanderau ML (2006) Science at any cost: the ineffectiveness and underenforcement of the animal welfare act. Penn State Environ Law Rev 14:721
Vapnek J, Chapman MS (2010) Legislative and regulatory options for animal welfare. FAO legislative study, 104: Available at SSRN: https://ssrn.com/abstract=2898362
Vitti JJ (2013) Cephalopod cognition in an evolutionary context: implications for ethology. Biosemiotics 6:393–401
Winders DJ (2006) Combining reflexive law and false advertising law to standardize "cruelty free" labeling on cosmetics. New York Univ Law Rev 81:454
Yamada KH, Xu L (2001) Properties, and applications of an organic fertilizer inoculated with effective microorganisms. J Crop Prod 3:255–268
Zhang B, Korolj A, Lai BFL, Radisic M (2018) Advances in organ-on-a-chip engineering. Nat Rev Mater 3:257–278

Producing Species

12

12.1 Nature/Society: A Permanent Conflict

Species' "production" includes both the reproduction of existing species and the creation of new ones.

As we know, the reproduction of plants and animals represents a fundamental human activity, that emerged already at the beginning of the Neolithic era, an activity that enabled human societies to cover essential needs in nutrition, clothing, and production of energy. However, from that very early period, humans also achieved the creation of new species by combining natural living organisms in order to serve various needs better. From Antiquity already, the progress of the scientific study of Nature, encompassing the more precise identification of living organisms' properties, and the detailed classification of them, drove traditional Biotechnology to applications yielding new species. These species originated through human technological interventions, represent a large part of the existing fauna and flora nowadays.

With the development of genetic engineering in the early 1970s, the prospect of creating new species has opened enormously.[1] Two decades earlier, the DNA structure discovery revolutionized Genetics and became the crucial factor for testing species' modifications at their molecular level since it has indicated a common biological ground characterizing all living organisms. Since then, technology focuses on species' genotype, which allows much more precise comparisons and, thus, a better understanding of differences amongst them than those revealed by phenotypic characteristics.

The obvious next step could not be other than the possibility of designed interventions targeting directly genomes, which is based on the knowledge of specific correlations between genes and respective phenotypes. The results of such molecular interventions are new genomes, not existing in Nature, products of natural

[1] Cf. Reiss and Straughan (1996), pp. 96 seq.

T. Vidalis, *The Emergence of Biolaw*, Springer Textbooks in Law,
https://doi.org/10.1007/978-3-031-02359-0_12

species' genetic combinations even if these may belong to diverging biological classification (transgenic plants or animals). These new species of genetic engineering would be impossible to occur, following the foreseen course of Evolution, and practically they would never be created by traditional intercrossing methods.

Both the reproduction of existing species and the creation of new ones determine the current situation of biodiversity. In that sense, biodiversity cannot be addressed as a natural fact "outside" human civilization with an isolated and independent development. With industrialization in agriculture and stock-breeding, human societies became a crucial factor that influences biodiversity, a factor usually expressing interests and values in conflict with its conservation. This is the real reason for which biodiversity needed to be recognized as a distinct ethical and legal value, with the already mentioned normative consequences.

The crucial coexistence of Nature and Society in a field of conflicting values and interests occurred only in the era of the Industrial Revolution. Up to that period, human societies have been perfectly harmonized with Nature (more or less like the animal societies), submitted to its norms, with minimal or null possibility to affect the existing fauna or flora. Hunter-gatherer activities of the distant past, as well as the elementary farming activities of Antiquity and Middle-Age (even including their traditional methods of biotechnology), for a long period until the Modern era were insufficient for making human civilization a crucial factor for the Evolution process; at these times, Society was part of the entire Nature, not standing against the latter.

The introduction of machinery processing in agriculture and stock-breeding can be identified as the primary cause of change. The possibility of undertaking intensive productive activities enlarged the farming market immensely. On the one hand, these activities fulfilled the basic needs of nutrition for entire populations. On the other, they engaged wild natural territories for farming purposes by converting them into suitable land after draining, deforestation, and other land reclamation works.[2] They entail thus considerable changes in ecosystems, deeply affecting populations of animal and plant species and leading to the extinction of many of them.[3]

The extensive progress of scientific study about factors influencing agriculture activities is also relevant to that process. The involvement of the chemical industry seeking soil improvement and the protection of cultivations with the use of fertilizers and pesticides has been a crucial parameter here. On the other hand, strategies of monoculture plantations in farming (cotton, potatoes, etc.) were proved catastrophic for local ecosystems. The next step towards a complete dominance of science and technology, with the aim of massive and intensive production, is nowadays that of modern biotechnology, through applications of genetic engineering in plant and animal species, a step influenced by research uncertainty, though.

[2]Often resulting in citizens' reactions, like in the known example of judicial actions against deforestation in the Amazon area. See Bustos et al. (2020).

[3]The example of rain forests' devastation is characteristic. Cf. Janzen (1988), pp. 130 seq. About the influence on local communities, see Taylor (1988), pp. 138 seq.

Most of all, it was the industrial era, as such, that demanded a huge volume of energy, needing the exploitation of the mineral wealth at a planet-scale, which concluded in massive pollution. This development has dramatically changed the environmental conditions, influencing profoundly the sustainability of ecosystems and the conservation of species living there. Chain consequences of that phenomenon, like the greenhouse effect or the ozone hole in the atmosphere, entail significant changes in the planet's ecology, with the rise of temperature, melting of ice in the poles, extensive drought in large territories, the rise of extreme weather events, etc. Nowadays, these changes represent the most critical threat to biodiversity since the extent of their impact on existing ecosystems remains mostly unforeseen. For instance, a permanent drought of soils fertile in the past changes radically the micro-environment for significant populations of plant and animal species, the fate of which is to be extinct. Eventually, this fact influences the food chain and, thus, species' survival in much larger territories.

Finally, the above description needs to be associated with a critical change at the symbolic level. The industrial era illustrates a dominant instrumentalistic approach to Nature, an approach suggesting the unlimited use of natural components to meet the constantly multiplied human needs; Nature is considered a mere instrument for serving humans, with any corresponding value having only a utilitarian purpose. This instrumentalistic approach to Nature, part of the more general instrumentalistic transformation of Reason in Modernity,[4] represents now an essential feature of common social perceptions, also reflected in modern legal systems. This symbolic factor supports a competitive relationship between Society and Nature and triggers its manifestation in many fields.

For Biolaw, that competitive relationship means the need for balancing other legal values with the value of biodiversity, not when the latter is affected by environmental changes in general (this belongs to the study of Environmental Law), but specifically when biodiversity is affected by the creation or reproduction of living organisms or microorganisms, that is, with the appearance of new forms of life.

In that context, we can distinguish three topics of interest; a first, relevant to the conventional agricultural production, a second, concerning enrichment of biodiversity with new species deriving from genetic engineering, and finally, a third topic, referred to future applications with methods of "synthetic biology," which seem to open a tremendous possibility of artificial creation of life.

12.2 Biodiversity and Agriculture

The way we organize agricultural activities affects ecosystems directly and, therefore, the condition of biodiversity.

Agricultural production involves

[4] See the Frankfurt School's classic analysis, in Horkheimer (1947).

a) The determination of land for agricultural exploitation, which may alter the situation of micro-environments leading to the potential extinction of local plant or animal populations
b) Methods in use for cultivations, which may also affect the local biodiversity
c) Changes in biodiversity due to changes in cultivations

12.2.1 Land Use

The European law does not provide any standards in order to determine lands to be used for agricultural activities. With the exceptions of forests and other protected lands (Natura, etc., where these activities are limited, if not prohibited), national legislators regulate this area freely, shaping largely the general conditions of ecosystems and biodiversity in their country.

We can distinguish three major categories of land uses

1. Forest land, governed by strict regulation, excluding in principle the exploitation for farming purposes. That regulation covers, therefore, conditions of forest biodiversity.
2. Zones of special environmental protection, determined by the EU and international legislation, like the Natura and Ramsar protected ecosystems, are governed by similar restrictive national regulations, permitting limited productive activities (organic agriculture, cultivations of aquatic organisms, etc.). Compared to the first category of forest land, these areas have minimal extent and thus negligible significance for systematic agricultural exploitation.
3. A third category comprises the much larger part of lands freely available for agriculture and livestock farming. In these lands, changes in ecosystems may always occur, entailing a direct influence on the conditions of local fauna or flora.

The distinction between the third category and the other two should not be considered as permanently fixed due to changing facts or legal reasons. Indeed, continuous transformations of environmental conditions even without human intervention result in the extinction of old and the emergence of new ecosystems. This is a common, natural phenomenon generated by the expansion of housing or by extensive periods of drought, natural disasters, the rise of the average temperature, etc. This makes necessary a redetermination of land use, according to standards provided by legislation. Besides, vague legal provisions or legislative gaps may also entail similar differentiation in land use determination. For instance, the lack of official forest mappings or insufficient registries leads to the emergence of grey areas with no clearly determined use. As evident, such a situation of legal uncertainty cannot support, if not obstructs, the protection of local biodiversity.

In addition, it is worth mentioning that limitations to agriculture exploitation, referring to the above categories of land, correspond to restrictions on the fundamental right of property,[5] regarding private lands. Yet, such restrictions cannot violate the essence of that right, according to the general principle of the European law.[6] A problem emerges when the state attributes a status of "absolute protection" of Nature to certain private lands (private forests, Natura habitats, etc.) since minimal exploitation of these by their owner cannot be legally excluded. In these cases, the only option for ensuring the conservation of local biodiversity is the expropriation by the state, under the terms of the owner's due compensation.

In conclusion, the above presentation illustrates a very unsafe status of legal protection for biodiversity that directly affects living conditions for plant and animal populations or even entire species when a national policy determines the use of lands. The European law comprises very few means of preventive control regarding national decision-making compliance with that protection. Initiatives on the part of national legislators and administrative authorities, for determining zones of protected lands and mostly for ensuring permanent monitoring of protection remain practically uncontrolled by a mechanism at the European level. Sanctions are provided only when specific environmental harms occur; these harms may be irreversible, though, undermining an efficient protection of biodiversity.

12.2.1.1 Biofuels

Current technologies on the cultivation of plants for the production of biofuels have opened a new chapter in land use and the protection of biodiversity. Thanks to these technologies, we are now able to exploit cultivations of maze, soy, sunflower, etc., for yielding substantial amounts of energy.

The primary advantage of biofuel production is the substitution of clean and renewable energy in the place of conventional energy deriving from non-renewable sources resulting in environmental pollution. There are two disadvantages, though: a) that biofuels require the use of fertile lands, which alternatively would be used for cultivations of plants destined to cover food and feed needs, and b) that this practice encourages the expanding of agricultural exploitation to free lands, including forests and other habitats of wildlife, which may be detrimental for local biodiversity.

Therefore, biofuel cultivations may conflict with alternative land uses equally important for crucial human needs or the protection of environmental values. A balancing of interests needs to be observed here. For instance, by restraining the use of lands that might be available for food and feed production, biofuel cultivations may influence supply or market pricing. Regarding essential goods that are absolutely necessary for consumption in a certain community, it is apparent that we need to accept a limit to the production of biofuels. Moreover, this limit is justified, taking into account that, at present, biofuels play no significant role in environmental protection, being a minimal source of the total supply of energy.

[5] See art. 17 of the EU Charter of F. Rights, art. 1 of the 1st Additional Protocol of the ECtHR.

[6] See, e.g., art. 52 of the EU Charter of F. Rights.

At the level of the EU legislation, the use of biofuels is regulated mostly by Directives 2018/2001, and 2009/30, determining a system of monitoring for biofuel sustainability and environmental impact, based on each member state's relevant competence.[7] It is also recognized, as a mandatory minimal goal for all member states, a determined percentage of biofuel participation in the content of gas and diesel used in transports. Directive 2009/30 contains a specific reference to biofuel cultivations, by excluding lands with "high biodiversity value" from such cultivations (art. 7 b para 3).

12.2.1.2 The Exploitation of Marine Biodiversity

A noteworthy example of land use concerns the exploitation of marine areas, particularly for nutrition purposes. This is a characteristic relic of the ancient food-collecting human activities in our days, which is not based on the cultivation and reproduction of plant and animal species, as in agriculture activities. Certainly, some marine species' cultivations represent a developing industrial activity, yet fishing continues to cover the most significant part of the market on a global scale.

That extensive work of collecting marine species' populations is supported by technological applications that intensify the exploitation pace, with no guarantees of reproduction though, contrary to what happens with land cultivations and farm animal breeding. The lack of a similar "inherent" mechanism in fishing activities inevitably raises risks for the protection of marine species.[8] Here, protection is only a matter of "external" legal control mechanisms, that is, a matter of regulatory measures imposing restrictions and prohibitions. The efficiency of such mechanisms with the purpose of preventing activities implying risk for marine biodiversity remains highly uncertain due to market priorities and differentiation of measures taken at a national level.

More specifically, threats to marine biodiversity include

a) Overfishing practice that overcomes the sustainability threshold for a concrete population of marine species due to intensive marketing. Overfishing consequences manifest mainly in seas with limited local populations of marine species, like the Mediterranean Sea.
b) "Unrevealed" fishing occurred due to the continuous destruction of marine organisms for considerable periods by abandoned fishing tools (old nets, etc.).
c) Unintentional fishing, namely fishing of other marine organisms besides the population targeted, with no commercial interest, which eventually are rejected. This is due mostly to the use of unsuitable fishing methods and tools.
d) Destruction of marine species' natural habitats, caused again by unsuitable methods of fishing.

---

[7] These Directives replaced the previous Dir. 2003/30, 2009/28.

[8] On marine biodiversity from the aspect of international law and policies, see Greer and Harvey (2004). At the EU level, see Directive 2005/35/EK.

e) The "competition" of fishing with the survival of certain marine species (dolphins, whales, etc.), the feeding of which is dependent on the targeted population (small fishes, etc.).[9]

For preventing the above risks, there are legal instruments in relevance, at the level both of international and EU law, along with national legislation.

The UN Convention for the Law of Sea provides a general framework for the protection and conservation of the marine environment (art. 21 para 1 d, art. 56 para 1 b III, art. 145, 192, etc.) and the exploitation of living marine species, under the states' sovereignty (art. 61, etc.) or beyond this (art. 116, etc.).[10]

The EU law in relevance comprises mostly:

a) Regulations 1380/2013, 1224/2009, and 1954/2003 related to the application of the common fishing practice rules, determining a system of controls in all instances of fishing activities. This legislation affects to a certain extent the conservation of marine biodiversity, providing (among others), rules on fishing methods, percentages of fishing products, control of unrevealed fishing, etc.
b) Regulations 1967/2006 and 1343/2011 which govern fishing in the Mediterranean Sea, with detailed provisions regarding the protection of marine habitats, the location, and mapping of ecosystems, the determination of protected areas, prohibited methods and tools of fishing, mandatory characteristics for those permitted (although legitimate deviations are also mentioned).[11] These instruments also refer to the appropriate size of collected fishes, the enrichment of marine areas with additional marine populations, and terms of amateur fishing. These two Regulations bind directly and concretely the national legislator regarding both the objective and procedures of fishing activities.
c) Finally, Regulation 1005/2008 relevant to illegal, unreported, and unregulated fishing establishes a complex control system (applicable mostly to marketing procedures) that includes certification of fishing products (art. 12).

Other EU Regulations also govern the protection of immigrant species' populations, for which national measures are proved inefficient.[12]

12.2.2 Farming Methods

The devastation of lands exploited for farming purposes, that is, the gradual restraint of their sustainability, is related directly or indirectly to the destruction of local

[9] In general, Cole (1998), pp. 168 seq.
[10] On restrictions to fishing activities, see Louka (2006), pp. 244 seq.
[11] See in relevance, Regulation 517/2008.
[12] Reg. 1936/2001, 520/2007, 302/2009, 640/2010, and Dec. 207/2011.

biodiversity. Usually, this is the consequence of the pressure that market factors exercise for adopting intensive methods of agricultural production to meet increasing consumer demands.[13] The generalized use of chemical compounds (pesticides, etc.) necessary for intensive productive activities has been identified as the main reason for that damaging process.[14] A shift occurred during the last decades towards organic agriculture redefined to some extent the market demands, but this has no impact on the total volume of products needed for consumption, that is, on the dominant trend for intensive farming activities, and safety issues are not excluded also.[15] Cultivations of genetically modified plants have not changed that trend.

It is worth noting that the protection of land sustainability and the subsequent conservation of local biodiversity are dependent mostly on the exercise of private property rights. Contrary to the determination of land use, public policies intervene only indirectly, usually by defining quality standards for chemical compounds or by controlling the pricing for agricultural hydration, etc.[16] However, the methodology of land exploitation is decided freely by the owner of lands, and this certainly involves the possibility of the land's potential abuse that may affect the situation of biodiversity.

In conclusion, the conservation of local biodiversity is influenced by priorities that private interests set, even if biodiversity is a matter of public interest. Therefore, animal and plant local populations that face the risk of extinction and live in private farms remain unprotected, or at least they are dependent on private care. Here, the only exception is that of private lands identified as protected areas, which are subject to direct state control involving property rights restrictions.

12.2.3 Changes in Cultivations

Changes in agricultural strategies, adopted by farmers for achieving better marketing results, may have a similar influence on local biodiversity. Such changes are often encouraged by the public policy for rationalizing agricultural activities. In some cases, a change may ameliorate the conditions for local biodiversity, particularly when traditional practices of monocultures are abandoned for adopting more differentiated exploitation of the land.[17]

As in farming methods, changes in cultivations are decided freely by the farmer, who exercises property rights according to priorities that the private interest sets. In that framework, farmers acting in the same geographical area may prefer cultivations of completely different plants demanding also different methods of land management, use of chemicals, hydration, etc. This involves the possibility of a significant

[13] Cf. OECD (2010), pp. 82 seq.

[14] Cf. Jans and Veder (2008), pp. 402 seq.

[15] Cf. Hansen et al. (2002).

[16] See in relevance, Reg. 306/2005, 1107/2009, Dir 2009/128, at the EU level.

[17] Cf. Adam (1994), p. 103.

change in local biodiversity, including the extinction of entire populations of animal or plant species. That possibility remains legally uncontrolled, given the strong status of property rights in the law that enable the farmer to prioritize economic interests over environmental protection.

12.3 Balancing the Value of Biodiversity

Given the current trends and policies in agricultural production, this weak status of the animal and plant species' protection poses a general problem. If modern law recognizes biodiversity as a necessary condition for the existence and survival of our own species, isn't it also necessary a new approach regarding its conservation? Don't we need to accept that this crucial dimension of the common interest a) should, in principle, prevail over other aspects of that interest (of economic nature, in particular) when the state makes relevant decisions, and b) should also prevail over private interests expressed by legitimate property rights?

To provide an answer, we need to clarify in more detail the value of biodiversity beyond its vague recognition as a necessary condition for the protection of human life. The fact that human life is dependent biologically (in various aspects and in an absolute manner) on functions performed by other living organisms or microorganisms does not indicate a concrete connection of humans with each and every one of these organisms. For instance, it is certain that the loss of entire populations of animals and plants, even the extinction of certain species, did not influence the survival of human communities during various historical periods, even if it changed, sometimes significantly, the conditions of their life. The same is true for the emergence or extinction of plant and animal species that changed the biodiversity content, without any concrete contribution of humans. This natural process of change represents something different from the results of interventions violating the status of biodiversity; the emergence or extinction of species or populations of them, as such, is not necessarily harmful to biodiversity as a legal value.

Biodiversity as a legal value is violated when its change entails an evident degradation of living conditions for human communities. Extensive destruction of plants and animals, if occurred, may deprive an isolated human community of necessary means of nutrition; also, the sudden appearance of insects or parasites in a local environment may produce a similar result or provoke serious epidemics. On the other hand, the extinction of the last few individuals of a rare endangered animal species may deprive a local human community of tourism resources. Are all these cases similar in terms of influence on the life of humans?

In fact, the answer is negative because, in the first two cases, the degradation of life for human communities refers directly to conditions of biological existence of humans (as nutrition and health are), whereas this does not happen in the last case. We cannot equate a change in biodiversity that obliges humans to reconsider their social and economic terms of living to the direct influence on biological conditions of living. The first change does not affect humans as biological organisms; therefore, it cannot be considered a necessary condition of human survival in a strictly biological sense of interdependence between species.

If we treated all these cases similarly, we would consider biodiversity as a sort of "super-value," an absolute limit for any human activity, ultimately as a value *per se*. Ethically speaking, this is accepted under a deep-ecology approach, yet as we already mentioned, for modern law, biodiversity remains a value supporting legitimate human interests, inevitably related to them. We must admit, then, a "core" in the concept of biodiversity that links the biological existence of humans with that of other living organisms. Without a doubt, the law ensures the absolute protection of that core, considering it inviolable by human activities of any sort because it is directly related to the human organism's existence and function, in purely biological terms. Some clarifications are, here, useful.

a) The content of that core of biodiversity does not encompass the total of living species; it cannot be defined at a global scale, but only in relation to concrete ecosystems larger or smaller. As such, it may be composed of entire species or populations. We cannot determine the extent of that content in general terms; the question is always whether potential changes to the concrete environment, involving the destruction or extinction of living organisms, may influence the survival or health (that is, the biological condition) of humans directly.
b) At the narrow level of its core, the legal status of biodiversity is strong, in the sense that no balancing with other legitimate interests or values of public interest is accepted. This represents a limit for any human activity and always prevails, even over activities serving widely accepted common economic interests, housing, etc. That conclusion derives from the direct connection of the core content with the human biological existence as such.
c) On the contrary, in its "periphery," biodiversity may be balanced with other expressions of common interest, economic, social, cultural, etc., at a certain time and place. Again, we need to suppose a necessary limit for ensuring its protection, even if no direct danger for human life and health occurs, a limit pertinent to extensive environmental changes that overcome national borders. Public policies of national states in the territory of which exist habitats of vital importance for the survival of the human species (like the Amazon or Central African rain forests, the big rivers and lakes, etc.) need to comply with the conservation of these habitats, even if such policies serve legitimate aspects of social interest. Current tendencies in international law accept that limit, no matter if a problem of implementation certainly exists, due to the entailed constraint of national sovereignty.[18] Therefore, beyond the core of biodiversity lies a wide area that encompasses the total of species at the planet scale. In that area, the value of biodiversity is not associated with the biological existence of humans in a certain community but with more general interests, like the survival of our species or our care for future generations. In legal terms, that area is governed by the already

[18] Cf. Taylor (1998), esp. pp. 295 seq., 323 seq.

mentioned principle, according to which individual plants or animals are indifferent if no risk for the whole species occurs, and interventions affecting the existence of certain species are legitimate if the whole biodiversity remains safe.

The above analysis leads to some conclusions:

- The protection of biodiversity and supporting habitats prevail absolutely over other values of common interest if a direct danger for the life or health of humans living in these ecosystems emerges. Practically, this happens if, for instance, the use of chemical compounds in cultivations involves a direct danger to public health due to land and water pollution, or if the development of cultivations, radically influencing the conditions of biodiversity, lead to the manifestation and spread of infectious diseases. This is a minimal precaution, serving both public health and the conservation of local biodiversity.
- Public policies concerning large-scale changes in ecosystems (i.e., deviations of rivers, installations of industrial zones near forests, etc.) presuppose an assessment concerning the potential impact on biodiversity, in order to avoid risks to human life and health, due to radical changes that may occur in living conditions of local species.
- When changes in habitats result from public policies, citizens are entitled to protect the local biodiversity, not only on the ground of environmental rights (as recognized by national law) but also on the ground of individual rights to health, if a direct danger to public health emerges. Moreover, these rights are the most suitable legal ground for relevant claims against private interventions, jeopardizing the local status of biodiversity under the title of property rights. Therefore, methods of cultivation adopted by a landowner, which may damage the local biodiversity seriously, and involve risks for epidemics, constitute property rights abuse and violate the rights to other citizens' health.

12.4 Products of Genetic Engineering

The first applications of genetic engineering in the early 70s resulted in the creation of genetically modified microorganisms. As already mentioned, the methods of DNA recombination represent the beginning of the modern Biotechnology era. In this era, science is in place to create new forms of life, impossible to emerge through traditional cross-breeding methods and even not expected to occur naturally following the foreseeing future of Evolution and natural selection of species.

The basis of these new applications is either the artificial removal of genes from concrete regions of DNA ("knock-out"), which is a genetic modification in the strict sense, or the intended transfer of genes from another species to the DNA of a particular organism or microorganism ("knock-in"), intending to develop unique phenotypic characteristics to the offspring, which is the case of transgenic species.[19]

[19] Cf. Reiss and Straughan (1996), pp. 1 seq., 11 seq.

In microorganisms, modification of genomes may produce new species capable of functioning as means for environmental cleaning, as pesticides, or as catalysts for the production of food ingredients or pharmaceuticals.[20]

In plants, genetic modifications intend to create species free from certain disease infections, or less water consumption, as well as species more effective in productivity, with increased nutritional properties, or even with better physical appearance.[21]

In farm animals, the aim is to create new races capable of providing more products (milk, wool, etc.) or even products with pharmaceutical properties (like milk containing insulin, etc.).[22] Even in laboratory animals, as we mentioned, applications of genetic engineering make it possible to create genetically modified or transgenic "animal models" for more accurate pre-clinical research in new medicinal products (like the creation of mice models developing cancer tumors rapidly, for facilitating pre-clinical research in oncology[23]).

It is important to note that these interventions in the genome of microorganisms, plants, and animals involve a significant degree of uncertainty regarding the expected results. This is due to the insufficient knowledge we still have regarding the functions of genes, even if the precise mapping of various genomes (the human genome included) is already concluded.[24] This lack of knowledge is expected to last for the next decades, although the progress in Genetics is rapid and continuous. Therefore, removal of a specific gene from a particular genome, or inclusion of an alien gene into that genome, although suitable for influencing concrete phenotypes, cannot be considered as safe regarding potential undesirable implications to other phenotypes, insofar as sufficient statistic data on the properties and behavior of the newly created species are still missing.

12.4.1 The Precautionary Principle

This last element characterizes the rationale of law since the legal regulation of modern Biotechnology became an issue.

Generally speaking, the uncertainty concerning the potential results of genetic interventions causes reservations when enacting legislation at any level. Legislators need to balance, on the one hand, expected benefits for the economy, which include more robust and productive cultivations or other applications of genetic engineering

---

[20] *Ibid*, pp. 95 seq.

[21] Cf. WHO/FAO (2009), Reiss and Straughan (1996), pp. 131 seq. Regarding expectations on addressing the famine problem in developing countries, see Nuffield Council on Bioethics (1999), pp. 59 seq.

[22] Cf. Houdebine (2009), pp. 107 seq. Also, Reiss and Straughan (1996), pp. 173 seq.

[23] See the example mentioned above of "Harvard oncomouse." On GM animals, in general, EGE (1996).

[24] Cf. Collins et al. (2009), pp. 721 seq.

to the industry of food, pharmaceuticals, etc., and on the other, dangers for the environment and the public health that may emerge, due to our lack of knowledge regarding implications of that technology. The contemporary international law has adopted, as a guide for this work of balancing, the "precautionary principle".[25]

According to that principle, in conditions of uncertainty about possible risks that a certain technological application may create, reasonable measures of caution must be developed. Uncertainty exists when no concrete evidence of potential implications has been scientifically identified for designing precise prevention measures.

A too conservative approach to the precautionary principle would suggest abstaining from any intentional application of new technology to the open environment since risks can never be excluded outside a controlled laboratory of experimentation. Nevertheless, this approach is legally misleading because it concludes with an entire alienation of scientific research from any practical exploitation, contrary to the fundamental rights of Research and Economic Freedom.[26] A realistic and legally sound approach does not exclude applications to the open environment on conditions of their reasonable control; this presupposes continuous research for the detection of concrete dangers and the development of respective prevention measures.[27] Under this view, the objective of a technological novelty should not be considered a sort of "taboo" prohibiting by definition any potential activity; on the contrary, it represents a field of intensive research effort for revealing risks and achieving the highest degree of certainty, in order to avoid relevant implications.[28]

---

[25] Besides the preamble of the CBD, see art. 174 of the EU Treaty, and art. 1 of Dir. 2001/18 on GMOs. See in relevance, Andorno (2004), pp. 11 seq., and (2013), sec. I, 3, Wynne (1994), pp. 179 seq. From the EU case-law, see references to the principle in T- 13/99 Pfizer Animal Health v. Council, T- 76/00 Bruno Farmaceutici and Others v. Commission, T- 84/00 Roussel and Roussel Diamant v. Commission, T- 85/00 Roussel and Roussel Iberica v. Commission, T- 132/00 Trenker v. Commission, C- 24/00 Commission v. France, C- 3/00 Denmark v. Commission, C-192/01 Commission v. Denmark, C- 236/01 Monsanto Agricoltura Italia SpA and Others v. Presidenza del Consiglio dei Ministri and Others. A criticism, from a free trading view, see in Victor (2001), pp. 295 seq. The principle is also relevant to non-biological procedures, like the use of dangerous chemical compounds (see the REACH Reg. 1907/2006).

[26] EU Charter of F. Rights, art. 13, 16.

[27] Regarding scientific uncertainty that influences the application of the precautionary principle in decision-making, see Lemons (1998), pp. 75 seq., 77 seq. Also, Hilbeck et al. (2015). In the U.S. case-law, see Daubert v. U.S. [113 S. Ct. 2786 (1993)], regarding standards on the principle's application in concrete fields. A mathematic analysis on risk assessment, see in Basili and Franzini (2006), pp. 219 seq.

[28] See also Hansson (2020), arguing on necessary conditions for the principle's implementation. On the other hand, a sceptical view, see in Stefánsson (2019).

12.4.2 Biosafety: The International Legal Framework

The precautionary principle is mentioned explicitly in the Declaration of Rio de Janeiro on Environment and Development (Principle 15) and in article 1 of the First Protocol to the CBD on Biosafety (Cartagena Protocol), which is the basic binding instrument of international law on modern Biotechnology.

According to the Declaration:

> In order to protect the environment, the precautionary approach shall be widely applied by States according to their capabilities. Where there are threats of serious or irreversible damage, lack of full scientific certainty shall not be used as a reason for postponing cost-effective measures to prevent environmental degradation.

This is a rather moderate precaution approach since the principle's application depends on each state's possibilities, and relevant measures should be cost-effective. Both factors give priority to economic assessments. Thus, if no necessary infrastructure, equipment, or resources exist at the national level, or if a specific measure is considered excessively costly, a relevant technical application may be promoted, despite the uncertainty about potential risks. *A contrario*, we can conclude that, if "full scientific certainty" regarding potential threats is achieved, preventive measures need to be taken, regardless of relevant costs or the state's possibilities; this could also mean respective support by other states or institutions of the international community.

In the above framework, the Cartagena Protocol refers to *"an adequate level of protection in the field of the safe transfer, handling and use of living modified organisms"* (article 1). The terms "transfer," "handling," and "use" cover, in fact, the total of possible applications involving GMOs, that is, research, seeds marketing, cultivations, use food and feed, or other products, marketing of these, etc.

However, the Protocol includes also reservations regarding its implementation, namely:

i) Following other international law's provisions, national sovereignty is prioritized over the territorial sea, economic zones, and continental shelves (art. 2 para 3). This may favor GMOs' transfer and marketing within the area of national sovereignty, even without compliance with the protections on safety of the Protocol.
ii) The exception of GMOs for pharmaceutical use from its area of interest (article 5).
iii) Similar exceptions occur regarding "transit" and "contained use" of GMOs. The first refers to already existing free marketing before the Protocol's enforcement, while the second is related to procedures of controlled production in terms of safety (article 6).

The management and particularly the transboundary transfer of GMOs are governed by a formal procedure of information exchange between importers and exporters, and decision-making in determining notices, for certainty reasons. The risk assessment stipulations (art. 15) are an essential element of the Protocol since they favor relevant research to identify real risks, considering that confirmation of "uncertainty" is insufficient.

Special provisions also exist to protect information confidentiality, given the rights on industrial and intellectual property (art. 21), and for the rights of indigenous populations on local biodiversity, which could be damaged by the presence of imported GMOs (art. 26). Biosafety is also approached as a special culture, which is encouraged with measures aiming at the promotion of capacity-building in developing countries (art. 22), as well as with public awareness and educational initiatives or measures for participation of the public in relevant decision-making (art. 23).

These are general provisions, needing specific implementation and further development at the national level, although a system of international monitoring has been established that includes a central "Biosafety Clearing-House" and national focal points (art. 19, 20).[29]

12.4.3 The EU Law

In the framework of the Cartagena Protocol's above provisions, the EU law contains more specific legislation. The most important relevant instruments are:

- Directive 2001/18 on genetically modified organisms *(GMOs)*[30]
- Directive 2009/41 on genetically modified microorganisms *(GMMs)*
- Regulation 1829/2003 on GM products of food and feed[31]
- Regulation 1830/2003 on genetic modifications' traceability and the labeling of products containing GM ingredients
- Regulation 1946/2003 on the transboundary transfer of GMOs (incorporating the Cartagena Protocol).

In addition, Regulation 178/2002 concerning food safety, as well as the EU general legislation concerning the safety of cultivations (use of chemical compounds, etc.), are relevant here, particularly with respect to the meaning and assessment of risks.[32]

[29] On the Protocol's relevance with the WTO law, see Böckenforde (2004), pp. 118 seq.

[30] After its amendments by the Transparency Regulation (1381/2019), and mostly, by Directive 2015/412, which accepts national discretion to ban cultivations of GMOs approved at the EU level, following also the general principle of subsidiarity.

[31] After its amendment by the Transparency Regulation (1381/2019).

[32] On the EU legislation, see in general, Bruetschy (2019). Still, essential gaps in legislation occur due to the ongoing scientific progress. This is the example of nanoscale biotechnological products, given that no specific legislation on nanomaterials exists. Cf. Smets and Rudelscheim (2018). Another example is the current controversy regarding the applicability of GMOs legislation to

12.4.3.1 Genetically Modified Organisms

What is of legal interest in the case of GMOs, is the deliberate release of them into the environment. Here, we mean any possible use except for the restraint use in conditions of safety, for research purposes.

The release of GMOs needs to be addressed, notably when it aims at the commercial disposal of GM products or products containing GM ingredients.[33] That use may involve risks for the environment and the public health,[34] the prevention of which constitutes the objective of the EU legislation. The law provides a system of relevant procedures of technical nature, based upon implicit balances of legal values, as we will see in the following.

– *The concept of "risk"*

The first issue is the concept of "risks" to be prevented. The wording of Directive 2001/18 (art. 4) makes clear that "risk" means potential harm to the environment or public health.[35] Similarly, as described in the Cartagena Protocol, the precautionary principle refers to "threats" of "environmental degradation".
Evidence for risk is undoubtedly a scientific question. Still, in article 9, the Directive refers to the non-expert public's participation in procedures regarding decision-making for the deliberate release of GMOs. This involves the possibility of a common belief about risks, not based upon scientific evidence to prevail over scientific opinions. More generally, a question is whether it would be legally valid to enlarge the concept of risk by including certain crucial socio-economic factors and data related to the agricultural economy and the protection of consumers. In such a case, the public's non-expert opinion reflects a rational option by the legislator, even if this creates a potential conflict with the scientific reasoning about risks.

– *Reservations for the protection of the environment and public health*

The intentional release of GMOs is associated mostly with their use in food and feed products. In that sense, it is of interest for companies, farmers, and traders acting in the field of production and transfer of seeds, food, feed, or their ingredients. This constitutes a legitimate dimension of their economic freedom, which nevertheless is protected by law in conformity with the protection of the environment and public

---

products of genome editing (CrisprCas9 etc.), as ruled by the ECJ (C-528/16). See Wasmer (2019), arguing in favor of specific legislation for clarifying the issue. See also, Zimny and Eriksson (2020).

[33] Art. 2 para 3 of the Dir.

[34] Although objections to GM products may also be grounded in fundamental values and integrity of consumers, requiring, for example, appropriate information and non-maleficence. See Pascalev (2003), pp. 583–594.

[35] Cf. Holl et al. (2004), pp. 263 seq.; Domingo and Bordonaba (2011), pp. 734 seq.; European Commission (2010). On the activism from both sides of opponents and defendants, see Kloor (2013) and White (2013).

health.[36] In that context, an issue emerges regarding the extent of economic freedom's limitations, mainly if we consider the element of risk uncertainty that we mentioned before; is it legally justified the prohibition of GMOs release when risks are only a possible implication with no firm evidence?
There is no doubt that the precautionary principle, as recognized by the EU law, suggests an affirmative answer. However, this does not support a denial of economic freedom for those interested. Indeed, on the one hand, as we already underlined, the precautionary principle is implemented, taking into account economic considerations as well. On the other, its implementation is limited only to "serious" and "irreversible" harms, which means that minimal harms are accepted when an economic (productive or commercial) activity is to be developed.

Therefore, we need to trace a clear line separating "serious" from "minimal" harms for suggesting an accurate implementation of the precautionary principle.

i) Regarding the environment, potential harms threaten the conditions of biodiversity.[37] Genetically modified species may influence natural species by subverting the latter's reproduction process or affecting conventional cultivations (with the appearance of resistant weeds), or violating the food chain with vast implications for the local fauna.
   The concept of "serious" harm encompasses not only the extinction of a rare species but also the loss of a certain population, which entails a future threat to the species. On exceptional occasions, even the extinction of individuals belonging to a rare species may be considered serious harm if the species is threatened (like in animals of limited numbers that may be deprived of their feed due to changes that occurred in the food chain). On the other hand, minimal harms affect specific local animal or plant populations with no further influence on the flora or fauna they belong to.
ii) Regarding public health, potential serious harms may occur directly (i.e., with the unforeseen appearance of pathogen microbes that may be favored in a surrounding environment containing GMOs) or indirectly (i.e., if consumption of food containing GMOs results in an unforeseen impact on human health). Here, minimal harms should be considered those that may be compared with potential implications from the consumption of conventional food products.

This classification of potential harms to the environment and public health requires, certainly, the performance of specific research. This is a crucial legal presupposition for implementing the precautionary principle and not a mere option related to the freedom of Research in this particular topic. Indeed, if a basic assessment at least on the potential nature of harm (if not on the certainty of risks) does not exist, the precautionary principle as a legal rule is weakening in front of claims based on economic rights that favor an uncontrolled development of GM applications.

---

[36] See, art. 114, 169 of the EU Treaty.

[37] Cf. Nuffield Council on Bioethics (1999), pp. 103 seq.

– *Reservations supporting "social perception"*

If protection of the environment and public health is considered a permanent reason for restricting the GMOs' development, the opinion of non-expert citizens is an additional, although tacit, reason in relevance.
Indeed, the EU Directive provides a process of public consultation that competent national organs need to organize.[38] The question, here, as we already mentioned, refers to the real significance of that non-expert opinion; is there a possibility for that opinion to prevail over scientific assessments about potential risks that involve the deliberate release of GMOs?

A closer examination of that question indicates that a dissenting opinion of non-experts is due either to specific disagreement with the concrete scientific assessment of risks (arguing, for instance, that the latter is not impartial but reflects economic interests[39]), or to a general mistrust over radical technological interventions in Nature. The first approach may cover rational objections, either scientific (if, for instance, based on alternative risk assessments produced by equally reliable scientific teams or organizations, etc.), or political ones (if a conflict of interest is proved involving scientists that have produced the original report). In contrast, the second approach (that of general mistrust over technological interventions) reflects a "technophobic" attitude only, which cannot be judged with reference to objective standards.[40]

Objections that may be expressed in public debate cannot be appraised but within a rational framework, according to objective standards, if we intend to a consistent legal interpretation of that Directive's provision. It is important to notice that, for the law, objections of the general public concerning the use of GM products, despite a positive scientific assessment by legally competent experts, contravenes the economic freedom of GM producers, which is supported by that positive assessment. Consequently, only if these objections would be subject to a rational review (like in the examples mentioned above) and proved convincing, it is meaningful to override the risk assessment report of competent organs and determine the final decisions. Otherwise, if we considered as valid non-expert beliefs, even those reflecting technophobic and irrational attitudes, we would exclude modern biotechnology from the area of economic freedom. Indeed, ultimately, we would accept prohibitions in cultivations and trading of GMOs on the basis of alleged "risks" with no scientific evidence, based merely on common beliefs.

Bearing in mind the intense economic and political controversies on the GMOs issue that occurred worldwide during the last three decades, it is evident that the dimension of legal certainty is, here, of crucial importance. Approaches that demonize any biotechnological innovation, on the one hand, and others that tend to undermine real dangers, on the other, divide the scientific community itself; this

---

[38] On a relevant experience, see Aasen and Vatn (2013), pp. 461 seq.

[39] See Diels et al. (2011), pp. 197 seq.

[40] On "technophobia" see Gilbert et al. (2003), pp. 253 seq.

problem indicates the sensitive role that those who implement that particular legal provision, need to undertake, whether they are politicians, administrative officers or judges.

In conclusion, a decision that takes into account the opinion of non-experts dissenting to the official scientific evaluation of a particular GM product should be based upon rational evidence indicating specifically the reasons for which that dissenting opinion must prevail. General objections regarding the intentional release of GMOs, or political statements claiming that the public opinion has been considered, with no reference to an organized process of debate, are in legal terms invalid, even if they may influence *de facto* economic or political attitudes on the matter.

– *Reservations in favor of consumers*

The use of GMOs in food products needs to observe an additional legal requirement, that of "labeling" for ensuring clear information to consumers.[41]
Labeling represents an example of the precautionary principle implementation, as a preventive measure corresponding to uncertainty regarding potential implications to health due to consumption of GM products. That measure affects the economic freedom of producers directly since a certain product appears in the market with the advertisement for such negative implications, namely with a relative disadvantage compared to similar GM-free products.

Known examples of labeling indicating negative implications include, of course, numerous conventional products. Still, there is a difference with the GM example because implications in these products are evident and detected specifically. In contrast, labeling GM products reflects a scientific reservation on potential risks, which remain still unproven. Nevertheless, this is sufficient for generating a general impression of real risks to the non-expert public. One may argue that this fact eventually violates the economic freedom of producers since GM products are stigmatized on the basis of "alleged" and not real risks, which creates a serious disadvantage in market competition.

Indeed, a drastic intervention of legislation in market conditions requiring labeling cannot be based on the sole element of scientific uncertainty. It is necessary to insist on specific research for identifying real risks, mostly because it would be unfair for the non-expert consumers to carry the responsibility of consumption instead of the expert scientists. Indeed, if based on mere scientific doubts, labeling could be considered a disproportional measure of restriction on economic freedom. This conclusion is relevant mainly when a specific indication on the product's package exists, separately from the other ingredients; such a practice cannot be considered a

[41] Art. 21 of the Directive and art. 4 B of Regul. 1830/2003. On consumer rights, see the general recognition by art. 38 of the EU Charter of Rights. Specifically, cf. Nuffield Council on Bioethics (1999), pp. 9 seq. Also, Buckens and Changeur (2000).

"reasonable measure," under the view of the precautionary principle, since it may be misleading for consumers, as stigmatizing the product.[42]

– *Reservations in favor of farmers*

A final issue related to reservations on GMOs' development, particularly regarding plants' cultivations, refers to the "farmers' privilege." By this term, we mean the possibility of control of cultivated plants' breeding by the farmers. This is based on their discretion to collect and exploit the plant's seeds freely, a traditional practice accepted in conventional cultivations.[43]

Criticism of the GMOs industry's development focuses on the fact that this "farmers' privilege" is canceled since the applied technology creates sterile plants. This obliges the farmer to permanent supply with new seeds from the industry for each new term of cultivation, which results in close dependence on the producing company. A similar problem is already known, as it occurs in conventional farming insofar as productivity is dependent almost absolutely on the use of chemical compounds, that is, on the necessary support of the chemical industry. Moreover, the "farmers' privilege" has little significance nowadays, even in conventional farming, since the use of seeds that have been produced after a first harvest is expected usually to provide poor quality and quantity of plants.

In any case, the legal question in relevance is whether the loss of control over plants' reproduction infringes the economic freedom of farmers. The current law does not regulate this problem with explicit provisions. Nevertheless, relevant litigation may arise, particularly if the use of GM plants proved harmful to conventional cultivations, making them economically disadvantageous. On this occasion, conventional plant farmers would be deprived of a primary option regarding their activity, resulting in their absolute dependence on the GMOs industry. It is worth noting that a supposed harmful influence of GM plants does not necessarily mean danger for the environment or public health; it just results in economic damage for traditional farmers.[44] This is sufficient to imply a serious overthrow of local economic activity and even the local social life of a traditional agricultural community.

Obviously, the emergence of any novel technology involves big changes in traditional economic activities. Therefore, it would be unreasonable to argue for rejecting technological applications on that ground solely, since innovations contribute significantly to the entire economic development and, as such, they are protected legally through economic rights; in principle, potential changes resultant from new technology are not considered as a legal reason justifying its restriction.

[42] See in relevance, Markie (2007), pp. 88 seq.; American Medical Association House of Delegates (2012); Wolf (2007), pp. 178 seq. A critical approach of the recent federal legislation in the U.S., see in Bovay and Alston (2018) and Mosier et al. (2020).

[43] Cf. Philips (2007), pp. 49 seq.; Lee (2009), pp. 179 seq.

[44] On the "coexistence" of GM with conventional plants, see Devos et al. (2009).

Still, we should trace a limit here. If an innovation involves serious damage in economic and social terms, which could be the future exclusion of certain persons from local economic life, it is not protected by its owner's economic freedom. In that case, restricting the use of that innovation is meaningful because its damaging results are irreparable in social terms.

In the example of GM plants, traditional farmers are obliged to use GM seeds due to the harm that occurred in conventional cultivations, which entails the loss not only of the "farmers' privilege" but also of other possible advantages in the market (i.e., loss of customers preferring conventional products, etc.).[45] This forced re-orientation of farmers' economic activity, profoundly affecting their economic freedom, constitutes the exact reason for restricting that innovation.

In institutional terms, relevant measures can be taken following the process of public consultation, according to the relevant provision of the EU Directive. Indeed, the opinion of non-experts could have a significant impact on decisions regarding concrete cultivations of GM plants if providing evidence for serious overthrows in local economic and social life. Still, a general rejection of GM plants' cultivations should not be observed but as an exceptional measure of protection. A coexistence of conventional with GM plants in farms, with the implementation of appropriate protection measures to avoid harmful influence, needs to be examined in any case. This presupposes the prioritization of the proportionality principle for seeking a balanced protection of potentially conflicting economic interests.

12.4.3.2 Genetically Modified Microorganisms

Contrary to the GMOs' deliberate release to the environment and the relevant risks that may occur, genetically modified microorganisms (GMMs) are of legal interest regarding their potential use in laboratory conditions for research purposes or in controlled spaces like greenhouses, seedbeds, animal cages, reservoirs, etc.

Directive 2009/41 refers to measures of safe management after risk assessment. By "microorganisms", the Directive means microbes, cellular or not, capable of reproducing or transferring genetic material. Viruses, viroids, and animal or plant cells cultures also belong to that concept (art. 2). Genetic modification produces changes to the DNA of microorganisms due to biotechnological methods, which cannot occur naturally.

Since the use of microorganisms is controlled in order to restrain influence on humans and the environment, measures of protection are simpler; they are relevant mostly to possible accidents (namely unintentional release of GMMs) and waste management. Again, an essential legal prerequisite is the appropriate implementation of the precautionary principle, concerning, in particular, the following points.

1. The Directive establishes four distinct levels of risk, corresponding to specific protection measures that need to be considered by GMM's users. Risks are related

[45] Regarding the effects of GMOs development on traditional trading, see Nuffield Council on Bioethics (1999), pp. 66–67.

to the environment and human health. Users may undertake activities of a) null or trivial risk, b) low risk, c) medium risk, and d) high risk (art. 4 para 3).
2. Any doubt about the classification of a specific activity should be addressed, in principle, at the relatively stricter level, following the precautionary principle (art. 4 para 4).
3. Risk assessment is a dynamic process. Measures of protection are subject to permanent updating, on the basis of new scientific evidence, to ensure their efficiency (art. 5 para 2).
4. The Directive sets up a network of regular communication between users and competent authorities, at both the EU and national levels, for exchanging notifications on new facilities, incidents, specific measures for addressing accidents, scientific data, etc. (art. 6 seq).

The Directive also includes analytical annexes, determining in detail the levels of risk and the corresponding protective measures.

That legislation illustrates, mostly, specific limitations to freedom of Research. Although the use of GMMs in laboratories and the control of their reactions in experimental conditions may lead to future nutritional or pharmaceutical products available for consumption, the objective of Directive 2009/41 does not cover this dimension. Therefore, issues related to economic rights, similar to those discussed regarding GMOs, are subject to the general EU legislation in relevance.

12.5 Synthetic Biology

12.5.1 Concept and Applications

By "synthetic biology," we mean the artificial creation of genomes to be merged with natural cells in order to develop new forms of life.[46]

This technology imitates natural procedures at the molecular level of DNA structuring and its functional capacity. The difference from genetic engineering is that the latter is limited to targeted interventions in a particular organism's natural genome by inserting new, or for deactivate existing genes to achieve phenotypic outcomes. These interventions need to adjust themselves to natural genetic contexts for becoming functional. In contrast, with the artificial creation of genomes from simple chemical nucleobases, we expect to overcome this natural limitation, even if, so far, we just "imitate" the natural prototype genome.

To be functional, the artificial genome needs to be inserted as a new nucleus into a natural cell, from which the original nucleus has been removed. The contribution of the natural cytoplasm's organoids to the new form's functionality (the emergence of a new original form of life) is crucial, although still largely unknown. In that sense, synthetic biology does not create entirely "artificial life"; it just annexes an

[46]Cf. EGE (2009), pp. 13–14.

artificially created fundamental element of life (the genome) to a natural element (the cell).[47] Compared to conventional methods of genetic engineering, this development allows more creativity through technology than the known targeted interventions in the composition of natural genomes. Still, natural limitations are always present, regarding, in particular, the need for using natural cells (which cannot be replicated artificially); we have not yet developed a technology permitting the creation of prototype genomes with original combinations of their nucleobases not occurred in Nature.

Expected applications from synthetic biology may cover an extensive area, including the development of synthetic bio-pharmaceuticals (the example of artemisinin for treating malaria is typical) or biofuels for producing clean energy and synthetic microorganisms for environmental cleaning. Also, it covers the creation of synthetic cells capable of targeting and destroying cancer cells, or the construction of biomaterials, including biosensors or even tissues, that may contribute to advanced diagnostic and therapeutic practices (transplantations), etc.[48]

12.5.2 The Law in Relevance

So far, the European law or national laws do not include specific provisions regulating synthetic biology. The legal issues that we need to address are similar to those emerging in GMMs and GMOs. There is no doubt that the uncertainty for potential implications of future applications is vast; therefore, the significance of the precautionary principle is vital here.[49]

General legislation pertinent to the regulation of synthetic biology includes mostly:

1. The Rio de Janeiro CBD, which does not exclude applications of natural species' artificial replications, aiming at the creation of living microorganisms or organisms. In particular, the term "genetic material" has a broad meaning in the CBD, since it does not exclude an artificial origin of "functional units of heredity" (art. 2). As it happens with other biotechnology products, the CBD determines a general limit for the protection of biodiversity, pertinent to future applications of synthetic biology as well.
2. The Cartagena Protocol on Biosafety is also applicable since it refers to "living modified organisms" deriving from "modern biotechnology" (art. 3 g). Indeed, forms of life produced artificially are considered "modified organisms," given that they involve an intervention in natural cells. Moreover, the method itself

[47] Cf. Gibson et al. (2010), pp. 329 seq., regarding the creation of "Synthia" by the J. C. Venter's team.

[48] Nevertheless, a futuristic scenario involving even the creation of "synthetic" humans, and raising stimulating ethical questions is also discussed. See, Anomaly (2020), chapt. 5.

[49] Regarding the dual-use danger (exploitation of technology for military or terroristic purposes), see EGE (2009), pp. 43–44. On ethical issues emerged, see O'Neill et al. (2012).

refers to an "in vitro technique of nucleonic acids," as appears in the definition of "modern biotechnology" (art. 3 h). In that context, the Protocol is relevant, particularly regarding its detailed requirements based upon the precautionary principle, regulating the transportation and controlled use of synthetic bio-products.

3. The EU legislation on GMOs and GMMs is applicable for the same reasons. Indeed, both Directive 2009/41 and Directive 2001/18 define, respectively, the "GM" microorganisms and organisms, by referring only indicatively to various techniques of genetic modification; they leave thus margin for an artificial development of genomes as well. Therefore, that legislation will cover future synthetic biology applications, involving the requirements already mentioned, relevant now to conventional GM products.
4. Finally, the legislation on patenting biological material and living organisms and microorganisms (TRIPS Convention and Directive 98/44) needs to be considered also. Provisions of that legislation do not exclude synthetic biology applications, on the condition that these are found in conformity with the "ordre public" and "morality," clauses that indeed leave an extensive margin of interpretation, as we will explain in the relevant Chap. 13.

At the level of fundamental rights, the freedom of Research, as recognized by the EU Charter of Fundamental Rights (art. 13) and by the EU Treaties (art. 179–180), covers basic research in synthetic biology, when the purpose of the relevant experimentation is to enrich our knowledge for artificial forms of life. The freedom of Research also protects activities of applied research aiming at the development of synthetic bio-products for commercial use. Nevertheless, in the latter case, conflict with other rights may occur, in particular with the right to Health and environmental rights, which means that a balance between them *in concreto* needs to be considered under the standard of the precautionary principle.

12.5.3 Future Applications to Living Organisms

So far, applications of synthetic biology are limited to unicellular microorganisms. The creation of synthetic pluricellular organisms is only a prospect, given that genomes, even the simpler of them, present enormous complexity. Theoretically, however, we can imagine such applications concerning the formation of artificial genomes of plants, animals, or even humans.

The current law accepts that prospect in plants at the basic research level. Concerning deliberate release to the environment through cultivations, the legal regulation on GMOs, as exposed previously, also allows activities under control. The same we should accept for possible applications to non-vertebrate animals.

In the case of vertebrates, the legislation on GM laboratory animals would also be applicable at the stage of experimentation, and the legislation on patenting biological applications regarding commercial exploitation. The latter excludes, in particular, patents for applications entailing unnecessary suffering and pain to animals.

Finally, in the case of humans, the futuristic manufacturing of an artificial human genome, which would be inserted into a human oocyte, currently contravenes the absolute prohibition of article 18 para 2 of the Oviedo Convention, referring to the creation of embryos for research purposes. Indeed, even if the final aim, here, might be an innovative technique in assisted reproduction, basic research on "synthetic embryos" would be a necessary prerequisite, which at present is not acceptable by the European law.

12.6 Further Studying...

ECJ: CASE C-236/01 Monsanto Agricoltura Italia SpA and Others V Presidenza del Consiglio dei Ministri and Others, 9 September 2003
(Novel foods and novel food ingredients — Placing on the market — Simplified procedure — Substantial equivalence to existing foods— Use of that procedure notwithstanding the presence of residues of transgenic protein in novel foods — Whether permissible — Limits — Potential risks to human health — Detailed rules sufficient to ensure a high level of protection for human health and the environment — Compliance with the precautionary principle and the principle of proportionality)

Question

Comment on the following paragraphs under the light of the ECHR, the CBD, and the EU Charter of Fundamental Rights also:

75. The concept of substantial equivalence should be placed in the context of the work carried out by the international scientific institutions where it was elaborated, as set out inter alia in Recommendation 97/618.

76. It is true, as follows from its legal basis, namely Article 4(4) of Regulation No 258/97, that that recommendation was adopted for the purpose of clarifying the normal procedure. That explains why the requirement for a conventional toxicological evaluation referred to in Part I, Section 5, point IV of the Annex to the Recommendation (read in conjunction with points 3.3 and 3.7 of Section 3 thereof), to which the national court refers, is not relevant in the present case. The concept of substantial equivalence is there applied in the specific context of a risk analysis such as that provided for under the normal procedure

77. The Recommendation is nevertheless useful for the purpose of defining the concept of substantial equivalence as referred to in the first subparagraph of Article 3(4) of Regulation No 258/97. It is clear from the first and second subparagraphs of Part I, Section 3, point 3.3 of the Annex that the concept does not in itself involve a safety assessment, but rather constitutes an approach for comparing the novel food with its conventional counterpart in order to determine whether it should be subject to a risk assessment as regards, in particular, its unique composition and properties. It also follows therefrom that the absence of substantial equivalence does not necessarily imply that the food in question is unsafe, but simply that it should be subject to an assessment of its potential risks.

78. In order further to define the concept of substantial equivalence, it must also be placed in the context of the process of risk analysis as commonly defined at international and Community level. As in the present case, a concept which is applied by specialised scientific bodies charged with assessing the risks inherent in novel foods is involved.

79. That concept must, more precisely, be understood as a specific method concerning novel foods, relating to the identification of hazards which comprises the first stage in scientific risk assessment, namely the identification of the biological, chemical and physical agents liable to give rise to adverse health effects which may be present in a given food or group of foods and which call for scientific assessment in order better to understand them (see to that effect, inter alia, the procedure manual of the Codex Alimentarius Commission of the United Nations Food and Agriculture Organisation (FAO), 12th edition, pages 51 and 52, and Annex III to the provisional communication by the Codex Alimentarius Commission of the FAO and of the World Health Organisation (WHO) CX 4/10, CL 2000/12 — GP, April 2000; Article 3(9) to (14) of Regulation (EC) No 178/2002 of the European Parliament and of the Council of 28 January 2002 laying down the general principles and requirements of food law, establishing the European Food Safety Authority and laying down procedures in matters of food safety (OJ 2002 L 31, p. 1) and points 5.1.1 and 5.1.2 of and Annex III to the Communication from the Commission on the precautionary principle of 2 February 2000 (COM (2000)1); see also Case T-13/99 Pfizer Animal Health v Council [2002] ECR II-3305, paragraph 156, and Case T-70/99 Alpharma v Council [2002] ECR II-3495, paragraph 169).

80. Since the protection of public health is a fundamental objective of Regulation No 258/97, the concept of substantial equivalence cannot be interpreted in such a way that the simplified procedure, which according to the wording of the first subparagraph of Article 3(4) of that regulation is in the nature of a derogation, amounts to a relaxation of the safety requirements which must be met by novel foods (see to that effect, in the area of proprietary medicinal products, Case C-368/96 Generics (UK) and Others [1998] ECR I-7967, paragraph 22).

81. As to the unpredictable effects on human health which the insertion of foreign genes may produce, which are referred to by the Norwegian Government in particular, if such effects were identifiable as a danger to human health according to available scientific evidence at the time of the initial examination by the competent body, they would have to be subject to a risk assessment, and a finding of substantial equivalence would therefore be excluded.

82. Another element of the regulatory context of the concept of substantial equivalence in the first subparagraph of Article 3(4) of Regulation No 258/97 which reinforces the interpretation according to which that concept does not preclude differences in composition that are not relevant to public health is apparent on reading the second paragraph of Article 5 and Article 8 of that regulation.

83. It follows from those provisions that certain differences, inter alia as regards the composition of novel foods, do not prevent those foods from being deemed substantially equivalent in accordance with the first subparagraph of Article 3(4) of Regulation No 258/97, since Article 8 of that regulation, on the contrary, provides that such differences must be specifically referred to on the labelling.

84. Therefore, the answer to the first question must be that the first subparagraph of Article 3(4) of Regulation No 258/97 must be interpreted as meaning that the mere presence in novel foods of residues of transgenic protein at certain levels does not preclude those foods from being considered substantially equivalent to existing foods and, consequently, use of the simplified procedure for placing those novel foods on the market. However, that is not the

case where the existence of a risk of potentially dangerous effects on human health can be identified on the basis of the scientific knowledge available at the time of the initial assessment. It is for the national court to determine whether that condition is satisfied.

...

106. If the twofold objective of Regulation No 258/97, namely ensuring the functioning of the internal market in novel foods and protecting public health against the risks to which those foods may give rise, is not to be adversely affected, protective measures adopted under the safeguard clause may not properly be based on a purely hypothetical approach to risk, founded on mere suppositions which are not yet scientifically verified (see to that effect, as regards a non-harmonised field, the judgment of the EFTA Court in Case E-3/00 EFTA Surveillance Authority v Norway, EFTA Court Reports 2000-2001, p. 73, paragraphs 36 to 38).

107. Such protective measures, notwithstanding their temporary character and even if they are preventive in nature, can be adopted only if they are based on a risk assessment which is as complete as possible in the particular circumstances of an individual case, which indicate that those measures are necessary in order to ensure that novel foods do not present a danger for the consumer, in accordance with the first indent of Article 3(1) of Regulation No 258/97.

...

111. According to the case-law of the Court, it follows from the precautionary principle that where there is uncertainty as to the existence or extent of risks to human health, protective measures may be taken without having to wait until the reality and seriousness of those risks become fully apparent (see Case C-157/96 National Farmers' Onion and Others [1998] ECR I-2211, paragraph 63, and Case C-180/96 United Kingdom v Commission [1998] ECR I-2265, paragraph 99).

112. Therefore, protective measures may be taken pursuant to Article 12 of Regulation No 258/97 interpreted in the light of the precautionary principle even if it proves impossible to carry out as full a risk assessment as possible in the particular circumstances of a given case because of the inadequate nature of the available scientific data (see to that effect Pfizer Animal Health v Council, cited above, paragraphs 160 and 162, and Alpharma v Council, cited above, paragraphs 173 and 175).

113. Such measures presuppose, in particular, that the risk assessment available to the national authorities provides specific evidence which, without precluding scientific uncertainty, makes it possible reasonably to conclude on the basis of the most reliable scientific evidence available and the most recent results of international research that the implementation of those measures is necessary in order to avoid novel foods which pose potential risks to human health being offered on the market.

ECJ: C-528/16 Confédération paysanne and Others v Premier ministre and Ministre de l'Agriculture, de l'Agroalimentaire et de la Forêt, 25 July 2018

(Deliberate release of genetically modified organisms into the environment — Mutagenesis — Directive 2001/18/EC — Articles 2 and 3 — Annexes I A and I B — Concept of 'genetically modified organism' — Techniques/methods of genetic modification conventionally used and deemed to be safe — New techniques/methods of mutagenesis — Risks for human health and the environment — Discretion of the Member States when transposing the directive — Directive 2002/53/EC — Common catalogue of varieties of agricultural plant species — Herbicide-tolerant plant varieties — Article 4 — Acceptability of genetically modified varieties obtained by mutagenesis for inclusion in the common catalogue — Human health and environmental protection requirement — Exemption)

Question

Comment on the following paragraphs under the light of the ECHR, the CBD, and the EU Charter of Fundamental Rights also:

27 Article 2(2) of Directive 2001/18 defines a GMO as an organism, with the exception of human beings, in which the genetic material has been altered in a way that does not occur naturally by mating and/or natural recombination.

28 Account being taken of the information provided by the referring court, it must be noted, first, that the mutations brought about by techniques/methods of mutagenesis such as those at issue in the main proceedings, the implementation of which is intended to produce herbicide-resistant varieties of plant species, constitute alterations made to the genetic material of an organism, for the purposes of Article 2(2) of Directive 2001/18.

29 Secondly, since, as is apparent from the order for reference, certain of those techniques/methods involve the use of chemical or physical mutageneous agents, and others involve the use of genetic engineering, those techniques/methods alter the genetic material of an organism in a way that does not occur naturally, within the meaning of that provision.

30 It follows that organisms obtained by means of techniques/methods of mutagenesis must be considered to be GMOs within the meaning of Article 2(2) of Directive 2001/18.

31 That interpretation is supported by the general scheme of that directive, which is one of the factors to be taken into account for the purpose of its interpretation.

32 It should be noted that the definition of a GMO in Article 2(2) of Directive 2001/18 is made clear by a distinction between techniques the use of which results in genetic modification and techniques which are not considered to result in such genetic modification.

33 In that regard, Article 2(2)(a) of Directive 2001/18 states that, for the purpose of the definition of a GMO, genetic modification occurs at least through the use of the techniques listed in part 1 of Annex I A to that directive.

34 Although part 1 of Annex I A to that directive does not explicitly refer to techniques/methods of mutagenesis, that fact is not such as to exclude organisms obtained by means of those techniques/methods from coming under the definition of a GMO in Article 2(2) of the directive.

35 It should be noted, first, that, as follows from the expression 'inter alia' in the first sentence of part 1 of Annex I A to Directive 2001/18, the list of genetic modification techniques in that part is not exhaustive. Consequently, that list cannot be regarded as excluding genetic modification techniques other than those to which it specifically refers.

36 Secondly, it must be noted that the EU legislature has not included mutagenesis in the exhaustive list of techniques not resulting in a genetic modification, referred to in Article 2(2)(b) of Directive 2001/18, read in conjunction with part 2 of Annex I A to that directive.

37 On the contrary, mutagenesis is expressly cited, in Annex I B to that directive, as one of the techniques/methods of 'genetic modification' referred to in Article 3(1) of that directive, relating to organisms that have to be excluded from the scope of the directive.

38 In the light of the foregoing considerations, Article 2(2) of Directive 2001/18 must be interpreted as meaning that organisms obtained by means of techniques/methods of mutagenesis constitute GMOs within the meaning of that provision.

...

39 It is apparent from Article 3(1) of Directive 2001/18, relating to exemptions, that that directive does not apply to organisms obtained through the techniques of genetic modification listed in Annex I B to that directive.

40 In that regard, Annex I B lists the techniques/methods of genetic modification yielding organisms which, on condition that they do not involve the use of recombinant nucleic acid molecules or GMOs other than those produced by one or more of the techniques/methods listed in that annex, are to be excluded from the scope of that directive. Among those techniques/methods, point 1 of that annex refers to mutagenesis.

41 At the outset, it should be pointed out that, as a provision derogating from the requirement to subject GMOs to the obligations laid down in Directive 2001/18, Article 3(1) thereof, read in conjunction with point 1 of Annex I B to that directive, must be interpreted strictly (see, by analogy, judgment of 17 April 2018, Commission v Poland (Białowieża Forest), C-441/17, EU:C:2018:255, paragraph 189 and the case-law cited).

42 Furthermore, for the purpose of interpreting a provision of EU law, it is necessary to consider not only its wording but also the context in which it occurs and the objectives pursued by the rules of which it is part (judgment of 27 April 2017, Pinckernelle, C-535/15, EU:C:2017:315, paragraph 31).

...

48 As the referring court states in essence, the risks linked to the use of those new techniques/methods of mutagenesis might prove to be similar to those which result from the production and release of a GMO through transgenesis. It thus follows from the material before the Court, first, that the direct modification of the genetic material of an organism through mutagenesis makes it possible to obtain the same effects as the introduction of a foreign gene into that organism and, secondly, that the development of those new techniques/methods makes it possible to produce genetically modified varieties at a rate and in quantities quite unlike those resulting from the application of conventional methods of random mutagenesis.

49 Moreover, as stated in recital 4 of Directive 2001/18, living organisms, whether released into the environment in large or small amounts for experimental purposes or as commercial products, may reproduce in the environment and crossnational frontiers, thereby affecting other Member States. The effects of such releases on the environment may be irreversible. In the same vein, recital 5 of that directive states that the protection of human health and the environment requires that due attention be given to controlling risks from such releases.

50 Furthermore, it has been emphasised, in recital 8 of that directive, that the precautionary principle was taken into account in the drafting of the directive and must also be taken into account in its implementation. Emphasis is also placed, in recital 55 of Directive 2001/18, on the need to follow closely the development and use of GMOs.

51 In those circumstances, Article 3(1) of Directive 2001/18, read in conjunction with point 1 of Annex I B to that directive, cannot be interpreted as excluding, from the scope of the directive, organisms obtained by means of new techniques/methods of mutagenesis which have appeared or have been mostly developed since Directive 2001/18 was adopted. Such an interpretation would fail to have regard to the intention of the EU legislature, reflected in recital 17 of the directive, to exclude from the scope of the directive only organisms obtained by means of techniques/methods which have conventionally been used in a number of applications and have a long safety record.

Provide legally appropriate solutions

1. A dairy company intends to market ice creams produced with the milk of GM cows. The company wishes to mention on the ice creams labels and the advertising spots that regular consumption in each daily meal reduces cholesterol levels by up to 50%. Also, consumption is recommended particularly for children, and the product provides vitamins and other nutritional elements in quantities that cannot be ensured by any conventional diet. A member of the competent licensing authority claimed that these ice creams should be considered "pharmaceuticals"; therefore, they need to be tested in clinical trials, otherwise they may be dangerous for the consumers.

Comment on the company's application and the above suggestion, in the light of the current European legislation.

2. A municipal department rejects the application of food company C concerning cultivation licensing for a type of GM maize. The department justified its decision with reference to an "objection of the citizens' majority." The company asks for revision based on an expert report demonstrating safety evidence.

Comment on the legal validity

- Of the department's decision
- Of the company's argumentation

3. G is a substance used for the treatment of neurodegenerative diseases. It is the natural product of a plant that yields G in minimal quantities. A research team makes possible the artificial creation of a bacterium's genome, yielding pure G in significant quantities at low cost. Under the current European law

- Is the artificial development of the new bacterium legally accepted, particularly under the view of the "precautionary principle"?
- Is it legally accepted a possible use for developing anti-aging substances instead of therapeutic drugs?

References

Aasen M, Vatn A (2013) Deliberation on GMOs: a study of how a citizens' jury affects the citizens' attitudes. Environ Values 22:461–481

Adam B (1994) Running out of time: global crisis and human engagement. In: Redclift M, Benton T (eds) Social theory and the global environment. Routledge, London – New York, pp 92–112

American Medical Association House of Delegates (2012) Labeling of bioengineered foods. Council on Science and Public Health Report 2, in http://www.amaassn.org/resources/doc/csaph/a12-csaph2-bioengineeredfoods.pdf

Andorno R (2004) The precautionary principle: a new legal standard for a technological age. J Int Biotechnol Law 1:11–19. https://doi.org/10.1515/jibl.2004.1.1.11

Andorno R (2013) Principles of international biolaw. Seeking common ground at the intersection of bioethics and human rights. Bruylant, Brussels

Anomaly J (2020) Creating future people: the ethics of genetic enhancement. Routledge, London

Basili M, Franzini M (2006) Decision-making under uncertainty and irreversibility: a rational approach to the precautionary principle. In: Basili M, Franzini M, Vercelli A (eds) Environment, inequality, and collective action. Routledge, London, New York, pp 219–244

Böckenforde M (2004) Grüne Gentechnik und Welthandel: Das Biosafety-Protokoll und seine Auswirkungen auf das Regime der WTO. Springer, Heidelberg

Bovay J, Alston JM (2018) GMO food labels in the United States: economic implications of the new law. Food Policy 78:14–25

Bruetschy C (2019) The EU regulatory framework on genetically modified organisms (GMOs). Transgenic Res 28:169–174. https://doi.org/10.1007/s11248-019-00149-y

Buckens MM, Changeur L (2000) GMOs and food safety. Europe Information Service, Brussels

Bustos C, Rozo-Ángel V, Eslava-Bejarano G (2020) Protecting the rights of future generations through climate litigation: lessons from the struggle against deforestation in the Colombian Amazon. In: Henry C, Rockström J, Stern N (eds) Standing up for a sustainable world. Edward Elgar Publishing, UK, USA, pp 163–170

Cole CF (1998) Sustaining world fisheries resources: an ethical dilemma. In: Lemons J, Westra L, Goodland R (eds) Sustainability and integrity: concepts and approaches. Springer, Dordrecht, Boston, London, pp 168–181

Collins SR, Weissman JS, Krogan NJ (2009) From information to knowledge: new technologies for defining gene function. Nat Methods 6:721–723

Devos Y, Demont M, Dillen K, Reheul D, Kaiser M, Sanvido O (2009) Coexistence of genetically modified (GM) and non-GM crops in the European Union. A review. Agronomy for Sustainable Development, Springer Verlag/EDP Sciences/INRA, 29: 11–30, ffhal-00886478f

Diels J, Cunha M, Manaia C, Sabugosa-Madeira B, Silva M (2011) Association of financial or professional conflict of interest to research outcomes on health risks or nutritional assessment studies of genetically modified products. Food Policy 36:197–203
Domingo JL, Bordonaba JG (2011) A literature review on the safety assessment of genetically modified plants. Environ Int 37:734–742
EGE (1996) Opinion no 7 on Ethical Aspects of Genetic Modification of Animals
EGE (2009) Opinion no 25 on the Ethics of Synthetic Biology
European Commission (2010) A decade of EU-funded GMO research (2001–2010), Brussels, in http://ec.europa.eu/research/biosociety/pdf/a_decade_of_eu-funded_gmo_research.pdf
Gibson DG, Glass JI, Lartigue C, Noskov VN, Chuang R-Y, Algire MA, Benders GA, Montague MG, Ma L, Moodie MM, Merryman C, Vashee S, Krishnakumar R, Assad-Garcia N, Andrews-Pfannkoch C, Denisova EA, Young L, Qi Z-Q, Segall-Shapiro TH, Calvey CH, Parmar PP, Hutchison CA III, Smith HO, Venter JC (2010) Creation of a bacterial cell controlled by a chemically synthesized genome. Science 329:52–56
Gilbert D, Lee-Kelley L, Barton M (2003) Technophobia, gender influences and consumer decision-making for technology-related products. Eur J Innov Manag 6:253–263
Greer D, Harvey B (2004) Blue genes: sharing and conserving the World's Aquatic Biodiversity. Earthscan, IDRC, London – Sterling US – Ottawa
Hansen B, Fjelsted Alrøe H, Steen Kristensen E, Wier M (2002) Assessment of food safety in organic farming, DARCOF Working Papers 52, in http://orgprints.org/206/1/Hansen_organic_food_safety.pdf
Hansson SO (2020) How extreme is the precautionary principle? NanoEthics 14:245–257
Hilbeck A, Binimelis R, Defarge N, Steinbrecher R, Székács A, Wickson F, Antoniou M, Bereano PhL, Clark EA, Hansen M, Novotny E, Heinemann J, Meyer H, Shiva V, Wynne B (2015) No scientific consensus on GMO safety. Environmental Sciences Europe 27: 1–6, in http://www.enveurope.com/content/pdf/s12302-014-0034-1.pdf
Holl R, Johnston S, Sendashonga C (2004) Risk assessment and precaution in the Biosafety Protocol. Rev Eur Community Int Environ Law 13:263–269
Horkheimer M (1947) The eclipse of reason, Oxford U.P
Houdebine LM (2009) Production of pharmaceutical proteins by transgenic animals. Comp Immunol Microbiol Infect Dis 32:107–121
Jans J, Veder H (2008) H.H.B., European environmental law, 3rd edn. Europa Law Publ., Groningen
Janzen DH (1988) Tropical dry forests: the most endangered major tropical ecosystem. In: Wilson EO (ed) Biodiversity. National Academy Press, Washington D. C., pp 130–137
Kloor K (2013) Greens on the run in debate over genetically modified food. Bloomberg, 7 January, in http://www.bloomberg.com/news/2013-01-07/green-activist-reverses-stance-on-genetically-modified-food.html
Lee M (2009) EU Regulation of GMOs. Law and decision making for a new technology. Elgar Publ., Cheltenham UK, Northampton MA
Lemons J (1998) Burden of proof requirements and environmental sustainability: science, public policy and ethics. In: Lemons J, Westra L, Goodland R (eds) Sustainability and integrity: concepts and approaches. Springer, Dordrecht, Boston, London, pp 75–103
Louka E (2006) International environmental law. Fairness, effectiveness and world order. Cambridge U. P., Cambridge, New York
Markie P (2007) Mandatory genetic engineering labels and consumer autonomy. In: Weirich P (ed) Labeling genetically modified food. The philosophical and legal debate. Oxford UP, Oxford, pp 88–105
Mosier SL, Rimal A, Ruxton MM (2020) A song of policy incongruence: the missing choir of consumer preferences in GMO-labeling policy outcomes. Rev Policy Res 37:511–534
Nuffield Council on Bioethics (1999) Genetically modified crops: the ethical and social issues. London

O'Neill J, Baertschi B, Schark M, Sandler R, Deplazes-Zemp A, Attfield R (2012) Special Issue: synthetic biology. Environ Values 21:1–94

OECD (2010) Challenges for agricultural research. OECD Publ., Paris

Pascalev A (2003) You are what you eat: genetically modified foods, integrity and society. J Agric Environ Ethics 16:583–594

Philips PWB (2007) Farmers' privilege and patented seeds. In: Philips PWB, Onwuekwe CB (eds) Accessing and sharing the benefits of the genomics revolution. Springer, Dordrecht, pp 49–64

Reiss MJ, Straughan R (1996) Improving nature? The science and ethics of genetic engineering. Cambridge U. P., Cambridge, New York

Smets G, Rudelscheim P (2018) Biotechnologically produced chitosan for nanoscale products: a legal analysis. New Biotechnol 42:42–47. https://doi.org/10.1016/j.nbt.2018.02.005

Stefánsson HO (2019) On the limits of the precautionary principle. Risk Anal 39:1204–1222. https://doi.org/10.1111/risa.13265

Taylor KI (1988) Deforestation and Indians in Brazilian Amazonia. In: Wilson EO (ed) Biodiversity. National Academy Press, Washington D. C., pp 138–144

Taylor P (1998) An ecological approach to international law. Responding to challenges of climate change. Routledge, Oxon, New York

Victor M (2001) Precaution, or protectionism? The precautionary principle, genetically modified organisms, and allowing unfounded fear to undermine free trade. Transnatl Lawyer 14:295–321

Wasmer M (2019) Roads forward for European GMO Policy—uncertainties in wake of ECJ judgment have to be mitigated by regulatory reform. Front Bioeng Biotechnol 7:132. https://doi.org/10.3389/fbioe.2019.00132

White M (2013) The scientific debate about GM foods is over: they're safe. Pacific Standard Magazine, 24 September, in http://www.psmag.com/health/scientific-debate-gm-foods-theyre-safe-66711/

Wolf C (2007) Labeling genetically engineered foods: rights, risks, interests, and institutional options. In: Weirich P (ed) Labeling genetically modified food. The philosophical and legal debate. Oxford U.P., Oxford, p 178. seq

World Health Organization (WHO)/ Food and Agriculture Organization (FAO) of the United Nations (2009) Codex Alimentarius: foods derived from modern biotechnology, 2nd edn. Rome, in ftp://ftp.fao.org/codex/Publications/Booklets/Biotech/Biotech_2009e.pdf

Wynne B (1994) Scientific knowledge, and the global environment. In: Redclift M, Benton T (eds) Social theory and the global environment. Routledge, London – New York, pp 179–200

Zimny T, Eriksson D (2020) Exclusion or exemption from risk regulation? A comparative analysis of proposals to amend the EU GMO legislation. EMBO Rep 21. https://doi.org/10.15252/embr.202051061

13 Life as a Commodity

The significant progress of biotechnology raises issues related to the commodification of living organisms or parts of them, from microorganisms to humans themselves. Here, the fundamental question is to what extent commodity as a concept might be compatible with the value of life in its various forms.

We have already examined this question specifically regarding the commercialization of products or parts of the human body under the light of human dignity. Nevertheless, life as a commodity needs to be investigated in the broader meaning covering, firstly, the commercialization of plants and animals and, secondly, the commercialization of non-living products, which are necessary for the conservation and development of life (products of food, feed, pharmaceuticals, etc.). A third issue is that of patenting living organisms or elements of them (including elements of the human organism), for promoting their commercial exploitation.

13.1 Commercialization of Plants and Animals

Commercialization is closely related to the availability of plants and animals for meeting various human needs. This refers mostly to their direct use as food or feed products. There are also needs for clothing or entertainment or (in a few cases) for working support justifying the placing on the open market of living organisms or their products.

Legally speaking, the commercial exploitation of non-human species seems obvious, given their status as objects of law that makes them available for transactional activities, according to the relevant provisions of civil and commercial law.

Still, that general approach is not without exceptions. For particular reasons, the commercial exploitation of certain species is either restricted or even prohibited. This happens, as we have said already, regarding rare species of flora and fauna (mostly the wild ones), according to article 3 para 1 of the EU Regulation 338/97, the CITES instrument, and relevant legislation, and also regarding the use of vertebrate animals for entertainment purposes.

T. Vidalis, *The Emergence of Biolaw*, Springer Textbooks in Law,
https://doi.org/10.1007/978-3-031-02359-0_13

In the first case, restrictions and prohibitions are grounded on the protection of biodiversity, that is, the prevention of species extinction. Thus, although accepting the use of rare plants and animals for serving human needs, the legislator considers that the motivation of profit is likely to create unsafe conditions for their conservation. For instance, the removal of wild animals from their natural habitats, where they live freely, and their transport in conditions of captivity for commercial exploitation, does not ensure their breeding to a sufficient degree for the conservation of the species. To exclude these living organisms from marketing procedures is legally justified because the market's supply and demand mechanism does not guarantee their existence, which is dependent closely on living in specific natural habitats.

In the second case, prohibition is related to the danger that emerges for individual vertebrates' welfare when involved in entertainment activities. By excluding animals' use in shows and exhibitions, any motivation for commercial exploitation of these animals is annulled. The animal welfare prevails in that case over the economic freedom of those interested in such activities.

Still, as a general rule, plants and animals are considered by the law as commodities, subjects of free commercialization, insofar as two legal standards are ensured: The conservation of biodiversity, as a programmatic principle pertaining to all non-human species, and the protection of the animal welfare, regarding, in particular, the individual vertebrates.

13.2 Products Supporting Life

The commercialization of products indispensable for the conservation and reproduction of human life and other species' lives also needs to be examined.

Under the particular view of Biolaw, food, and feed products, as well as pharmaceuticals or other chemical products, deserve special attention, Compared to other consumable goods, since a lack of them influences the conditions of life directly. We can then argue that, if these products are essential for supporting the value of life, their disposal should not be subject exclusively to market priorities.

Following this approach, the law accepts interventions in their free marketing that differ in their sort, following the nature of the products concerned.

a) Food and feed marketing regulation is limited, in principle, to basic goods, establishing mechanisms of price control for ensuring sufficient supply. That regulation is part of the national law in each country.
b) Regarding pharmaceuticals, which are subjected to marketing on a global scale, national policies have only little impact. The EU relevant legislation is the basis, here, for the adoption of national policies on pricing.[1] In pharmaceuticals, supply depends on public insurance systems' economic conditions and other factors, like

[1] See, in particular, the Transparency Directive 89/105/EEC.

the national health systems operation, etc. This is crucial, particularly in periods of economic crisis when state interventions are frequent and more intensive.

c) The management of water, particularly potable water, as a minimal condition for the conservation of life is of great importance also. Most national laws adopt a regime that combines free access to natural water resources, a general supply provided by public companies, and access to potable water commercially disposed.

The management of water is dependent on sufficiency, given that climate change causes significant restrictions on natural resources. As already mentioned, the international law contains provisions for managing water resources among states for securing sufficiency; however, the problem is a long-run one. There are opinions in relevance suggesting a fundamental "right to water" also emerging in specific elaborations in international fora.[2] Given the lack of binding international or EU specific legal instruments, at the national legislation level the main question is whether an opening to the markets of the water supply, through its privatization, would be legally possible, following the model adopted for energy and telecommunications.

In terms of Biolaw, the answer does not differ from that concerning the other products that support life. Contrary to what happens with energy and telecommunications, where a practically unlimited availability of resources is possible, the water represents a limited natural resource, not excluding potential scarcity. Therefore, a guarantee of general and equal access to water resources is necessary for the conservation of life, including human life. Moreover, this access does not refer only to the direct consumption of water but also to its appropriate use in sufficient quantities and adequate quality for agricultural and industrial activities related to food and feed production, also crucial for life-supporting. This necessity impedes, in principle, an absolute opening of the water management to market priorities.

13.3 Patents in Biological Applications

The main legal issue concerning the commercialization of life refers to the recognition of patent rights on research findings. Like any sector of advanced technologies, biotechnology is based on profit motivation attracting mostly private funding, and also involves a relatively high risk of failures. By ensuring a privilege to the commercial exploitation of novel products, patents represent the crucial mechanism for attracting economic investment in biotechnological research since funding in that field is associated with significant risks. This privilege means a monopoly granted to the patent owner for a limited period. During that period, the owner prospects to

[2] See esp. Resolution 64/292 of the UN (2010) and art. 5 of the European Charter on Water Resources (Council of Europe, 2001).

cover the concrete investment or previous research costs of failed attempts and to gain commercial profits, without the risk of exposure to commercial competition.[3]

Biotechnological products include living organisms or parts of them (like GM or transgenic organisms, cells, and tissues, not existing in Nature); they also involve discoveries of biological properties through the application of new research methods (like the identification of genes responsible for the manifestation of serious diseases[4]), or they may be artificially created organisms or microorganisms identical to natural original models, like the products of synthetic biology. The question is whether patenting of all these products is allowed under the existing legislation.

In Europe, the basic rule for granting patent rights is the characterization of a product as an "invention." This condition is met by products emerging for the first time in the natural world, thanks to the creative work of researchers and developers. Therefore, existing natural products, even unrevealed or unknown, cannot be considered "inventions"; they may be findings of the research, but only as "discoveries" not subject to patenting. Nonetheless, the methodology and newly developed tools enabling researchers to reach a discovery may be subject to patenting since, as such, they are "inventions."

Apart from that general rule, the existing law in Europe excludes from patentability some inventions in order to protect generally acknowledged moral values. These inventions are considered to oppose the legal concepts of "ordre public" and "morality," and, for that reason, the law intends to discourage their marketing. "Moral" clauses appear in particular in the field of biotechnology, where the commodification of life in general (mostly, of human life) remains an open issue. Hence, for patents specifically related to biological applications, the legal framework is consisted of

a) The general legislation on patenting,
b) The EU specific legislation, and its most important instrument, Directive 98/44 for the legal protection of biotechnological inventions, and
c) Specific provisions of Biolaw prohibiting the commercialization of certain objects explicitly and, consequently, rejecting commercial rights like patenting.

Given that legal framework, we can conclude that patenting of a specific object presupposes two conditions; its characterization as an "invention" and its compliance with the ethics clauses mentioned by the law.

13.3.1 Inventions and Discoveries

According to the first of these two conditions, the patent object must be recognized as an "invention." This does not exclude the involvement of natural biological

[3]Cf. Kelves (2002).

[4]See the case of BRCA 1 and 2 genes on breast cancer.

mechanisms to create new living organisms or microorganisms. Indeed, the artificial creation of a new form of life is also based on the natural biological process of reproduction. Eventually, "creation" occurs when we mobilize properties already existing in the natural world. Yet, this activity may be seen as a "discovery" of their potential, unrevealed so far. The question, then, is why we accept that "discovery" as being "patentable"?

Firstly, because we are logically obliged to; even in the non-living world, the artificial creation of novel objects means nothing but a discovery of existing material properties; this does not influence the originality of these objects and their characterization as "inventions." Secondly, the concept of "originality" concerning the creation of living beings focuses on their appearance in the natural world for the first time, or in other words, on the fact that this appearance would be not possible, or at least not foreseen, given our current knowledge for the process of Evolution. Therefore, the patent law assumes that the "contribution to technical knowledge" is decisive for distinguishing a mere discovery from an invention.

In that view, the creation of microorganisms, plants, or animals with new phenotypes, resulting from artificial changes of the genome in concrete species through genetic engineering, meets the legal criteria of "inventions." Likewise, original organisms produced through conventional methods of reproduction with no involvement of molecular interventions (like in the case of crossbreeds between different races or even species of plants or animals) represent also "inventions" of living objects, subjected to patenting.[5]

Moreover, "inventions" should be considered microorganisms or future organisms that may occur through techniques imitating natural genomes, as it happens in synthetic biology. In such cases, it would be false to claim that it is not the object itself but only the particular technique that should be identified as an "invention" given that the object exists already in the natural world. Indeed, we need to distinguish, here, between the natural organism, which is resulted through the natural process of reproduction with the presence of life, and the artificial one which is created through the mere exploitation of biochemical reactions, in principle without the presence of life; the latter, even if phenotypically identical, is a different organism that may develop unknown properties also.

We conclude that the concept of "invention" can cover the phenomenon of life as well; there is no characteristic of "sanctity" or any specific element in the legal value of living objects that would exclude life from patentability. The above argumentation clarifies the explicit recognition of "inventions" in biotechnological applications by the law and the possibility of granting patent rights. A practical consequence of this is that, as a matter of principle, we cannot challenge the patentability of life on the basis of its special legal value. It is another issue whether we can exclude particular life forms from patentability due to certain ethical reasons, as we will see when examining the ethics clauses of the current legislation.

---

[5] *Contra*, Sagoff (1996), pp. 15 seq., arguing that GM animals are not products of designing; therefore, they cannot be considered as "inventions."

Nevertheless, apart from clear cases of inventions, the experience of contemporary technological applications indicates that patenting rights are claimed even for elements of natural organisms, after isolating them artificially, elements encompassing valuable biological information. Moreover, patent rights are discussed even for entire organisms or microorganisms having unique properties.

A characteristic example is that of genes in the human genome, the study of which has shown their correlation with serious diseases.[6] As such, genes cannot exist isolated from their natural context, consisting of proteins or other biological elements that form a complex biochemical unity. Still, claims on their patentability are based on their isolation, namely on their appearance for the first time in the natural world under that particular form.

A similar example is that of plants revealing significant nutritional or therapeutic properties, known traditionally by indigenous communities of certain developing countries. The molecular composition of these plants was investigated by pharmaceutical and biotechnological companies, which developed new products based on that original knowledge. For marketing reasons, companies are interested in obtaining patent rights over the plants as such, despite the lack of any original contribution to their existence in Nature, namely the absence of a characteristic element for "inventions." The acknowledgment of patent rights over the biological wealth of developing countries, permitting exclusive exploitation in the markets, deprives local communities of the traditional use of these precious organisms; this phenomenon is identified as a new manifestation of colonial exploitation, known by the term "biopiracy".[7]

For the protection of indigenous communities' rights over local genetic resources and the benefit-sharing international policy in relevance, the CBD's Nagoya Protocol on *"Access to Genetic Resources and the Fair and Equitable Sharing of Benefits Arising from their Utilization to the Convention on Biological Diversity"* (2010), is the legal instrument of reference. Especially in its articles 5 para 5, 6 para 2, 7, 12, 22, and 23, the Protocol entitles countries providing local biological resources to enact legislation for the fair and equitable sharing of benefits (economic or others) arising from the exploitation of the acquired genetic knowledge. The Protocol accepts the legal rights of local indigenous communities over these genetic resources that need to be taken into account on the basis of "mutual agreed terms" with companies developing relevant technology and applications. In that context, the prerequisites of "informed consent or approval and involvement" of indigenous communities are necessary for such agreements. Furthermore, provisions on state

[6]BRCA 1, 2, genes, already mentioned.

[7]Cf. Mgbeoji (2006) and Robinson (2010). See also art. 8 I of the CBD, regarding respect and protection of indigenous communities' traditional knowledge related to the use of biological resources. The introduction of agricultural biotechnology in these countries creates wider problems concerning the substitution of traditional products that they export, which negatively influences their economy. Cf. Kumar (1998), pp. 47–48. On the distinct term of "bioprospecting" (meaning "the exploration of biodiversity for commercially valuable genetic resources and bio chemicals"), see Mukku et al. (2021).

cooperation for promoting capacity-building and technology transfer to developing countries (notably the least developed of them) are expected to facilitate the implementation of the benefit-sharing strategy.[8]

Legally speaking, either we examine entire organisms (like plants), or their elements (like genes), we cannot characterize them as "inventions" since they exist already in the natural world; therefore, in principle, they are not patentable objects.

In the example of genes, what is indeed an "invention" is the method of isolation or "purification" from their natural biological context, but not genes as such, since they represent an existing, although "hidden" element of the human organism. The practice of patent offices' looked ambivalent on that issue, as expressed regarding patenting claims over the breast cancer genes (BRCA 1 and 2).[9] Moreover, the relevant legal provisions are somewhat vague, even those of the EU law (Directive 98/44), which remains stricter than the law in the U.S.[10] Yet, after these patents' rejection by the U.S. Supreme Court, the issue becomes more evident in terms of legal certainty.[11]

In the example of plant patenting, the elaboration of existing properties with scientific means for developing new pharmaceuticals etc. is not sufficient to conclude that the original plant has been converted into something "new" in the natural world for being subject to patents. Indeed, its natural properties already encompass the potential of such a development, which is not "invented" by technology. Certainly, in this case, also, methods guiding the discovery of these essential properties and the development of new products could be patentable.

The commercial interest in discoveries in Genetics and Molecular Biology is indeed intense, as the future of Medicine, Pharmacology, and Agriculture is considered directly dependent on these. Therefore, there is an obvious concern for "locking" exclusive sectors when developing research and potential applications through patenting even known elements of living organisms, including elements of the human life, given that the patent owner gains a crucial advantage in the market. However, this industrial tendency remains legally problematic since it challenges the importance of original creation that characterizes inventions, which is fundamental for the European patent law. Moreover, the prerequisite of the original creation, for recognizing a market advantage, has a wider significance, since it enables the development of free research in any field of the natural world, without the restrictions imposed by patented "sectors"; this contributes to discovering particular

[8]On benefit-sharing, see also art. 15 of the UNESCO's Bioethics Declaration.

[9]On gene 1, see the initial EP 705902 and the definitive decision 699754 of the European Patent Office, which accepted appeals by European political parties, NGOs, research institutes, scientific associations, etc.

[10]See, esp. the "permissive" art. 5 para 2 of Dir. 98/44, according to which: *"An element isolated from the human body or otherwise produced by means of a technical process, including the sequence or partial sequence of a gene, may constitute a patentable invention, even if the structure of that element is identical to that of a natural element".*

[11]Cf. Association for Molecular Pathology v. Myriad Genetics, 569 U.S. (2013). Cf. Kesselheim et al. (2013), pp. 869 et seq.

manifestations and elements of the Nature that can be elaborated by technology to produce real patentable "inventions." This vital cycle that connects research with industrial exploitation, through free communication and cooperation between research teams, would be disturbed drastically if the freedom of research as a fundamental right were limited within patented "reserved domains" of the natural world.

13.3.2 "Morality" Clauses

However, regarding products of biological applications, the "invention" standards are not sufficient requirements for patenting. The EU law describes an additional standard, not of technical but of ethical nature, precisely because these applications refer to life, namely an objective with a special value. Thus, according to Directive 98/44 (art. 6):

1. Inventions shall be considered unpatentable where their commercial exploitation would be contrary to ordre public or morality; however, exploitation shall not be deemed to be so contrary merely because it is prohibited by law or regulation.
2. On the basis of paragraph 1, the following, in particular, shall be considered unpatentable:
   (a) processes for cloning human beings;
   (b) processes for modifying the germ line genetic identity of human beings;
   (c) uses of human embryos for industrial or commercial purposes;
   (d) processes for modifying the genetic identity of animals, which are likely to cause them suffering without any substantial medical benefit to man or animal, and also animals resulting from such processes.

These provisions of the Directive have the meaning of "morality" clauses for the recognition of patents in the field of life sciences specifically; in that sense, they elaborate the concept of "morality," as referred to the general European legislation on patenting (pertinent to inventions regardless of their objective) and obstruct patentability.

We can distinguish two sorts of morality clauses in the field of biological applications. First, those referring to particular inventions that are considered as non-patentable by definition. These are:

(a) processes for cloning human beings;
(b) processes for modifying the germ line genetic identity of human beings;
(c) uses of human embryos for industrial or commercial purposes;
(d) processes for modifying the genetic identity of animals, which are likely to cause them suffering without any substantial medical benefit to man or animal, and also animals resulting from such processes.

In the law's view, these four cases represent typical, although indicative, examples of unpatentable inventions.

Second, beyond these examples, the inclusion of others remains open, following the general clause of paragraph 1, if "*their commercial exploitation would be contrary to ordre public or morality*".

Directive 98/44 clarifies a preliminary point by considering unpatentable inventions opposing the "ordre public" or "morality". This means a) that this particular legal instrument does not exclude commercial exploitation without patent rights (no matter if profitable or not), and b) that, contrary to its patentability, the invention, as such, is not characterized as illegal.[12]

Thus, even unpatentable, a biological invention remains ethically neutral outside the scope of the Directive. Although with no commercial interest, that invention may be used for other non-commercial purposes, in research, education, etc. This concerns all inventions falling within the above provision, namely the four explicitly mentioned and any other characterized as contrary to the ordre public or morality.

Besides that point, the ethics clauses' interpretation regarding patentability needs to be examined separately for each case mentioned in the Directive.

1. The reference to "methods of human cloning" is general, with no distinction between reproductive and therapeutic cloning. Unpatentability, here, conforms with the explicit prohibition of reproductive cloning by both the EU Charter of Fundamental Rights, the relevant Additional Protocol of the Oviedo Convention, and the prohibition of creating embryos for research purposes, according to article 18 para 2 of the Convention, which implicitly prohibits therapeutic cloning as well. Therefore, the European law addresses cloning not only as an unpatentable invention but also as an illicit one, excluding all potential uses, commercial or not. Yet, this approach refers only to cloning *stricto sensu* (the incorporation of an adult human cell's nucleus into a human oocyte). It does not cover other methods, like for instance, the research use of totipotent stem cells obtained by a human blastocyst, or in general any research in human stem cells. This also includes their "reprogramming" or their genetic modification or other intervention in their biological composition (on the condition that the ultimate research purpose is not human reproduction or commercial or industrial use, as we will see later).
   The patentability of organisms produced after combining the human genome with genomes from different species (hybrids, chimeras) is another issue. The methodology for creating such an organism may be similar to that of cloning when it involves the transfer of a human cell's nucleus to an animal oocyte. Still, it is not human cloning *stricto sensu*; therefore, we need to address patentability here under the light of the general ethics clause of the Directive's Article 6 para 1. In certain European legislations, the creation of hybrids or chimeras is explicitly prohibited, which means that patentability is excluded at their national level since

[12]The limits of these generally accepted patent law standards were wrongly blurred in the Brüstle decision of the ECJ (Case C-34/10, Oliver Brüstle v. Greenpeace e.V., Judgement of the Court (Grand Chamber) of 18 October 2011) against the clear wording of the Directive. See, comment by Spranger (2012), pp. 1197 seq.

that prohibition represents evidence regarding the characterization of these inventions as "contrary to morality". At the European level, though, such evidence does not exist. Assuming that a relevant case could be brought before the European Patent Office, patentability would be dependent on the potential uses of hybrids or chimeras. If these uses are restricted to the exploitation of biological material (stem cells as an alternative source of transplants, for instance), and reproductive purposes are clearly excluded, patentability cannot be contested on the ground of "ordre public" or "morality."

2. By "*processes for modifying the germ line genetic identity of human beings,*" we mean genetic engineering interventions in the human genome, intending to modify gametes (sperm and oocytes) or cells that generate them.
   In the first part of the study, we mentioned that article 13 of the Oviedo Convention prohibits explicitly such interventions, given that their consequences are not limited to the person involved, but they affect the genome of future descendants as well. Even if we assume that some of these modifications may be beneficial for the offspring (as preventing genetic predispositions responsible for severe diseases), an application of the precautionary principle justifies that prohibition due to the lack of knowledge about functions and interactions between genes, which creates uncertainty about possible risks of genetic implications.[13]
   Directive 98/44 shares this attitude of the Oviedo Convention, like in the case of cloning. If, in the future, indications of germline cells' safe modifications occur, the law will need to change accordingly, in order to enable both the development of technical issues (mostly through designing of clinical trials) and patentability, as a commercial motivation. However, the rationale of these prohibitions suggests that the interventions concerned are only those related to persons and not merely to human biological material (isolated cells or tissues). This means that the basic research in that field is legally accepted since it does not negatively influence a human organism; the patentability of that basic research results is equally acceptable for the same reason.
3. The reservation regarding "*uses of human embryos for industrial or commercial purposes*" raises questions of interpretation. It covers, without a doubt, possible uses of embryonic tissue or cells for fabricating products of common consumption (cosmetics, etc.[14]). On the contrary, uses serving basic research or future therapeutic means (like the development of tissue or even solid organs from embryonic stem cells suitable for transplantations) cannot be characterized as "industrial" or "commercial," even if they involve economic considerations. On the other hand, if the law aimed at excluding the use (and, moreover, the creation) of embryos for purposes other than reproduction, it would be expressed explicitly, mostly because of the attitudes of national legislations in Europe

[13] Genome editing techniques (particularly the CRISPR/Cas9 method) facilitate targeted interventions even in human gametes and embryos. On ethical issues in relevance, see *supra*.

[14] See article in the Washington Times by Richardson (2009).

regarding that issue present significant differences.[15] Indeed, while some national laws prohibit entirely any research on embryos, others recognize their use for research purposes, on the condition that these embryos, although initially created for reproductive purposes, eventually remained "spare." Given that controversial normative environment, it would be peculiar to interpret the Directive in a way that essentially would deny the attitude of these latter national legislations.[16]

4. The fourth reservation refers to methods of animal genetic modifications, guiding to the emergence of new species, the individuals of which, though, suffer because of that modification, with no specific scientific usefulness. That case is rather of theoretical significance. It might apply in laboratory animals used for medical pharmacological or veterinary research, often requiring "model" organisms manifesting particular properties not existing in natural species. These organisms result either from the genetic modification of natural species' individuals or from combined genomes, and occasionally they suffer from birth, presenting pathological symptoms intentionally designed for research reasons. The use of these animals, based on scientific protocols, aims by definition at the progress of science. Thus, only on extreme occasions of irresponsible behavior by researchers, we could demonstrate a use "*without any substantial medical benefit*," making that provision applicable. In any case, the prohibition of patentability of such methods, or the animals created, conforms to the protection of the animals' corporal integrity, which we repeatedly meet in the law as a general rule in the context of animal welfare.

These four explicit reservations of the Directive are only indicative. Prohibition of patentability is relevant to any biotechnological invention that may be considered contrary to the "ordre public" or "morality." The *in concreto* interpretation of these (generally known in the European law) clauses is a very sensitive issue, precisely because the very object of the different life forms' management has a strong ethical weight. Therefore, we need to thoroughly study the necessary balance between the freedom of research as a fundamental right and the relevance of the Directive's reservations for preventing absurd interpretations.

In that respect, we could assume a basic framework for interpretation, formed by certain elements deriving from the European Biolaw. This framework would be based on a rule, according to which whatever Biolaw's provisions do not prohibit could be eligible as an object of patenting, on the condition that it is recognized as an invention. This may cover, for instance,

[15] In the EU countries, relevant laws, e.g., in Germany, Austria, and Italy, prohibit any research on embryos. Other national legislations allow research only on "spare" embryos initially created for fertility treatment. In the U.K. and Sweden, it is allowed the creation of embryos for research purposes.

[16] On the similar issue of embryonic stem cells' patentability, see Spranger (2004), pp. 247 et seq. Against an open approach to the different attitudes of national laws in relevance, see in particular, the ECJ Brüstle case (*supra*), recognizing human embryos as individual subjects of dignity.

- Future technical improvements of assisted reproduction
- New methods of genetic testing, revealing correlations between genetic characteristics and phenotypes not necessarily related to health matters
- Development of artificially improved food, feed, additives, cosmetics, etc.
- Methods for creation of genetically modified or transgenic animals or animal hybrids and chimeras, as well as these animals as such, if not entailing animal suffering or pain

For all the above and similar inventions, we can conclude that a favorable or at least neutral will of the European legislator, as expressed in the entire corpus of Biolaw, reflects an objective ethical standard; therefore, no issue regarding violations of "ordre public" or "morality" occurs. On the other hand, we must also accept that, where national laws prohibit certain inventions on ethical grounds, patentability may be excluded only at the national level, namely the level of the respective national legal system, not affecting the European system. Under this view, the European Patent Office may grant a patent, which would be enforceable only in countries that accept this invention, in terms of "ordre public" and "morality."

13.4 Further Studying...

ECJ: Case C-192/01 Commission of the European Communities v Kingdom of Denmark, 23 September 2003
(Free movement of goods — Quantitative restrictions — Measures having equivalent effect — National legislation subjecting the marketing of foodstuffs enriched with nutrients to proof of the population's nutritional need — Not permissible — Justification — Protection of public health — No justification — Breach of the ¡principle of proportionality (Arts 28 EC and 30 EC))

Question

Comment on the following paragraphs under the light of the EU Charter of Fundamental Rights also:

44. Community law does not therefore, in principle, preclude a Member State from prohibiting, save for prior authorisation, the marketing of foodstuffs incorporating nutrients, such as vitamins or minerals other than those whose use is lawful under Community legislation.

45. However, in exercising their discretion relating to the protection of public health, the Member States must comply with the principle of proportionality. The means which they choose must therefore be confined to what is actually necessary to ensure the safeguarding of public health; they must be proportional to the objective thus pursued, which could not have

been attained by measures which are less restrictive of intra-Community trade (see Sandoz, paragraph 18; Bellon, paragraph 14; and Harpegnies, paragraph 34).

46. Furthermore, since Article 30 EC provides for an exception, to be interpreted strictly, to the rule of free movement of goods within the Community, it is for the national authorities which invoke it to show in each case, in the light of national nutritional habits and in the light of the results of international scientific research, that their rules are necessary to give effective protection to the interests referred to in that provision and, in particular, that the marketing of the products in question poses a real risk to public health (see, to that effect, Sandoz, paragraph 22; Case 227/82 Van Bennekom [1983] ECR 3883, paragraph 40; Beer purity law, cited above, paragraph 46; and Case C-228/91 Commission v Italy [1993] ECR I-2701, paragraph 27).

47. A prohibition on the marketing of foodstuffs to which nutrients have been added must therefore be based on a detailed assessment of the risk alleged by the Member State invoking Article 30 EC (see, to that effect, EFTA Surveillance Authority v Norway, cited above, paragraph 30).

48. A decision to prohibit marketing, which indeed constitutes the most restrictive obstacle to trade in products lawfully manufactured and marketed in other Member States, can only be adopted if the real risk alleged for public health appears sufficiently established on the basis of the latest scientific data available at the date of the adoption of such decision. In such a context, the object of the risk assessment to be carried out by the Member State is to appraise the degree of probability of harmful effects on human health from the addition of certain nutrients to foodstuffs and the seriousness of those potential effects.

49. It is clear that such an assessment of the risk could reveal that scientific uncertainty persists as regards the existence or extent of real risks to human health. In such circumstances, it must be accepted that a Member State may, in accordance with the precautionary principle, take protective measures without having to wait until the reality and seriousness of those risks are fully demonstrated (see, to that effect, Case C-157/96 National Farmers' Onion and Others [1998] ECR 1-2211, paragraph 63). However, the risk assessment cannot be based on purely hypothetical considerations (see, to that effect, EFTA Surveillance Authority v Norway, paragraph 29, and Case C-236/01 Monsanto Agricoltura Italia and Others [2003] ECR 1-8105, paragraph 106).

50. In assessing the risk in question, it is not only the particular effects of the marketing of an individual product containing a definite quantity of nutrients which are relevant. It could be appropriate to take into consideration the cumulative effect of the presence on the market of several sources, natural or artificial, of a particular nutrient and of the possible existence in the future of additional sources which can reasonably be foreseen (see EFTA Surveillance Authority v Norway, paragraph 29).

51. In many cases, the assessment of those factors will demonstrate that there is a high degree of scientific and practical uncertainty in that regard. A proper application of the precautionary principle presupposes, in the first place, the identification of the potentially negative consequences for health of the proposed addition of nutrients, and, secondly, a comprehensive assessment of the risk to health based on the most reliable scientific data available and the most recent results of international research (see, to that effect, EFTA Surveillance Authority v Norway, paragraph 30, and Monsanto Agricoltura Italia and Others, paragraph 113).

52. Where it proves to be impossible to determine with certainty the existence or extent of the alleged risk because of the insufficiency, inconclusiveness or imprecision of the results of studies conducted, but the likelihood of real harm to public health persists should the risk materialise, the precautionary principle justifies the adoption of restrictive measures (see, to that effect, EFTA Surveillance Authority v Norway, paragraph 31).

53. Such measures must not be allowed unless they are non-discriminatory and objective (see, to that effect, EFTA Surveillance Authority v Norway, paragraph 32).

54. It must be added that, in such a context, the criterion of the nutritional need of the population of a Member State can play a role in its detailed assessment of the risk which the addition of nutrients to foodstuffs may pose for public health. However, contrary to the interpretation of the Sandoz judgment suggested by the Danish Government, the absence of such a need cannot, by itself, justify a total prohibition, on the basis of Article 30 EC, of the marketing of foodstuffs lawfully manufactured and/or marketed in other Member States.

55. In the present case, the Danish administrative practice is disproportionate since, apart from the four restrictively defined cases of what is considered to constitute a nutritional need and which are mentioned in paragraph 1 1 of this judgment, it systematically prohibits the marketing of all foodstuffs to which vitamins and minerals have been added, without distinguishing according to the different vitamins and minerals added or according to the level of risk which their addition may possibly pose to public health.

56. Indeed, the systematic prohibition under the Danish administrative practice on the marketing of enriched products which do not meet a nutritional need of the population does not enable Community law to be observed in regard to the identification and assessment of a real risk to public health, which requires a detailed assessment, case-by-case, of the effects which the addition of the minerals and vitamins in question could entail.

ECJ: C-34/10, O. Brüstle v. Greenpeace e.V., REFERENCE for a preliminary ruling under Article 267 TFEU from the Bundesgerichtshof (Germany), 18 October 2011

(Directive 98/44/EC - Article 6(2)(c) - Legal protection of biotechnological inventions - Extraction of precursor cells from human embryonic stem cells - Patentability - Exclusion of 'uses of human embryos for industrial or commercial purposes' - Concepts of 'human embryo' and 'use for industrial or commercial purposes')

Question

Comment on the following paragraphs under the light of the ECHR, the Oviedo Convention, and the EU Charter of Fundamental Rights also:

28. The lack of a uniform definition of the concept of human embryo would create a risk of the authors of certain biotechnological inventions being tempted to seek their patentability in the Member States which have the narrowest concept of human embryo and are accordingly the most liberal as regards possible patentability, because those inventions would not be patentable in the other Member States. Such a situation would adversely affect the smooth functioning of the internal market which is the aim of the Directive.

29. That conclusion is also supported by the scope of the listing, in Article 6(2) of the Directive, of the processes and uses excluded from patentability. It is apparent from the case-law of the Court that, unlike Article 6(1) of the Directive, which allows the administrative authorities and courts of the Member States a wide discretion in applying the exclusion from patentability of inventions whose commercial exploitation would be contrary to ordre public and morality, Article 6(2) allows the Member States no discretion with regard to the unpatentability of the processes and uses which it sets out, since the very purpose of this provision is to delimit the exclusion laid down in Article 6(1). It follows that, by expressly excluding from patentability the processes and uses to which it refers, Article 6(2) of the Directive seeks to grant specific rights in this regard (see Commission v Italy, paragraphs 78 and 79).

...

32. In that regard, the preamble to the Directive states that although it seeks to promote investment in the field of biotechnology, use of biological material originating from humans must be consistent with regard for fundamental rights and, in particular, the dignity of the person. Recital 16 in the preamble to the Directive, in particular, emphasises that 'patent law must be applied so as to respect the fundamental principles safeguarding the dignity and integrity of the person'.

33. To that effect, as the Court has already held, Article 5(1) of the Directive provides that the human body at the various stages of its formation and development cannot constitute a patentable invention. Additional security is offered by Article 6 of the Directive, which lists as contrary to ordre public or morality, and therefore excluded from patentability, processes for cloning human beings, processes for modifying the germ line genetic identity of human beings and uses of human embryos for industrial or commercial purposes. Recital 38 in the preamble to the Directive states that this list is not exhaustive and that all processes the use of which offends against human dignity are also excluded from patentability (see Netherlands v Parliament and Council, paragraphs 71 and 76).

34. The context and aim of the Directive thus show that the European Union legislature intended to exclude any possibility of patentability where respect for human dignity could thereby be affected. It follows that the concept of 'human embryo' within the meaning of Article 6(2)(c) of the Directive must be understood in a wide sense.

35. Accordingly, any human ovum must, as soon as fertilised, be regarded as a 'human embryo' within the meaning and for the purposes of the application of Article 6(2)(c) of the Directive, since that fertilisation is such as to commence the process of development of a human being.

36. That classification must also apply to a non-fertilised human ovum into which the cell nucleus from a mature human cell has been transplanted and a non-fertilised human ovum whose division and further development have been stimulated by parthenogenesis. Although those organisms have not, strictly speaking, been the object of fertilisation, due to the effect of the technique used to obtain them they are, as is apparent from the written observations

presented to the Court, capable of commencing the process of development of a human being just as an embryo created by fertilisation of an ovum can do so.

. . .

43. Although the aim of scientific research must be distinguished from industrial or commercial purposes, the use of human embryos for the purposes of research which constitutes the subject-matter of a patent application cannot be separated from the patent itself and the rights attaching to it.

. . .

46. The answer to the second question is therefore that the exclusion from patentability concerning the use of human embryos for industrial or commercial purposes in Article 6(2)(c) of the Directive also covers use for purposes of scientific research, only use for therapeutic or diagnostic purposes which is applied to the human embryo and is useful to it being patentable

. . . .

49. Accordingly, on the same grounds as those set out in paragraphs 32 to 35 above, an invention must be regarded as unpatentable, even if the claims of the patent do not concern the use of human embryos, where the implementation of the invention requires the destruction of human embryos. In that case too, the view must be taken that there is use of human embryos within the meaning of Article 6(2)(c) of the Directive. The fact that destruction may occur at a stage long before the implementation of the invention, as in the case of the production of embryonic stem cells from a lineage of stem cells the mere production of which implied the destruction of human embryos is, in that regard, irrelevant.

ECJ: C-636/11, Karl Berger v. Freistaat Bayern, Request for a preliminary ruling under Article 267 TFEU from the Landgericht München I (Germany), 11 April 2013
(Regulation (EC) No 178/2002 — Consumer protection — Food safety — Public information — Placing on the market of food unfit for human consumption, but not constituting a health risk.)

Question

Comment on the following paragraphs under the light of the EU Charter of Fundamental Rights also:

28. The referring court proceeds on the assumption that, in the case in the main proceedings, the foodstuffs which were the subject of warnings pursuant to Article 40(1), second sentence, point (4), of the LFGB during January 2006 could not be regarded as injurious to human health. Therefore, by its two questions, which it is appropriate to consider together, the referring court asks in essence if Article 10 of Regulation No 178/2002 must be

interpreted as precluding national legislation allowing information to be issued to the public, mentioning the name of a food and the name or trade name of the food manufacturer, processor or distributor, in a case where that food, though not injurious to health, is unfit for human consumption.

29. Article 10 of that regulation, which is the subject of the request for a preliminary ruling, simply places public authorities under an obligation to inform, where there are reasonable grounds to suspect that a food or feed may present a risk to human or animal health.

30. Accordingly, that provision as such does not prohibit public authorities from informing the public where a food is unfit for human consumption whilst not injurious to health.

31. In order to provide the referring court with a helpful answer to allow it to resolve the dispute before it, the Court considers that it is also necessary to give an interpretation of the second subparagraph of Article 17(2) of Regulation No 178/2002 even if that provision is not specifically mentioned in the request for a preliminary ruling before it (see to that effect, inter alia, Case C-243/09 Fuß [2010] ECR I-9849, paragraph 39 and case-law cited).

32. Pursuant to the second subparagraph of Article 17(2) of Regulation No 178/2002, which, under the second indent of Article 65 thereof, is applicable from 1 January 2005, Member States shall maintain a system of official controls and other activities as appropriate to the circumstances, including public communication on food safety and risk.

33. Article 7 of Regulation No 882/2004, which is applicable from 1 January 2006, provides inter alia that, firstly, in general, the public shall have access to information on the control activities of the competent authorities and their effectiveness, and, secondly, that the competent authority shall take steps to ensure that members of their staff are required not to disclose information acquired when undertaking their official control duties which by its nature is covered by professional secrecy in duly justified cases.

34. Article 14 of Regulation No 178/2002, applicable under the second paragraph of Article 65 thereof from 1 January 2005, sets out food safety requirements. Under Article 14(2), a food which is unfit for human consumption is said to be 'unsafe'.

35. In so far as a foodstuff is unacceptable for human consumption and accordingly unfit therefor, it does not fulfil the food safety requirements under Article 14(5) of Regulation No 178/2002, and is, in any event, such as to prejudice the interests of consumers, the protection of whom, as stated in Article 5 of that regulation, is one of the objectives of food law.

36. It follows from the above that, where food, though not injurious to human health, does not comply with the aforementioned food safety requirements because it is unfit for human consumption, national authorities may, as provided under the second subparagraph of Article 17(2) of Regulation No 178/2002, inform the public thereof in accordance with the requirements of Article 7 of Regulation No 882/2004.

37. The answer to the questions referred must therefore be that Article 10 of Regulation No 178/2002 must be interpreted as not precluding national legislation allowing information to be issued to the public mentioning the name of a food and the name or trade name of the food manufacturer, processor or distributor, in a case where that food, though not injurious to health, is unfit for human consumption. The second subparagraph of Article 17(2) of that regulation must be interpreted as allowing, in circumstances such as those of the case in the main proceedings, national authorities to issue such information to the public in accordance with the requirements of Article 7 of Regulation No 882/2004.

Provide legally appropriate solutions

1. A Biotech Company developed a single diagnostic test, which can check the donated blood used in transfusions for 20 infectious agents at that same time (such as HIV, hepatitis B, and chlamydia). The test is quick and requires no special conditions for the transfer, and therefore is very useful for countries with poor infrastructure. The Biotech Company invested 80% of its budget on this test, which took 5 years to be developed. Is it ethically acceptable for the Biotech Company to file a Patent on the test regarding the method, the product, and its use for donated blood? Justify your answer.

2. A research team created the genome of a fly artificially, with the purpose to replace the nucleus of a human oocyte with the artificial cell nucleus.

a. Is that allowed by European law?
b. If the research team applies for patenting, what would be the European Patent Office's decision?

References

Kelves DJ (2002) A history of patenting life in the United States with comparative attention to Europe and Canada. European Group on Ethics in Science and New Technologies (EGE) – European Commission, Brussels

Kesselheim AS, Cook-Deegan RM, Winickoff DE, Mello MM (2013) Gene patenting — the Supreme Court finally speaks. N Engl J Med 369:869–875

Kumar HD (1998) Modern concepts of biotechnology. Vikas Publ., New Delhi

Mgbeoji I (2006) Global biopiracy: patents, plants, and indigenous knowledge. Cornell U. P., New York

Mukku S, Lalasa M, Nagalakshmamma K, Savitri G, Sujata P, Shanthi Sree KS (2021) The role of law and regulation on the intersection of bioprospecting and bio piracy. In: Proceedings of the 2nd International Conference on Computational and Bio Engineering, Springer, Singapore, pp 129–134

Richardson V (2009) Aborted fetus cells used in beauty creams. The Washington Times 3.11. 2009, in http://www.washingtontimes.com/news/2009/nov/3/aborted-fetus-cells-used-in-anti-aging-products/?page=all

Robinson D (2010) Confronting biopiracy: challenges, cases and international debates. Earthscan, Routledge, Oxon, New York

Sagoff M (1996) Animals as inventions: biotechnology and intellectual property rights. Philos Public Policy 16:15–19

Spranger TM (2004) Patentability of stem cell inventions with special emphasis to EC law. J Int Biotechnol Law 1:247–254

Spranger TM (2012) Comment on the ECJ Brüstle case. Common Mark Law Rev 49:1197–1209

14 Concluding Remarks

We are now in place to draw a general conclusion about the meaning of the value of life by the European Biolaw.

1. The key concepts for the legal value of life are those of "person" and "biodiversity." They are distinct, but still, they share a common ground permitting an integrated approach at the normative level.
2. The concept of "person" refers exclusively to the life of humans. Its characteristic is that it attributes the highest level of value to all individuals of our species by treating each and everyone as a *value-per-se*. This is the ground from which individual autonomy emanates; an essential dimension of individual autonomy is what we can call "biological autonomy", which refers to the direct control of our biological functions.
   To determine the "beginning" of the person's existence presupposes that we cope with some conceptual issues since, at the legal level, the various forms of human life before birth are not considered "persons." In most national legal systems, a person exists only from the moment of birth, when biological autonomy is meaningful. However, the European legal instruments do not clarify that issue.
   On the other hand, there is no standard view of whether the end of a person may be seen as a matter of biological autonomy. Most national legislations reject that perception, by prohibiting end-of-life decisions. Again, at the European level, the only relevant provision is the "right to life" of the ECHR and the EU Charter of Fundamental Rights. Its wording is too general for addressing the problem, although the Strasbourg Court denies an interpretation of that provision in favor of a right "to death"; beyond that verdict, death remains a largely unexplored legal issue, falling within the competence of each national legal system.
   Between the key moments of birth and death, throughout the whole lifecycle of a person, biological autonomy presupposes his/her informed consent for justifying interventions of others related to personal health or management of biological data. Exceptions in relevance are very few, particularly regarding

T. Vidalis, *The Emergence of Biolaw*, Springer Textbooks in Law,
https://doi.org/10.1007/978-3-031-02359-0_14

situations requiring urgent medical intervention or the processing of non-identifiable biological data.

3. The concept of "biodiversity" reflects the value of non-human species in contemporary law. The essential characteristic of that value is that, in principle, it does not focus on the protection of non-human individuals, covering the non-human life (of microorganisms, plants, and animals) as a unified whole. This means that the law's interest refers to the conservation of a given balance between species and not to the protection of their individuals. The rule, here, is that individuals matter only if the conservation of the entire species (and consequently that of biodiversity) is threatened. Moreover, human interventions influencing large populations of a certain species are legally accepted to the extent that they do not violate the existing balance of biodiversity.
   Nevertheless, the law establishes classifications to the legal treatment of species, associated with corresponding restrictions on potential human interventions. At the top of that hierarchical protection, we meet the wild and rare plant and animal species and the vertebrate animals also.
   Thus, for the law, non-human individuals, populations of them, or entire species, are not considered as having a value-per-se; they may be subjected to human interventions for serving our needs, being treated as mere instruments with only relative value.
4. At the level of global biodiversity, the approach is different though. It is true that, in practical terms, it is difficult to determine its "balance", particularly if we take into account that biodiversity has a dynamic nature, with continuous changes involving the extinction of species and the emergence of others, regardless of any human interventions.
   However, as a matter of principle, we must accept that *global biodiversity has a value-per-se, as well.* Indeed, a hypothetical condition of conflict between the latter and the human species seems meaningless since the human species could not even exist if an extensive catastrophe to the global biodiversity occurred.
   Most of all, this value-per-se is also reflected at an elementary level and makes sense for each and every separate human being. That is because the human value and the resulting property of personal autonomy are inextricably linked to the biological composition of the human organism, which in its turn is directly dependent upon the natural influence of other species on most human vital functions.
   It is precisely that organic connection of our biology with the biology of non-human species that establishes what we can define as an institutional "bridge" between the two fundamental concepts, the concept of Person and that of Biodiversity, equally considered as the supreme values of modern Biolaw.

Part III

Future Challenges for Biolaw

"Animal Rights" 15

15.1 The Problem

The current attitude of law regarding the status of animals reintroduces the debate between the human-oriented and the "deep ecology" approaches in more specific terms. Their conflict is expressed in legal reflections about the recognition of legal rights to animals. At the level of international soft-law, UNESCO has already adopted the Universal Declaration of Animal Rights (Paris, 1978),[1] and at least one national legal system in Europe (the Portuguese) uses the same terminology in a relevant law.[2] On the other hand, in the U.S., a court rejected the recognition of whales as subjects of the 13th Amendment of the Constitution. It was a claim supported by activists intending to the protection of that endangered species.[3]

Apart from these isolated initiatives, defenders of animal rights present theoretical arguments in legal journals as well.[4] Their controversy with the dominant human-oriented legal position definitely contributes to increasing attention to animal protection, regardless of the stance that we may adopt for its justification.

Nevertheless, this debate is legally meaningful for at least one reason. In extremis, the human-oriented approach concludes with different practical consequences than the opposite "deep ecology" approach. A supposed conflict between human interests and the survival of an animal population as a confrontation between subjects of "rights" may be Compared to the conflict of the distant past between masters and slaves within the human species. At that time, it should be proved that a slave was

---

[1] According to its art. 9 para 1, "The specific legal status of animals and their rights must be recognized by law."

[2] Cf. Protecção aos Animais, Lei n° 92/95.

[3] See on the U.S. case-law, Bjornkestam (2001). In newspapers, also: N Y Times, Enforcing the Legal Rights of Animals (The Opinion Pages, Room for Debate, 1.10.2014, in http://www.nytimes.com/roomfordebate/2014/10/01/enforcing-the-legal-rights-of-animals.

[4] See, for example, Kelch (1998), p. 531, Favre (2005), pp. 333 seq. Contra and Cupp (2007), pp. 18 seq.

T. Vidalis, *The Emergence of Biolaw*, Springer Textbooks in Law,
https://doi.org/10.1007/978-3-031-02359-0_15

indeed a human being for recognizing rights to slaves. Still, the attribution of rights meant that philanthropy (with the uncertainty of its expressions, since no guarantees for its supporters' commitment existed) was exclusively a matter of ethics, not a matter of law; a master was never legally obliged to recognize rights to the slave.

Are we facing the same situation with animal care nowadays? The real problem relates to institutional guarantees that we need to consider for the protection of animals, even if that means a certain retreat of human interests. An "ontological" comparison between humans and animals, and also the analysis of certain conceptual elements of the law, will demonstrate if such institutional guarantees are suitable and even necessary in terms of ethics.

15.2 The Legal Concept of "Rights": Some Clarifications

Legally, the concept of "rights" refers to powers that the law recognizes for persons to pursue or protect legitimate interests. This is a minimal essence that all rights share, regardless of their classification or specific nature.

Rights in the law do not necessarily have a corresponding legal "obligation" for persons interacting with their subject. The European Social Charter's social rights are a characteristic example of such "incomplete" rights, which nevertheless produce ethical and political commitments for their recipients.

Also, the exercise of rights does not necessarily mean a specific conscious "action" of their subject, even if that usually happens. Thus, some fundamental rights illustrate self-determination areas to which no intervention by the state or other persons is allowed, and they produce full legal force even if their subject does not "act" at all. These are "defensive" freedoms that include the right to Corporal Integrity, the freedom of Thought, and the right to Privacy, all related to situations not involving personal consciousness, that is, freedoms producing legal force even when their subject is unable to express will, lives in a coma, etc.

Another essential characteristic of the legal concept of rights is the distinction between the "title" and the ability to "exercise." The "title" of a right belongs, by definition, to its subject if it fulfills certain conditions prescribed by law. These vary from the simple state of a "person" (which is sufficient mostly for the title of human rights) to complex states required when a person participates in legal relationships like, for instance, it happens for the title of property rights or family rights.

On the other hand, the title of rights does not necessarily entail the ability to exercise by their subjects. Thus, although children or mentally ill persons, or persons temporarily unable to express valid will are subjects of rights, they cannot exercise those presupposing a conscious action; the exercise of these rights is, then, undertaken by legal representatives (usually family members), on behalf of the rights' subjects and for the best interest of them. However, all subjects are considered able to exercise freedoms not involving conscious actions, even if we are talking about a newborn or a patient living in a deep vegetative state.

15.3 The Equation "Human = Rights"

All these determinations of rights in the law indicate that the element of their subjects' "consciousness" is not as critical as usually assumed. Although in our common understanding, a right means the possibility to "act differently," which presupposes individual control through specific neurophysiologic human functions, for the law this is not a necessary element of the concept of right. However, the legal theory adopts, as an axiom, that subjects of rights are exclusively human beings, persons developing that specific neurophysiologic function.

That firm belief has been a given characteristic of past legal systems already but not unconditionally. We know that in pre-modern eras, the recognition of rights presupposed many more elements than the mere state of a human being. In the Antiquity and the medieval era, it was not sufficient for someone to be recognized merely as a conscious human being to enjoy or acquire rights. It also required a fundamental social condition, that is, that person's recognition as a free citizen or a member of certain social groups. In the Antiquity, in particular, that condition was so crucial that, if not existing, it meant that human beings were not considered as "persons"; they were "res" like animals, although they possessed the biological state of conscious humans.[5] Their liberation could undoubtedly change this, but without reference to a certain "right." A slave's liberation depended on the absolute discretion of his/her master.

Modernity modified this legal regime radically. With the recognition of Human Rights, that is, the fundamental individual freedoms, the equation "human = rights" was established without exceptions and additional social determinations; ever since, every human individual in the biological sense is recognized as a subject of rights. This produced a further shift in approach; if that equation is undisputable now, its justification does not refer to History or Sociology anymore, but rather it is a matter of Biology and Ethics.

What Biology indicates is the distinct neurophysiology of humans Compared to other animal species. We know that, in the distant past, among primates, some species (not a unique one) of "erect apes" had been developed gradually. The first findings identifying them were located in Africa and dated almost 3 million years back. Fighting desperately and continuously for their survival against hostile natural conditions, these animals progressively manifested certain pre-conscious functions of their central neural system, beyond instincts. For hundreds of millennia, these functions have been codified in the genome, building a hereditary basis for the descendants, distinct from that of other primate species.[6] In the most recent phases of that tremendous evolutionary process, one of these "advanced" species (Homo sapiens) dominated in their competition and continued to develop the exciting cognitive ability of conscious selection and pursuit of goals. It was a feature that overcame the characteristic of animals, instinctive activity, insofar as the brain is in

[5] Sohm (1970), pp. 165 seq.

[6] Cf. Diamond (1991), p. one and two.

place (to a limited degree first, more extensively then) to distinguish and appraise alternative options, in order to form "conscious" decisions.

It was precisely that primordial ability of will that separated the human species from the others. Indeed, the ability to "act differently," beyond instincts, caused a long period of humans' domination over Nature, already from the first permanent residences of the Neolithic era.[7] It was that ability that motivated conscious research acts to understand natural norms in order to mobilize them through the development of technology for serving humans' interests and needs.

Nevertheless, Evolution produced effects for the other animal species as well. The neurophysiologic functions of many of them were subjected to deep differentiations also codified in their genome, during long periods. Survival and extinction of animal species are natural processes associated mostly with that latter development. An approach that considers animals as passive "recipients" of transformations occurring in their natural environment, incapable of reacting because they do not possess advanced cognitive properties like humans, is deeply misleading. Current research in ethology (studying the social behavior of animals) demonstrates that many species, not only mammals but also insects, fishes, and frogs, produce a remarkable influence on their natural habitats when they create suitable conditions for social or individual living.[8] The impact of these animal species on the natural environment they live in is due to Evolution and cannot be attributed to some particular biological "surprise"; other species, already extinct or under threat of extinction, failed to adapt to natural conditions, exactly because they were unable to behave alike.

Thus, the spontaneous behavior that we observe in the animals' social or individual life (like the use of tools for collecting food, the nest building, and the organization of the bees' and ants' complex societies[9]) reflects a particular elementary form of "controlling" natural processes by these animals; even if that cognitive function is substantially different from human cognitive properties, its existence cannot be disputed. In other words, if we understand that action as a mere manifestation of instincts, we tend to adopt a too wide conception of these, profoundly false. If, for instance, we observe the complex strategy that a tiger develops when hunting, following the particular circumstances that the animal faces, we can realize that, to serve its instincts, it shows specific will properties, a sort of discretion to "act differently." In any case, it is difficult to characterize the hunting strategy as purely "instinctive," if with that word we mean a mere reflective "automatic" reaction to external incentives. Otherwise, we could also attribute to some manifestation of instincts, if not the total, but certainly the most part of the human technology; the critical intervention of our conscience to that process would not be that different if Compared to the animal action.

[7] See in relevance, Price (2000).

[8] On altruistic behavior occurring in many animal species, see Wilson (2000), pp. 120 seq., Dawkins (1989), pp. 166 seq.

[9] Cf. Shumaker et al. (2011).

If we examine the biological features of humans themselves, we conclude that biology plays a pivotal role in the legal concept of the subject of rights. The fact that the human organism undergoes a developmental process and also may experience unforeseen changes related to neurophysiologic characteristics throughout the lifespan produces significant differences. There are, thus, special conditions of human existence that may challenge the property of "subject," mostly from two aspects.

Firstly, we have seen that the unborn human life does not fall within the concept of "subject" of rights due to the state of dependence of its development on external conditions. However, the embryo and the fetus are forms of human life even without the property of persons. For that reason, as mentioned, the law needs to take into account their "potential" towards the condition of future persons. Thus, the developmental nature of human biology imposes a "transitional" regime that we can define as that of an "object-with-special-value." Still, this illustrates an exception to the equation "human = subject"; individuals of our species are recognized as legal subjects only after the moment of birth, not before.

Secondly, even after birth, although it is not questionable that persons are subjects of rights, these latter remain incomplete, insofar as the crucial element of conscience either is under a process of maturation or is entirely missing. Again then, an empirical element derived directly from human biology obscures the integrity of the concept of "subject." If conscience (that is, the ability of will) lacks completely or partially from a certain person, the property of "subject" becomes practically inactive, at least regarding rights the enjoyment of which presupposes conscious actions. Although a human, that person is recognized as a mere "titleholder" of rights, unable to control the actions of his/her legal representatives, a condition that contravenes the very meaning of "rights." This is the case for minors and even adults when experiencing permanent loss of conscience due to pathological reasons.

15.4 "Pure" and Empirical Perceptions of the Subject

Those studying the ethical status of humans usually do not pay attention to the above differentiations regarding the empirical condition of persons. A "pure" perception of the subject of rights, based mainly on the Kantian philosophy, considers the concept of rights as inseparable from human nature, stressing mostly that every right corresponds by definition to a relevant obligation, regardless of the empirical biological state of concrete human individuals. Briefly, the exercise of rights needs to comply with a rule of self-control that imposes responsibilities on the subject of rights for the sake of other persons. The discretion that a person has to "act differently" when exercising a specific right is judged on the basis of these responsibilities, not linked to the pursuit of egoistic interests or mere instincts, that is, of random needs deriving from empirical incentives.

Nevertheless, considering the "others" and their legitimate interests as necessary factors for enjoying our rights presupposes cognitive abilities that only humans and all humans possess. No matter if a person is or not in that particular cognitive state to exercise rights, he/she is recognized as an exceptional natural being, a member of a unique species that can behave in terms of responsibility. This is a privileged "title" inseparable from any human being, even those entirely determined by others in real life, a moral condition that deserves the highest respect making, at the same time, all humans equal; it is the title of the "moral subject" with the special quality of "human value" and "dignity" which by definition cannot be attributed to other species.[10]

Under this strict conceptual system, to suppose that individuals of an animal species may be recognized as titleholders of rights, even if their cognitive skills could be comparable to those of human individuals (like the cognitive skills of primates, that have been compared to those of 2 years old minors[11]) is absolutely absurd; because the term "right" is conceptually related to the mere term "human" regardless of any empirical evidence that may logically suggest something more. For that overly anthropocentric approach, any sort of care for animals can never overcome a generous sympathy for living "objects." This influences our social life's quality, which is based, nonetheless, on responsibilities towards our fellow humans and not towards animals themselves.

15.5 The Neurophysiology of Non-Human Species

However, it would be enlightening if we pay a little attention to the empirical evidence of science, stepping out from that closed universe of morality.

The neurophysiology of other living organisms regarding their capabilities to receive and process external stimuli is very different compared to humans. However, such capabilities exist in any form of life, or to put it better, life itself cannot be defined without them, namely without, on the one hand, the processing of information received from the surrounding world through a specific biological mechanism, on the other, the response of living organisms to that input, by modifying their form, particular movements, etc.. Even in microorganisms, a response to external stimuli can be identified by movements indicating a certain reaction to concrete environmental conditions. Apparently, the evolutionary process favoring the reproduction of species on the condition that they adapt to environmental changes leads also to genetic mutations that we can detect, for instance, in microbes causing infectious diseases.

Therefore, every life form is subjected to a continuous transformation process, at the level of either individuals or populations or entire species, being in firm contact with changing environmental conditions. This means that all life forms proceed to biochemical processing of external stimuli in their own molecular "laboratories" during an endless battle of adaptation and survival. This also illustrates the crucial difference even between elementary forms of life and non-living entities; the contact of non-living entities with the environmental conditions and changes remains

[10] Cf. Cohen (2001), pp. 36 seq.

[11] Wobber et al. (2013).

entirely "external," which means that although it produces a certain molecular motion, it does not result in characteristic phenomena of life (reproduction, self-movement, etc..).

In plants, the acquisition of external stimuli from the environment (when, for example, a reaction to light is manifested with phenotypic changes or even certain movements of the plant) is also due to simple biochemical changes in plant cells. In any case, there is no question here of any "senses" organized in a kind of neural system.

Animals, in contrast, are distinguished mostly by their capability to change positions in space that is due, precisely, to the existence of a particular neural structure, elementary or more developed. Sense organs cover the entire animal body, receiving input from the external world, and through the neuron network, environmental data is transferred to the brain. The processing of this data results in the respective reactions of the body. That function is common to all animals (vertebrate or not), and, in a sense, it represents the cause of their relative independence as living organisms from the environment in which they live.

Nevertheless, the capability of external data processing by the brain differs significantly among animal species. It is only elementary in the vast majority of them, namely in insects and other non-vertebrates. Reactions to external stimuli are automatic in these species, directly corresponding to the fulfillment of survival and reproduction instincts, which indicates that the brain does not perform substantial processing. This automatized behavior is characteristic even regarding the complex organization of several insect societies (ants, bees, etc..). The construction of the famous multi-room nest of ants or the organization of the bees' hive (with the strict allocation of roles between the queen bee, the worker bees, and the drones) are the result of predetermined moves of the population's members to the degree that the entire population functions like one individual, a single organism. There are no innovation elements or differentiation of individual behaviors in these artifacts, which indicates an almost absolute adaptation to natural demands.

The neurophysiologic condition is radically different in vertebrate animals. The brain of vertebrates possesses skills for more sophisticated processing of information deriving from the senses, to the degree that we can even observe a particular elementary "self-determination" of these animals. Indeed, in vertebrates, the cognitive functions of memory, of learning ability, even that of imagination,[12] occur in a specific state of development. These functions are visible, particularly in mammals, allowing them to distinguish between various choices for their needs' fulfillment, and to act thus "differently".[13]

However, it is important to stress that these needs correspond directly to animal instincts, and they are not mediated by "cultural" needs (of spiritual or emotional

[12] On the self-awareness of certain species, see Degrazia (2009), pp. 201 seq.

[13] Regarding differentiation in cognitive abilities, following the vertebrates' neurophysiology, see Salas et al. (2003), pp. 72 seq. A disagreement on the issue of advanced cognitive abilities in animals, see in Carruthers (1992), pp. 133 seq., 184.

nature, etc..). Thus, an animal makes "choices" between different actions, based on instincts of survival and reproduction, following an individual elementary skill of performing "appraisals" as expressed at the moment of that "choice"; animal "choices" do not involve incorporation of culturally determined values that, in humans, may prevail over the temporary fulfillment of instincts. That difference is undoubtedly crucial in terms of the animal moral status, as mentioned before, and though it has not have the same weight for the law.

15.6 Animal "Interests" and Human Responsibility

The above remark is fundamental, and it is related to the differentiation between ethics and law. The key element, here, is that contrary to what happens in ethics, for the law the motivation of an act does not matter but only in very exceptional cases. The law is merely interested in rule compliance regardless of the reason for which a subject decides to comply, because of fear, prudence, sense of duty, etc. Consequently, the motivation for the exercise of a legal right by its subject is, for the law, also irrelevant.[14] The abuse of a legal right cannot be characterized as an "irresponsible" exercise; it is not an "exercise" at all since it represents an act performed out of the field that the law recognizes for that right. Therefore, it is not strange that the law also comprises rights without respective legal obligations; other persons may legally ignore these rights, no matter if they may receive ethical or political criticism for their behavior.

However, if the motivation in exercising legal rights is indifferent, then their subject's cognitive ability to reflect on "responsibilities" towards the others is an element with no particular significance. Legally speaking, it is sufficient for the subject to determine purposes and consider alternative options for meeting them, that is, the cognitive ability to "act differently," for being identified as a titleholder of legal rights; it remains irrelevant whether these purposes and respective methods derive automatically from instincts or a specific cognitive elaboration involving an individual "conscience." That ability to "act differently," which is independent of any sort of responsibility and accountability, contrary to what happens with moral rights, might fit a non-human neurophysiologic condition also; not all of them, but certainly some animal species meet already that condition, which means that logically they could be recognized as subjects of legal rights.

[14]Regarding the issue of moral rights in animals, see Carruthers, pp. 194 seq. In favor of these rights, see the classic Regan (1983/2004), pp. 266 seq.

15.7 Positive Law Limitations and the Recognition of Animal Rights

There is no doubt that this last approach needs a more analytical explanation, particularly regarding two major issues; which animal species might be subjects of legal rights, and which rights could be these, also taking into account the legal rights of our own species?

15.7.1 Should We Recognize Rights to All Animals?

Reflecting on the animals' cognitive ability to "act differently" just for meeting instinctive needs, is critical regarding the recognition of animal rights. Indeed, if the neurophysiologic condition of an animal species is not quite advanced for "choosing" between various alternatives, it cannot also meet the minimal conceptual element of discretion that any right presupposes.

Therefore, the elementary neurophysiology of invertebrate animals, allowing only animal movements based on automatic reactions to external stimuli, with no specific elaboration by the neural system, does not meet the above minimal cognitive condition. In fact, these automatic movements of insects, for instance, are more similar to the model of simple biochemical, and cellular changes that we can observe in the movements of plants or even those of microbes, namely organisms and microorganisms with no neural system at all. This fundamental defect excludes invertebrates from the possibility to be identified as legal rights subjects, no matter if they also influence their environment through automatic "behavior." Some exceptions of particular invertebrates (cephalopods, mostly) should be observed indeed, still from the law's approach, given the current empirical data, the above distinction in the animal realm cannot be disputed.

In contrast, we can consider recognition of rights for the total of vertebrate animals since their neurophysiology, as mentioned, allows a minimal appraisal of choices for meeting their instincts, therefore, it is accustomed to the concept of rights. However, this minimal discretion should be referred only to their specific biological condition, not to a certain "similarity" to the human species. Here, there is no question about some enlargement of subjects of rights to cover species "similar" to humans, which may trigger human emotions of protection, affection, etc.., an approach that would restrict these animals to a few mammal species. In evolutionary terms, that neurophysiologic special quality represents a common background of all these species, including the human species. The unique nature of discretion as a cognitive ability is critical; therefore, its importance cannot be attributed to any sort of "affection," "protection" etc.., expressed by a biologically "privileged" human species.

15.7.2 Which Rights for Animals?

If we consider vertebrates as potential legal rights holders due to their particular neurophysiology, we need now to determine a special content for these rights.

15.7.2.1 Right to Life

The right to life for any vertebrate animal[15] is not only fundamental but also represents the logical condition for the recognition and exercise of all other rights.

In the context of human rights, we know that the right to life is recognized in its positive dimension, that is, the protection from threats against life, and it is also closely related to the human dignity principle. This strong status is generally acknowledged. Only in cases of legitimate defense, when a situation of conflict between the protection of two or more subjects of that right occurs, the life's removal is excused, at least in the European legal systems.

A pretty similar status could also be considered regarding the potential recognition of the right to life in vertebrate animals. Regardless of their species, that recognition would mean

(a) Protection against life-threatening acts, including those deriving from human actions
(b) Legitimate exceptions in situations similar to the "state of defense" as understood with reference to the respective human right

These exceptions raise an obvious problem since the use of vertebrates intentionally killed for serving human needs characterizes our civilization from its very dawn. If we decided to recognize an animal right to life with a legal status comparable to the human right, how could we justify that steady practice?

Here, the answer does not differ from the approach we adopt regarding the exceptional circumstances that may justify the intentional termination of human life. In the case of animals, the sole reason for justifying intentional killing could be a threat against human life, to the extent that we could describe it as a situation of "defense." That situation cannot be the case of animal killing for entertainment reasons (hunting, etc..),[16] or the case of killing for serving fashion or aesthetic preferences (use of natural leather, manufacturing of cosmetics, etc..). On the other hand, though, killing of animals for nutritional needs (representing the vast majority of relevant examples), under the assumption of a potential animal right to life, needs specific analysis.

In general, we cannot consider any nutritional need so important that could be understood within the meaning of a "state of defense" for protecting human life.[17] The classification of human nutritional needs and, in particular, their separation from

---

[15] See art. 2 of the UNESCO Declaration.

[16] See in relevance, Goodland (1998), pp. 248–249.

[17] *Ibid*, pp. 248 seq.

simple nutritional habits or "trends" would be necessary here. Nevertheless, even if we can distinguish our basic nutritional needs, this would not be sufficient for justifying animal killing. Before considering animal killing, we need to examine alternative means for serving these needs, deriving from other food sources that may include plants, lower animal species (non-vertebrates), or even synthetic food. Certainly, these alternatives would not cover sufficiently our basic needs; nonetheless, this is the only condition for deciding as inevitable the killing of vertebrate animals, in order to protect human survival individually or at the scale of a whole population, in other words, for confirming a legitimate "state of defense".[18]

The other basic category of human needs served nowadays by animal killing refers to the protection of health. Experimentation with laboratory animals is necessary for developing biomedical research and, eventually, for producing new medicinal means. A potential right to life of laboratory animals would require a similarly thorough examination of real needs in new medicines, as well as a respective study of alternative options for research, with no involvement of animals. The latter point is already acknowledged by the existing legislation, as mentioned. However, the real needs in medicines currently are not considered as an essential subject of relevant appraisal, for avoiding experimentation resulted in the elimination of excessive numbers of animals.

Following that remark, killing laboratory animals, if they were considered subjects of a right to life, would be justified only if they were used in research intending to cover needs for human health pertinent, again, to a "state of defense." To confirm this, we would require reasons for killing stronger than those emanating from the general state of animal protection that we accept currently, on the basis of animal welfare as the sole relevant legal value. The introduction of an animal "right to life" for every individual vertebrate restricts our discretion on animal use drastically, since it involves a comparison between human needs and a real legal right with a strong value status. Excessive current needs (determined exclusively by the pharmaceutical industry) would not stand to that comparison like they stand now prevailing over the relatively weaker value of animal welfare.

15.7.2.2 Right to Corporal Integrity

A right to corporal integrity for the animals would mean that no conditions of animal suffering or pain are acceptable if intentionally created or tolerated by humans.

This would be a right corresponding to the prohibition of tortures for humans, the core of "habeas corpus" related directly to the principle of human dignity that excludes any instrumentalization of individuals; it is precisely the question on the instrumentalization of animals that emerges here. Cases of animal suffering and pain are far more usual than those occurring in humans. They emerge in animal use serving a wide variety of human activities, from farm animals' confinement to the exploitation of animals in entertainment shows.

[18] Contrary to the argument that defends animals' use for our nutritional needs, on the sole condition of their humane treatment. See, for example, Sunstein (2002), pp. 387 seq., 395.

The recognition of a right to corporal integrity for any vertebrate would change again the level of animal protection that we accept currently requiring merely "humane treatment" to avoid corporal harm of any kind. At the level of humane treatment, protection does not balance the real significance of human needs, assuming that they always prevail over the animal welfare, and intending just to exclude an "unnecessary" and "useless" maltreatment of animals; a difference with the recognition of an animal right, here, would be precisely a required appraisal and balancing of our needs comparing to the advanced animal status.

In such a context, even a minimal violation of the animal corporal integrity could not be justified if alternative means for meeting our needs in relevance would not be observed. Again, human needs, like entertainment and clothing, would not justify interventions in the animal body; that is, they could not be prioritized over the right we examine. On the other hand, given that corporal harm threats do not necessarily result in life-threatening conditions, the standard of the "state of defense" for human needs that we suggested previously, as the only appropriate to assess the protection of the animal right to life, is not applicable here.

Cases in relevance might be, thus, only those not putting the life of an animal in a dangerous situation, like in examples of interventions for collecting animal products (milk, wool, etc..), experimentation not resulting in the animal killing, or collection of animal biological material for research purposes. In all these cases, we would first need to consider existing alternatives for avoiding any animal use, second, to evaluate the significance of our needs compared to the animal right and, third, to select the most appropriate means for painless animal treatment.

15.7.2.3 Rights to Safety, Free Movement, and Housing or Nest

Animal safety, free movement, and housing or nest correspond to the "habeas corpus" rights.

We already mentioned that captivity and restrictions to the animal movement (regarding wild animals in particular) are allowed by the current legal framework under special circumstances. Far more serious are restrictions imposed on pets and farm animals, where the determination of concrete conditions of housing should be suitable to these animals' nature for ensuring their welfare. Still, human needs that justify these conditions for the use of animals remain out of any assessment. In particular, pet animals destined to serve needs of pleasure, entertainment, and emotional fulfillment of their owners spend their entire lives in captivity.

A supposed change to the level of protection, through the recognition of the above animal rights, would raise a question of whether we can determine the objective of these rights accurately regarding the living conditions of different categories of animals (wild, pet, and farm animals).

In pet animals, permanent living in a restricted and controlled space guarantees the animal survival. The intentional release of these animals to unprotected environmental conditions involves dangers to their life, health, and corporal integrity, due to their nutritional habits and vulnerability to diseases. Thus, living in a controlled space cannot even be characterized as "captivity." However, this close relation of housing conditions to the animal nature also determines the concrete field of rights in relevance; acts of undue restriction to animal movements or disturbance of the animal calmness, and moreover acts of the owner that put in danger the life, health and corporal integrity of these animals cannot be accepted, as infringing these animal rights.

Regarding farm animals, the rights to free movement, safety, and housing are also dependent on specific human needs. The collection of animal products is a justified human need, but it should be fulfilled following the species' nature and the concrete habits of the individual animal. In any circumstances, a right-based approach would not be compatible with conditions of strict confinement, not allowing the animal movement, which may involve also stress, suffering, and pain, as it happens nowadays in various facilities of intensive production.[19]

The same rights have a different field of application in the case of wild animals when contacts with humans are observed.[20] These contacts may be accidental when human installations are situated close to wildlife habitats, or they may be intentional contacts for human pleasure and entertainment (visiting parks, hunting, fishing, etc..) or for professional reasons (professional fishing, research activities, journalism, etc..). In accidental contacts, restrictions on animal safety, free movement, or nest might be justified to ensure the regular life of human communities, on the condition that these acts do not involve captivity, health damages, or corporal harm to wild animals. Here, the only exception could be again a "state of defense" emerging, for instance, due to aggressive animal behavior.

As in the previous cases of farm and pet animals, legitimate restrictions on animal rights would be only those fitting the particular animal species and natural habits. In that framework, hunting and fishing for pleasure would not be accepted as legitimate human activities, threatening not only the life and corporal integrity of wild animals but also their safety, free movement, and nest. A similar conclusion should also be drawn for wild animal confinement and captivity, serving shows and exhibitions in circuses, zoos, etc.

15.7.2.4 Right to Serenity and to Reproduce

It would not be a strange hypothesis to consider, also for animals, rights pertaining to a state similar to the human "privacy."

A right to "serenity," either in the nest or within the wider animal habitat, represents a logical condition for animal welfare that is already acknowledged by current legislation in vertebrates, particularly the wild species of them. Serenity means, here, a state in which neither threats exist for the animal nor any disturbance from emissions, noises, etc. These conditions need to be observed even in pet, farm, or laboratory animals, where permanent contacts with humans are frequent. However, in these cases, specific measures are necessary, such as constructing suitable nests, sufficient protected spaces for animal convenience, etc.

---

[19] Cf. Van Horne and Achterbosch (2007).

[20] On the right of wild animals' free-living, see art. 4 of the UNESCO Declaration.

A right associated with serenity could be an animal right to reproduce and offspring care. Here, we need to point out that animal reproduction would not represent only a requirement for conserving biodiversity or fulfilling particular human needs, as considered under the current law approach. The right-based approach introduces a concept of reproduction suitable to serve individual animals themselves, meeting an animal biological need imposed by the sexual instinct. Therefore, any deprivation of adequate conditions that favor sexual intercourses during the reproductive cycles, or the feeding of newborns, results in harm to individual animals in their natural condition as mere biological entities, yielding pain, and suffering. This is a distinctive element not observed under the current legal regime.

Again, regarding the rights to serenity and to reproduce, human needs might impose relevant restrictions, but only after a balanced appraisal of their significance, taking into account the particular nature of the animal species and relevant habits, and not as representing a "privileged" legal status demanding priority by definition.

15.7.2.5 Right to Health

Finally, the animal corporal integrity is related to the recognition of a right to health that would also encompass access to water and appropriate feeding. This right would presuppose providing care by the human responsible (provision of feed, water, vaccinations, veterinary care, etc..).

The protection of animal health through interventional measures is not comfortable for wild animals that live in natural conditions. Initiatives for veterinary care taken by environmental organizations that involve the temporary confinement of wild animals are certainly exceptions here. The provision of feed and water in natural spots within the animal habitats is also a measure that serves the wild species' health care.

15.7.3 The Exercise of Animal Rights

We can illustrate, then, in general terms, the correspondence of vertebrate animals' basic needs with a supposed recognition of respective "rights" by future law. Still, a crucial issue emerges concerning the exercise of these rights, mainly if they conflict with human rights.

As we mentioned, the current law already endorses basic animal needs. What makes a radical difference with the adoption of a right-based animal status, is that this does not understand the needs of humans as "privileged" Compared to those of animals and thus prevailing over them by definition. Therefore, general references to the protection of public interests in the fields of economy, safety, health, research, etc.., would not be sufficient reasons, as such, for justifying restrictions to animal rights or, moreover, to annul them entirely. Specific reference to concrete human rights being in a state of conflict with an animal right would be such a reason, requiring balancing between them to determine priority in protection.

However, these appraisals would presuppose an explicit institutional provision for defending in practice the animal rights. Human "representatives", mostly, need to undertake that role, although a minimal instinctive defensive action is also expressed by animals themselves. The approach of legal representation for animals[21] would be appropriate, here, given that the model is already known in current law, in cases of representation either of unable to express will persons (minors, adults with mental disabilities, etc..) or of juridical persons (associations, companies, etc..).

Future legislation in relevance, adopting the institution of a certain kind of "animal representative" would be necessary and feasible in technical legal terms.[22] Animal representatives would be in charge of undertaking the legal defense of animal rights, either before or out of administrative and judicial authorities, for one or more individual animals, acting on their behalf and for their best interest. In that potential model, a representative could be identified as either the person who "cares" for the animal (the "owner" in current terminology), or the owner of the land where wild animals live (a private forest), or an environmental organization. Apparently, the details of that representative's role would need to be described by new legislation adapted to the right-based approach.

The right-based approach for animals requires nothing less than a fundamental change in modern law's structuring elements. That change would establish a wider concept of legal "subjects" by attributing the characterization of "person" to individual living beings beyond the human species. Until that change, any reference to the legal rights of animals remains unreasonable.

A supposed recognition as legal "persons" of animal individuals belonging to certain vertebrate species would be comparable to the consequences of slavery abolition in humans, although institutional "paths" that linked the legal statuses of free persons and slaves existed from the Antiquity already. These "paths" supported the passage from the status of slave/res into that of a free person. They were based on the common denominator of both statuses, that of the human condition. Yet, this is not the case for any animal species, even those biologically closer to humans; we do not share an identical biological nature with any animal species. Even the highly protective attitude for nonhuman primates (frequently expressed in ethics) looks static, not leading to a legal status higher than this of a valuable living "object"; this is because there are no institutional "paths" for bridging the normative gap, like in the old case of slaves.

Still, a strong reason that obliges us to reflect on that futuristic scenario of a potential recognition of vertebrates as subjects of legal rights does exist indeed; it is based on the findings of life sciences after Darwin and the subsequent constant confirmation of the Evolutionary approach. These findings pose persistently the crucial question of whether we need to adjust the basic structure of law accordingly. This step would mean that we are ready to abandon the axiom of that structure,

---

[21] Cf. Lilly (1976), pp. 67-68, and Favre, *supra.*

[22] On the U.S. existing law, see Hogan (2007), pp. 513 seq., 520, arguing that specific legislation would be not necessary, since legal representatives may defend animal interests.

affirming the human species' supremacy in the rank of legal values, an axiom of an essential theological origin. In its place, we would need to accept a system of values harmonized with the evidence of the common origin of species, indicating that there are no gaps in their comparative biological composition, including their cognitive functions.

However, we need to stress that to build the structure of a future legal system upon the findings of life sciences on the Evolution process, it is also a matter of better understanding the human species itself, particularly when focusing on the first stages of the embryo development. This is because, under the contemporary law, we do not recognize the full status of a "person" to the human embryo; this proves that we already have accepted an essential correspondence between the human biological development (that is, an *evolutionary fact* deployed in a relatively short time) with the generally recognized legal values. Perhaps, a further step that would accept a similar correspondence between the advanced biological status of vertebrate animals (again as an evolutionary fact deployed gradually over time within the conditions of biodiversity as a whole), and the legal status they deserve, looks not too distant.

References

Bjornkestam M (2001) Legal rights for animals. ASPCA Animal Watch, New York. https://www.petfinder.com/helping-pets/information-on-helping-pets/legal-rights-animals/

Carruthers P (1992) The animals issue. Moral theory in practice. Cambridge University Press, Cambridge

Cohen C (2001) In defense of the use of animals. In: Cohen C, Regan T (eds) The animal rights debate. Rowman & Littlefield Publ., Lanham

Cupp RL (2007) A dubious grail: seeking tort law expansion and limited personhood as stepping stones toward abolishing animals' property status. Southern Methodist Univ Law Rev 60:3–54

Dawkins R (1976/1989) The selfish gene, 2nd edn. Oxford University Press, Oxford

Degrazia D (2009) Self-awareness in animals. In: Lurz R (ed) The philosophy of animal minds. Cambridge University Press, Cambridge, pp 201–217

Diamond J (1991) The third chimpanzee. The evolution and future of the human animal. Hutchinson Radius, London

Favre DS (2005) Judicial recognition of the interests of animals-a new tort, Mich State Law Rev 333–367

Goodland R (1998) Environmental sustainability in agriculture: bioethical and religious arguments against carnivory. In: Lemons J, Westra L, Goodland R (eds) Ecological sustainability and integrity: concepts and approaches. Springer, Dordrecht, pp 235–265

Hogan M (2007) Standing for nonhuman animals: developing a guardianship model from the dissents in Sierra Club v. Morgan. California Law Rev 95:513–534

Kelch TG (1998) Toward a non-property status for animals. New York Univ Environ Law J 6:531–585

Lilly J (1976) The rights of cetaceans under human laws. Oceans 9:67–68

Price DT (ed) (2000) Europe's first farmers. Cambridge University Press, Cambridge

Regan T (1983/2004) The case for animal rights, 3rd edn. University of California Press, Berkeley

Salas C, Broglio C, Rodriguez F (2003) Evolution of forebrain and spatial cognition in vertebrates: conservation across diversity. Brain Behav Evol 62:72–82

Shumaker RW, Walkup KR Beck BB (2011), Animal tool behaviour: the use and manufacture of tools by animals. J. Hopkins University Press, Baltimore

Sohm R (1970) The institutes. A textbook of history and system of roman private law, Rothman Reprints Inc., S. Hackensack
Sunstein CR (2002) The rights of animals. A very short primer. Univ Chicago Law Rev 70:387–404
Van Horne PLM, Achterbosch TJ (2007) Animal welfare in poultry production systems: impact of European Union standards on world trade, Poultry in the 21st Century. FAO. http://www.fao.org/Ag/Againfo/home/events/bangkok2007/docs/part2/2_5.pdf
Wilson EO (1975/2000) Sociobiology. The new synthesis, 2nd edn. Harvard University Press, Cambridge
Wobber V, Herrmann E, Hare B, Wrangham R, Tomasello M (2013) Differences in the early cognitive development of children and great apes. Develop Psychobiol 56:547–573

Biolaw Beyond Biology: Artificial Intelligence and Smart Robots

16

16.1 General

Biolaw presupposes a well-determined concept of "life," which is given by life sciences. In general, "life" means organic entities, forming self-reproducing functional systems, with the presence of DNA; that concept embraces microorganisms, plants, and animals, including humans.

Nevertheless, current scientific developments make it necessary to consider an enlargement of that approach under the view of ethics and law. The development of artificial intelligence, and its applications in computers and robots, raises a key question in relevance: Does our ethical and legal interest in "life" emanate from the specific nature of living beings, their unique material and structuring, or what is important here, are the properties and functions they develop? If we focus on the functions of any object presenting properties characteristic for living organisms, regardless of its particular material and structuring, then an enlargement of the interest on the part of Bioethics and Biolaw should be observed.

For two reasons, this latter "functional" approach seems convincing, as more suitable for the purposes of a normative approach, either ethical or legal.

First, because, as we have seen with the examples of genetic engineering and synthetic biology in particular, the natural or artificial composition of entities manifesting functions of life is not considered as an important element. Although these technologies elaborate on biological material and exploit its natural properties, they form new entities through artificial means. Supposing that similar creations presenting functions pertinent to living organisms are possible through the processing of non-biological material, no reason occurs for excluding these from the field of normative interest. Second, and mostly, because Biolaw studies the legal regulation of life's manifestations to the measure that they influence our social relationships; Biolaw, thus, refers to the dynamic dimension of living organisms' functions, to their "potential," regardless of their nature, as biological entities. For instance, the reason for which we distinguish between the legal status of vertebrates and that of invertebrate animals is that the first manifest particular properties in their

T. Vidalis, *The Emergence of Biolaw*, Springer Textbooks in Law,
https://doi.org/10.1007/978-3-031-02359-0_16

life, of sense-expression (pain, stress, etc..), or of a cognitive sort (memory, etc..), to which we attribute a comparatively higher value; their functions are considered crucial for determining a respective legal status, precisely because these functions matter and produce effects on human relationships.

Under this view, we cannot exclude artificial intelligence and smart robots from the objective of this study, if we take into consideration the ultimate purpose of these applications, namely to substitute activities that now can be performed only by humans. Substitution, here, refers to specific cognitive functions, characteristic of the neurophysiology of our species.

We know that, to a certain degree, the process of human substitution reaches an advanced stage, thanks to the rapid progress of informatics. Main fields of research in artificial intelligence cover now development in machines of learning methods, communication with natural languages, logic processing, determination of goals and designing of their pursuit, and also the development of moving abilities and handling of various objects.[1] Apart from applications of algorithms for simulating the human ability to analyze and recompose data in a logical manner, since the late 1980s already, applications exist also for addressing problems with missing and uncertain data, based on Mathematics and Economics methods (probability theory, etc..). Thanks to the neural networks and the expert systems' technologies, which promoted the designing of models related to the structure of lower levels of the human brain, and the study of cognitive processing of symbols, the vision on machine simulation of human cognitive abilities became more realistic;[2] and this, even if we realize that there are barriers to further progress due to our limited knowledge about the human brain functions.

On the other hand, developmental robotics intend to create robots with advanced learning abilities based on specific "autonomous" data processing and their interaction with natural and social environmental conditions. That technology uses cutting-edge software applications that make possible automatic piloting, developmental language-learning, etc.[3] Already known applications permitting multipart processing of data, for tracking, piloting, or mapping, as well as properties identifying voices and sounds, characterize the current state-of-the-art in artificial intelligence that represents a starting point for reaching a level of "self-determined" machines.[4] In the long run, this comprises the development of social intelligence and creativity, based on research where the leading role play theories on decision-making, the Game Theory, but also neurosciences and Psychology, providing the necessary knowledge for imitating human senses and the respective human ability on information processing and reaction.

For some, this progress will make possible the creation of "strong" artificial intelligence that may result in the development of artificial conscience or even an

[1] Cf. Neapolitan and Jiang (2013), pp. 11 seq., 101 seq., 439 seq.

[2] Cf. Gurney (1997).

[3] Cf. Weng (2004), pp. 199 seq.

[4] Even developing some sort of "emotions" or "consciousness." See Scheutz (2014), pp. 247 seq.

artificial brain. Looking at that distant prospect, key problems remain to develop common knowledge and intuition abilities in robots. Common knowledge abilities require the collection and processing of massive data deriving from objects, relationships, and conceptual facts existing in the real world, which is not feasible given the current state of technology. For the achievement of intuition abilities in machines, the main difficulty refers mostly to the problem of speed in data processing that characterizes human intuition as a cognitive mechanism permitting immediate reactions to concrete stimuli from the external world.

16.2 Is Ethically Accepted the Development of Artificial Intelligence?

This technological progress presupposes that, in principle, artificial intelligence is developed in conformity with the fundamental values of a democratic society. To achieve that, needs a more detailed study in terms of ethics, though. Some hypothetical objections deserve, here, to be explored.

The first refers to the concept of human value. Suppose the superior value that we recognize for humans stems from their distinct biological nature. Are we justified to do experiments with machine applications under the prospect that these may reach a level of cognitive abilities (or other properties) comparable to that of humans? Is that vision compatible with human value?

Furthermore: A "smart" robot presenting abilities of "autonomous" function making possible self-programming actions for adapting to external conditions not passively but also by setting and pursuing goals, should be recognized as an agent of a certain moral status? And, if that cannot be the moral status of a human individual, could it be a status similar to that of animals? In other words, is it possible to attribute moral value to machines based on their functions?

Here, a positive answer looks reasonable, because as we argued previously, the material as such that forms any entity, including living beings, cannot be ethically significant. From microorganisms to humans, biological entities may be analyzed in a sum of organic chemical compounds, existing as such in non-biological objects, as well. What is crucial in terms of ethics, and stimulates our values, is the potential that develops any entity when interacting with its surrounding environment, not its static material form; that potential is translated into certain functions, and these make a natural or artificial entity ethically important or not.

Indeed, we have already noted that the moral significance of life as a natural phenomenon lies in the variety and interdependence of its forms (biodiversity) in their contribution to the overall quality of the natural environment. Moreover, it lies in the endorsement of vital functions of any existing species, including the human species. For humans, in particular, we noted that their biological nature is ethically significant because it functions as a necessary condition of autonomy, and this is the reason that explains why we distinguish, in terms of value, early embryos from viable fetuses or fetuses from newborns. The moral significance of functions that a biological organism develops, also explains why we consider nonhuman primates as ethically closer to our species than other categories of vertebrate animals.

Therefore, we do not have reasons to assume that the biological substance as such (what, in fact, includes DNA) is characterized by a specific ethical "preference" that makes it impossible to attribute ethical value to artificial non-biological entities as well.[5] Machines may also acquire a moral status, possibly comparable to that of animals or even of humans, if performing functions that we consider ethically significant; there is no question that the potential development of "autonomous" actions, even at an elementary stage of self-programming, would constitute such an ethically significant function.[6]

A further hypothetical objection could be similar to objections that we have already discussed and accepted in relation to positive eugenics, particularly regarding reproductive cloning in humans. Assuming that our technology makes possible the creation of "intelligent machines" at an advanced stage of conscience development, the fact that these machines would be products of intentional design would not undermine their supposed moral status? Would not these machines' "conscience" be already determined "by design," to such a degree that their creation should be considered fundamentally unethical?

There is no question that, even if for the moment this is only a futuristic scenario, logically the creation of machines with advanced cognitive abilities cannot be excluded, insofar as it belongs to visible purposes of artificial intelligence.

Compared to the reproductive cloning issue, we need to highlight a difference here. In cloning, the burden for autonomy results from the fact that, in humans, "designing" phenotypic characteristics contravenes the natural element of sexual reproduction that "attributes" phenotypes randomly; this is precisely the condition of individual uniqueness that supports autonomy. In machines, by definition, the element of reproduction is missing; designing and manufacturing are the only choices that we have for the appearance of an "intelligent machine" in the real world. Therefore, to suppose a "burden" for the potential moral status of that machine is meaningless; the "nature" of machines, as products of intentional designing and manufacturing, will determine their original ethical status. This also implies that, due to the lack of reproductive property (that supports autonomy in humans), even the more advanced form of "intelligent machines" would remain ethically inferior in terms of conscience development and self-programming abilities compared to humans. Nevertheless, certainly, this does not make unethical the creation of such machines.

---

[5]Moreover, although they include DNA, machines manufactured by biological material are nothing but "biological machines". Cf. Webster-Wood (2016).

[6]On the distinction between "computer ethics" (relevant to human behavior) and "machine ethics" (relevant to machine behavior), see Anderson and Anderson (2007), pp. 15 seq.

If objections, like the above, are not justified,[7] and the promotion of intelligent machines' technology is ethically acceptable, we need now to encounter two further problems; what could be the moral status of these manufactured products, and furthermore, the position of law in respect to that status?

16.3 The Moral Status of Intelligent Machines and the Future Role of Law

Given that the creation of machines is always dependent upon human initiatives and it is not the outcome of a natural process (like reproduction in biological entities), there are limits to what we can call machine "autonomy," that is, an advanced state of self-programming. This is because that state would be subjected to original design by the human manufacturer, setting up a basic machine functioning framework.

Theoretically, that framework allows accurate predictions of a machine's "behavior," and mostly the possibility of intentionally designed modifications, according to the manufacturer's preferences, including the interruption of all machine's functions. Certainly, we cannot exclude that absolute machine autonomy could be achieved in the future. On the other hand, purposing to entirely self-programming machines seems meaningless if we assume that this would entail the loss of human control over them. Any machine is created for serving aims posed by its manufacturer, not by itself, even if these aims require sophisticated calculations, development of advanced skills, or autonomous movements.

The fact that machines have by design a predetermined framework of functions should be taken into account when reflecting on their potential moral status. However, that reflection is unavoidable given the progress of technology. The prospect of appearing in the real world of artificial entities that might develop abilities of advanced processing and reaction to external stimuli, original combination of data, and discretion in movements is very close to what we call "self-determination," and even "conscience"; this cannot be treated as something ethically insignificant.[8]

Indeed, if "something" presents abilities to function similarly to animal or human behavior, this may influence our own life, possibly to a large extent. Relationships between machines and humans also manifesting emotional or other psychological reactions of the latter cannot be excluded, to give a simple example.[9] In that prospect, a sort of "respect" that we need to acknowledge towards intelligent machines seems not so strange. That respect may simply "reflect" credits that we

[7] On the relevance of the precautionary principle, as a general reason justifying objections, see a critical approach in Castro and McLaughlin (2019). However, the principle's model is in alignment with the recent proposal for the AI Regulation of the EU (AI Act) that introduces risk-levels for AI applications. See on that, Veale and Borgesius (2021).

[8] See the arguments presented by Santosuosso (2016).

[9] Cf. Scheutz, *supra,* Whitby (2014), pp. 233 seq. On the characteristic example of "love robots", that is, machines that may influence significantly the emotional state of their users, see Bendel (2021). The same we can say for caregiver robots, etc.

owe to their manufacturer. Nevertheless, it may also express recognition for the advanced abilities these machines develop by self-programming, bringing them even to some degree of self-determination.

From the moment that we adopt this fundamental ethical attitude towards artificial intelligence and smart robots, we need to consider the potential role of law.[10] Indeed, the confirmation of machines' moral status obliges us to think about their legal regulation, not as simple objects or industrial products, but under the light of this special status. Under this view, the value of machines should be perceived in a broader context of their coexistence with humans and other species. Their moral significance means that we need to regulate our relationships with them, accepting in principle new responsibilities for ourselves. More specifically, we need to ensure the machines' existence and functionality according to their potential, which could be unlimited following the progress of technology.

However, introducing intelligent machines under these terms into the legal system does not necessarily mean recognizing them as a sort of "legal subjects." A legal status similar to what we recognize currently for animals, as special "objects" deserving respect and protection, seems more suitable, insofar as machines develop skills of only limited self-determination, under controlled conditions by the manufacturer or their owner.[11]

This is the very general context, characterized by a major difference compared to living organisms, namely the lack in machines of reproduction ability, which results in their essential dependence upon the manufacturer's design and original programming. Within this context, it is up to the progress of technology to indicate the need for detailed regulation by future Biolaw. This prospect looks now absolutely realistic, as we can address "life" as something beyond a random combination of natural macromolecules.

References

Anderson M, Anderson SL (2007) Machine Ethics: creating an ethical intelligent agent. AI Magazine (Am Assoc Artif Intell) 28:15–26

Bendel O (2021) Love dolls and sex robots in unproven and unexplored fields of application. Paladyn J Behav Robot 12:1–12

Castro D, McLaughlin M (2019) Ten ways the precautionary principle undermines progress in artificial intelligence. Information Technology and Innovation Foundation. https://itif.org/publications/2019/02/04/ten-ways-precautionary-principle-undermines-progress-artificial-intelligence

Cofone IN (2018) Servers and waiters: what matters in the law of AI. Stan Technol Law Rev 21: 167–197

[10] See Cofone (2018), proposing a model for the legal status of artificial intelligence entities.

[11] On the draft motion presented at the EU Parliament, for the recognition of robots as "electronic persons" with "rights" and "obligations" (meeting, so far, objections by the industry, due to the expected boost of relevant costs if social security rights for robots are to be recognized), see Prodhan (2016).

Gurney K (1997) An introduction to neural networks. Routledge, London

Neapolitan RE, Jiang X (2013) Contemporary artificial intelligence. CRC Press, Boca Raton

Prodhan G (2016) Europe's robots to become 'electronic persons' under draft plan. Reuters 21(6): 2016. http://www.reuters.com/article/us-europe-robotics-lawmaking-idUSKCN0Z72AY

Santosuosso A (2016) The human rights of nonhuman artificial entities: an oxymoron? Jahrbuch für Wissenschaft und Ethik 19:203–238

Scheutz M (2014) Artificial emotions, and machine consciousness. In: Ramsey W, Frankish K (eds) The Cambridge handbook of artificial intelligence. Cambridge University Press, Cambridge, pp 247–268

Veale M, Borgesius FZ (2021) Demystifying the Draft EU Artificial Intelligence Act—Analysing the good, the bad, and the unclear elements of the proposed approach. Comput Law Rev Int 22: 97–112. https://doi.org/10.9785/cri-2021-220402

Webster-Wood V (2016) Biohybrid robots built from living tissue start to take shape. The Conversation. https://theconversation.com/biohybrid-robots-built-from-living-tissue-start-to-take-shape-62759

Weng J (2004) Developmental robotics: theory and experiments. Int J Human Robot 1:199–236

Whitby B (2014) Do you want a robot lover? The ethics of caring technologies. In: Lin P, Abney K, Bekey G (eds) Robot ethics. The ethical and social implications of robotics. The MIT Press, Cambridge, pp 233–248

Index

T. Vidalis, *The Emergence of Biolaw*, Springer Textbooks in Law,
https://doi.org/10.1007/978-3-031-02359-0

The manufacturer's authorised representative in the EU is Springer Nature Customer Service Centre GmbH, Europaplatz 3, 69115 Heidelberg, Germany. If you have any concerns regarding our products, please contact ProductSafety@springernature.com

Printed and bound by CPI Group (UK) Ltd, Croydon, CR0 4YY
30/09/2026
02225460-0018